PRINCIPLES OF DISCRETE TIME MECHANICS

Could time be discrete on some unimaginably small scale? Exploring the idea in depth, this unique introduction to discrete time mechanics systematically builds the theory up from scratch, beginning with the historical, physical and mathematical background to the chronon hypothesis.

Covering classical and quantum discrete time mechanics, this book presents all the tools needed to formulate and develop applications of discrete time mechanics in a number of areas, including spreadsheet mechanics, classical and quantum register mechanics, and classical and quantum mechanics and field theories. A consistent emphasis on contextuality and the observer–system relationship is maintained throughout.

George Jaroszkiewicz is an Associate Professor at the School of Mathematical Sciences, University of Nottingham, having formerly held positions at the University of Oxford and the University of Kent.

CAMBRIDGE MONOGRAPHS ON MATHEMATICAL PHYSICS

S. J. Aarseth *Gravitational N-Body Simulations: Tools and Algorithms*
J. Ambjørn, B. Durhuus and T. Jonsson *Quantum Geometry: A Statistical Field Theory Approach*
A. M. Anile *Relativistic Fluids and Magneto-fluids: With Applications in Astrophysics and Plasma Physics*
J. A. de Azcárraga and J. M. Izquierdo *Lie Groups, Lie Algebras, Cohomology and Some Applications in Physics*†
O. Babelon, D. Bernard and M. Talon *Introduction to Classical Integrable Systems*
F. Bastianelli and P. van Nieuwenhuizen *Path Integrals and Anomalies in Curved Space*
V. Belinski and E. Verdaguer *Gravitational Solitons*
J. Bernstein *Kinetic Theory in the Expanding Universe*
G. F. Bertsch and R. A. Broglia *Oscillations in Finite Quantum Systems*
N. D. Birrell and P. C. W. Davies *Quantum Fields in Curved Space*†
K. Bolejko, A. Krasiński, C. Hellaby and M–N. Célérier *Structures in the Universe by Exact Methods: Formation, Evolution, Interactions*
D. M. Brink *Semi-Classical Methods for Nucleus–Nucleus Scattering*†
M. Burgess *Classical Covariant Fields*
E. A. Calzetta and B.-L. B. Hu *Nonequilibrium Quantum Field Theory*
S. Carlip *Quantum Gravity in 2+1 Dimensions*†
P. Cartier and C. DeWitt-Morette *Functional Integration: Action and Symmetries*
J. C. Collins *Renormalization: An Introduction to Renormalization, the Renormalization Group and the Operator-Product Expansion*†
P. D. B. Collins *An Introduction to Regge Theory and High Energy Physics*†
M. Creutz *Quarks, Gluons and Lattices*†
P. D. D'Eath *Supersymmetric Quantum Cosmology*
J. Dereziński and C. Gérard *Mathematics of Quantization and Quantum Fields*
F. de Felice and D. Bini *Classical Measurements in Curved Space-Times*
F. de Felice and C. J. S Clarke *Relativity on Curved Manifolds*
B. DeWitt *Supermanifolds, 2nd edition*
P. G. O. Freund *Introduction to Supersymmetry*†
F. G. Friedlander *The Wave Equation on a Curved Space-Time*†
J. L. Friedman and N. Stergioulas *Rotating Relativistic Stars*
Y. Frishman and J. Sonnenschein *Non-Perturbative Field Theory: From Two Dimensional Conformal Field Theory to QCD in Four Dimensions*
J. A. Fuchs *Affine Lie Algebras and Quantum Groups: An Introduction, with Applications in Conformal Field Theory*†
J. Fuchs and C. Schweigert *Symmetries, Lie Algebras and Representations: A Graduate Course for Physicists*†
Y. Fujii and K. Maeda *The Scalar-Tensor Theory of Gravitation*
J. A. H. Futterman, F. A. Handler, R. A. Matzner *Scattering from Black Holes*†
A. S. Galperin, E. A. Ivanov, V. I. Ogievetsky and E. S. Sokatchev *Harmonic Superspace*
R. Gambini and J. Pullin *Loops, Knots, Gauge Theories and Quantum Gravity*†
T. Gannon *Moonshine beyond the Monster: The Bridge Connecting Algebra, Modular Forms and Physics*
M. Göckeler and T. Schücker *Differential Geometry, Gauge Theories, and Gravity*†
C. Gómez, M. Ruiz-Altaba and G. Sierra *Quantum Groups in Two-Dimensional Physics*
M. B. Green, J. H. Schwarz and E. Witten *Superstring Theory Volume 1: Introduction*
M. B. Green, J. H. Schwarz and E. Witten *Superstring Theory Volume 2: Loop Amplitudes, Anomalies and Phenomenology*
V. N. Gribov *The Theory of Complex Angular Momenta: Gribov Lectures on Theoretical Physics*
J. B. Griffiths and J. Podolský *Exact Space-Times in Einstein's General Relativity*†
S. W. Hawking and G. F. R. Ellis *The Large Scale Structure of Space-Time*†
F. Iachello and A. Arima *The Interacting Boson Model*
F. Iachello and P. van Isacker *The Interacting Boson–Fermion Model*
C. Itzykson and J.-M. Drouffe *Statistical Field Theory Volume 1: From Brownian Motion to Renormalization and Lattice Gauge Theory*†
C. Itzykson and J. M. Drouffe *Statistical Field Theory Volume 2: Strong Coupling, Monte Carlo Methods, Conformal Field Theory and Random Systems*†
G. Jaroszkiewicz, *Principles of Discrete Time Mechanics*
C. V. Johnson *D-Branes*
P. S. Joshi *Gravitational Collapse and Spacetime Singularities*†
J. I. Kapusta and C. Gale *Finite-Temperature Field Theory: Principles and Applications, 2nd edition*

V. E. Korepin, N. M. Bogoliubov and A. G. Izergin *Quantum Inverse Scattering Method and Correlation Functions*†
M. Le Bellac *Thermal Field Theory*†
Y. Makeenko *Methods of Contemporary Gauge Theory*
N. Manton and P. Sutcliffe *Topological Solitons*
N. H. March *Liquid Metals: Concepts and Theory*
I. Montvay and G. Münster *Quantum Fields on a Lattice*†
L. O'Raifeartaigh *Group Structure of Gauge Theories*†
T. Ortín *Gravity and Strings*
A. M. Ozorio de Almeida *Hamiltonian Systems: Chaos and Quantization*†
L. Parker and D. Toms *Quantum Field Theory in Curved Spacetime: Quantized Fields and Gravity*
R. Penrose and W. Rindler *Spinors and Space-Time Volume 1: Two-Spinor Calculus and Relativistic Fields*†
R. Penrose and W. Rindler *Spinors and Space-Time Volume 2: Spinor and Twistor Methods in Space-Time Geometry*†
S. Pokorski *Gauge Field Theories, 2nd edition*†
J. Polchinski *String Theory Volume 1: An Introduction to the Bosonic String*
J. Polchinski *String Theory Volume 2: Superstring Theory and Beyond*
J. C. Polkinghorne *Models of High Energy Processes*†
V. N. Popov *Functional Integrals and Collective Excitations*†
L. V. Prokhorov and S. V. Shabanov *Hamiltonian Mechanics of Gauge Systems*
A. Recknagel and V. Schomerus, *Boundary Conformal Field Theory and the Worldsheet Approach to D-Branes*
R. J. Rivers *Path Integral Methods in Quantum Field Theory*†
R. G. Roberts *The Structure of the Proton: Deep Inelastic Scattering*†
C. Rovelli *Quantum Gravity*†
W. C. Saslaw *Gravitational Physics of Stellar and Galactic Systems*†
R. N. Sen *Causality, Measurement Theory and the Differentiable Structure of Space-Time*
M. Shifman and A. Yung *Supersymmetric Solitons*
H. Stephani, D. Kramer, M. MacCallum, C. Hoenselaers and E. Herlt *Exact Solutions of Einstein's Field Equations, 2nd edition*†
J. Stewart *Advanced General Relativity*†
J. C. Taylor *Gauge Theories of Weak Interactions*†
T. Thiemann *Modern Canonical Quantum General Relativity*
D. J. Toms *The Schwinger Action Principle and Effective Action*†
A. Vilenkin and E. P. S. Shellard *Cosmic Strings and Other Topological Defects*
R. S. Ward and R. O. Wells, Jr *Twistor Geometry and Field Theory*
E. J. Weinberg *Classical Solutions in Quantum Field Theory: Solitons and Instantons in High Energy Physics*
J. R. Wilson and G. J. Mathews *Relativistic Numerical Hydrodynamics*

† Issued as a paperback

Principles of Discrete Time Mechanics

GEORGE JAROSZKIEWICZ
University of Nottingham

CAMBRIDGE
UNIVERSITY PRESS

CAMBRIDGE
UNIVERSITY PRESS

University Printing House, Cambridge CB2 8BS, United Kingdom

Published in the United States of America by Cambridge University Press, New York

Cambridge University Press is part of the University of Cambridge.

It furthers the University's mission by disseminating knowledge in the pursuit of education, learning, and research at the highest international levels of excellence.

www.cambridge.org
Information on this title: www.cambridge.org/9781107034297

First published 2014

Printed in the United Kingdom by CPI Group Ltd, Croydon CR0 4YY

A catalogue record for this publication is available from the British Library

ISBN 978-1-107-03429-7 Hardback

Contents

Preface

Long ago, great minds speculated on the nature of time. The following question was asked: could time be divided into ever smaller and smaller pieces, just like a length of wood? We know this for a historical fact, because some of Zeno's paradoxes have survived the ravages of time and these paradoxes discuss precisely this question.

Contrary to what might be believed, interest in Zeno's paradoxes has not been extinguished by the rigours of modern mathematics, although we are taught that it has. Yes, the paradox of Achilles and the tortoise can be explained away in terms of a convergent infinite sum. But the concept of an infinitesimal has not been killed off: far from it, for mathematicians have developed a rigorous, consistent mathematical framework called non-standard analysis that allows for such things.

What I believe this debate about time highlights is how conditioned humans can be. We learn from an early age to think in certain terms and, if we are not careful, we end up regarding them as the only possible framework for our thoughts. So it is with time, which has been regarded as continuous throughout the history of mathematics and physics. It is hard to imagine any physical theory without the concept of a time derivative, and that requires continuity in time.

However, it is the obligation of theorists not only to explore current theories to their natural horizons, but to look beyond those horizons and to step outside of them if that is possible. That's really what theorists are paid for, not for the confirmation of established paradigms. I started to be concerned about standard physics when I first encountered wavefunction renormalization, that notorious method of dealing with the divergences of quantum field theory. Now, many years later, I can see that this concern was a portent of what was to come, for a very large quantity can be regarded as the reciprocal of a very small quantity. Very large energies and momenta are related to very small timescales and intervals of space, as I shall discuss.

Can we resolve these problems? Is it possible to understand Zeno's paradoxes about the vanishingly small and understand the divergences in quantum field theory within the same framework?

I think the answer is *possibly*, but it will require a deeper examination of the role of the observer. The observer has long taken a back seat in scientific theory, because the focus in science has generally been on the systems under observation. It is my belief that the balance has to be redressed, particularly when it comes to time.

This book is not about approximations to continuous time models but an exploration of discrete time as a model in its own right. I am not interested in finding good discrete time approximations to continuous time equations or their solutions. So do not look here for advice about the latest and best convergent lattice discretizations in fluid mechanics, non-abelian gauge theory or gravitation. There are plenty of sources on those topics. I am exploring the following question: what would be the consequences of the conjecture that time is really discrete?

This book will necessarily be centred on my own experiences: what I have read, what I have written, and what has come to me through talking to others. So I will inevitably have missed some important topics and papers written by others, for which I apologize profusely in advance.

A preface is generally the place where an author expresses their unbounded gratitude to others. I do so now. I am indebted to all my teachers and lecturers, colleagues and students down the years who have given me far more of value than I have given them. In particular, I benefited from the wisdom and inspiration of Professors Nicholas Kemmer, Julian Schwinger and Peter Landshoff at various points in my career. I am indebted also to Andy Walker and Anne Lomax, and to Volodia and Rumy Nikolaev.

I thank also all my research students who took the risk of working with me on discrete time: Keith Norton, Jon Eakins, Jason Ridgway-Taylor and Fernando Aguayo.

My deepest thanks go to all members of my family, past and present: my wife Małgorzata and daughter Joanna for their endless patience and support, and to my parents for the priceless values they gave me.

George Jaroszkiewicz,
Walton on the Wolds

Part I

Discrete time concepts

1
Introduction

1.1 What is time?

This book explores the hypothesis that time is discrete rather than continuous. Time is an enigma, so we should expect some metaphysics and philosophy to creep into the discussion. Our inclination is to avoid those disciplines as much as possible, so let us deal with them right now.

Metaphysics and philosophy deal with statements and conjectures that cannot be empirically validated. In those disciplines there are constant references to absolutes such as existence, good and bad, and suchlike without further qualification, as if everyone accepted them as meaningful concepts. Absolutes are the key things we wish to avoid. For the record, we define an absolute statement as one that is considered to be true regardless of any caveats or criteria, i.e., context-free. In contrast, a contextual statement has a truth value that is meaningful only relative to its particular context.

The idea that physical truth can be contextual is an unfamiliar and uncomfortable one to physicists conditioned to believe that the laws of physics transcend the context of observation because they can be empirically validated. In fact, that is a circular line of reasoning. Every experiment is defined by its own context and experimentalists have to work hard to create that context: the search for the Higgs particle at the Large Hadron Collider did not happen overnight. Because it is impossible to actually perform all imaginable experiments, the known laws of physics have been validated only relative to a finite subset of all possible contexts. Therefore, the laws of physics are contextual, not absolute. It is metaphysics to think otherwise.

Despite that, there are numerous examples in physics of a conditioned metaphysical belief in an absolute. Lorentz covariance in special relativity (SR) is the principle that the laws of physics apart from gravitation take the same form in every inertial frame. In general relativity (GR) the corresponding concept is encoded into general covariance, the principle that the laws of physics are invariant with respect to arbitrary coordinate transformations. Are these absolute principles? Other examples come to mind: in thermodynamics, physicists make frequent references to the 'absolute temperature' of a system under observation (SUO), whilst in quantum mechanics (QM) they refer to 'the' probability of a quantum outcome. In fact, the temperature of an SUO is contextual on that SUO being in thermal equilibrium, whilst a probability in QM is always a conditional probability, i.e., contextual. As for Lorentz covariance and general covariance, these are more and more frequently these days being seen by cosmologists as useful guidelines in the construction of Lagrangians rather than absolute principles.

This issue impinges on us here because a potential criticism of discrete time (DT) mechanics is that it breaks Lorentz symmetry explicitly. That is true: we need to choose a preferred inertial frame in which to discretize time. We are not unduly concerned by this criticism, however, for several reasons. Three of these are as follows: (i) there is no empirical proof that time is continuous or otherwise; (ii) the aforementioned criticism does not take into account the empirical fact that we **can** use the laws of physics to identify a preferred local inertial frame anywhere in the Universe, the local frame relative to which the dipole anisotropy of the cosmic background radiation field vanishes (Cornell, 1989);[1] and (iii) conventional theories that are based on Lorentz symmetry are riddled with mathematical divergences, and DT may be a possible technique to grapple with them.

With the above in mind, we shall take as a guiding principle the view that there are no absolutes in physics: every concept or statement in physics should be accompanied by a statement of the context relative to which that statement's truth value makes sense.[2] Care should be taken to understand the opposite of relative truth: if a statement is not true relative to a given context, then it is false only relative to that context. Outside of that context, we should say *nothing.*

When we discuss any theory, the above principle of contextuality requires us to clarify the context in which our theory is to be discussed and held to be meaningful. In the case of DT, we should establish (i) *who* or *what* sort of observer is formulating the theory, (ii) for what purpose and to which ends the theory has been constructed, (iii) the principles underpinning the theory, including its limitations, and (iv) what might be done with the theory. We address these points in turn.

Point (i) In this book, time is studied from the perspective of the mathematical physicist, with no hidden agenda or philosophy. The reader will not

[1] This is a debatable point, in that it could be argued that the appearance of such a frame is a consequence of the laws of physics, not a fundamental feature in itself. But we would argue that the assertion that this frame was chosen at random by a quantum fluctuation is itself a metaphysical statement.

[2] Contextuality applies to mathematics as much as it does to physics. If it did not, why then do mathematicians spend time defining axioms and postulates? Theorems are true only relative to the relevant mathematical context.

be asked to believe either in continuous time (CT) or in DT. Since most of science deals with CT theories, it seems reasonable to redress the balance by investigating the consequences of DT mechanics. At our disposal will be mathematics backed up by intuition and supported by some empirical knowledge about time, such as its ordering property, its irreversibility and SR time dilation.

Point (ii) Many questions remain unanswered about the physical Universe, particularly the nature of space and time. We are not even sure how to classify time. Is it an object or a process? On the one hand, empty spacetime appears to have intrinsic physical properties such as curvature and vacuum polarizability, with particles being no more than quantum excitations of a basic state of empty space known as the vacuum. On the other hand, Mach's principle (Mach, 1912) and recent interpretations of QM (Rovelli, 1996) propose that space and time should be discussed in relational (contextual) terms. Which view is correct?

Even when spacetime is considered to be more than a relationship between objects, its structure remains debatable. Newtonian mechanics models space as a three-dimensional Euclidean manifold and time as a real line, whereas SR and GR merge space and time into a four-dimensional continuum known as spacetime. Although Einstein did acknowledge a debt to Mach's principle (Einstein, 1913), it is clear that GR spacetimes have intrinsic properties that can be measured, such as curvature. In GR, time is often identified with one of the four possible coordinates in a chosen coordinate patch and is continuous in that context.

On the other hand, some models of spacetime, such as Snyder's quantized spacetime (Snyder, 1947a, 1947b), suggest that continuous spacetime models may be too simplistic. Snyder's work motivated our particular interest in DT as an alternative to CT.

Point (iii) The principles we shall use are not controversial, apart from the single step of replacing the temporal continuum with a discrete set. All the standard principles of classical mechanics and QM adapted to DT are used in this book.

Point (iv) As for what DT can do for us, that remains to be seen. There are some nice things it can do for us, such as provide a natural (to the theory) cutoff in particle momentum. This may help cure some of the problems in CT quantum field theory, where the lack of any bound to linear momentum leads to divergences in loop integrals. This will be discussed in this book.

1.2 The architecture of time

We come now to a question central to this book: what is the architecture or structure of time? What sort of mathematical model best fits our intuitive notion of time?

This model or architecture should mirror the view of what we believe time represents and should incorporate into its rules whatever properties we believe time has. The architecture of time depends, therefore, on our beliefs about the Universe and how it runs. For instance, we might not believe that there is a single continuous strand of time such as Newton's absolute time (Newton, 1687). We might think there are many parallel strands of time each associated with a particular observer. Modelling such a 'multi-fingered architecture' would require the mathematics of parallel computer processing rather than the mathematics associated with single-processor computers.

In the following subsections we review several of the properties that the time concept should incorporate.

1.2.1 Events

Whatever model we decide on, it is a safe bet that we will incorporate into it the concept of an **event**.

Definition 1.1 An event is a well-defined, localized region of time and space, relative to a given observer.

Without the concept of an event, it would be impossible to discuss atoms and molecules, for instance. In general, events are assigned specific times and locations relative to a given observer. The existence of events is a supposition predicated on our world view. Whilst some quantum theorists view the Universe holistically as an enormous entangled state, quantum separability seems essential (Eakins and Jaroszkiewicz, 2003).

In particular, we should be aware of any hidden assumptions that we might be making about the nature of physical reality, as classically conditioned theorists find to their cost when they try to explain experiments such as the famous double-slit experiment (Tonomura and Ezawa, 1989). The quantum explanation of this experiment is at odds with the metaphysical classical explanation that an electron impacting on the final detecting screen had taken one or other of two possible paths on its journey from the source to that detecting screen. Quantum mechanically, we are not entitled to hold such a view. Therefore the following question arises: has a definite path been taken or not?

Classically we would have to believe that it had, because in our mind's eye we imagine a classical particle always follows a unique, continuous trajectory from source to screen. Quantum principles, however, require us to say nothing about this, *if* we have not attempted to detect anything about which path was taken.

For reasons such as this we have included the reference to an observer in Definition 1.1, for, if we did not, we would be implying that events could exist and things could happen regardless of who or what was observing them. That is a very classical, absolutist perspective on reality.

1.2.2 Temporal ordering

Suppose now that an observer had detected two or more events. How would that observation be modelled mathematically? In addition to any other attributes such as position, the observer would assign a time to each event, that is, a real number, which we call the assigned time (relative to that observer). Now, a crucial property of real numbers is that they are an ordered set. If we pick any two real numbers x and y, then only one of the following three statements can be true: x is less than y, or x equals y, or x is greater than y.

This means that there is temporal ordering relating any two observed events. If event A is assigned a time t_A and event B is assigned a time t_B, then $t_A < t_B$, or $t_A = t_B$, or $t_A > t_B$. These mathematical statements are interpreted physically as follows. In case (i) we say that A is earlier than B (or, equivalently, that B is later than A), whilst in case (ii) we say that A and B are simultaneous.

When SR emerged into the general consciousness of physicists, a significant conceptual problem for theorists conditioned to believe in Newtonian absolute time (Newton, 1687) was that simultaneity in SR is contextual. In contrast to Newtonian mechanics, where all classical observers agree on the relative temporal ordering of all events, SR asserts that A could be earlier than B relative to one observer and later than B relative to another.

The loss of absolute simultaneity in SR concerns two or more observers. We may bypass this issue by the simple method of restricting our attention to a single observer. In that context, all observed events have a well-defined temporal ordering relative to that observer.

1.2.3 Causality

A new factor now enters into the discussion: cause and effect. Suppose we have two events A and B, with A earlier than B according to some observer. That observer may have reason to believe in a causal link between A and B, in that there may be evidence in support of the notion that A caused B, or at least had some influence on B.

The notion of causality is notoriously difficult to pin down, principally because it requires us to contemplate counterfactuality, that is, valid logical conclusions that are based on premises known to be false. The 'mark-method' of Reichenbach (1958) demonstrates the point clearly (Whitrow, 1980). Reichenbach considers two events A and B, with A regarded as the cause of B. This relationship is denoted by AB, with the left–right ordering implying causal association. Now suppose that what happened at A had been slightly altered. We indicate this by marking the symbol A with an asterisk, i.e., A is replaced by A^*. Then one of two things could happen: either B is unchanged or else B is changed to B^*. Reichenbach asserts that the combinations AB, AB^* and A^*B^* are consistent with A being the cause of B, but A^*B is inconsistent with A being the cause of B.

The problem with this line of reasoning is that it is based on classical counterfactuality, which assumes counterfactual arguments are empirically meaningful. This is not the case in QM. A much quoted dictum attributed to John A. Wheeler states the quantum position elegantly: 'No elementary phenomenon is a phenomenon until it is an observed (or registered) phenomenon' (Wheeler, 1979).

1.2.4 The dimensions of time

Time is generally regarded as having a single dimension, but the question of physics based on two or more times has been discussed by experimentalists and theorists. We give an example of each.

Several decades ago, the astrophysicist Tifft measured the red shift of distant galaxies and came to the conclusion that redshifts were 'quantized'; that is, they appeared to be clustered into groups or bands. Subsequently, he developed an interpretation of his data that was based on a model of three-dimensional time. In his model, he asserted that 'each galaxy evolves along a 1-d timeline such that within a given standard galaxy standard 4-d space-physics is satisfied. The model deviates from ordinary physics by associating different galaxies with independent timelines within a general 3-d temporal space.' (Tifft, 1996). In the model, temporal quantization, involving photon exchange between galaxies and observers, was invoked to account for the discrete structures in his redshift data.

It would be unfair to criticize this approach since it is no more than an attempt to fit an unusual mathematical model to actual observations. Unfortunately, although Tifft's data were consistent with some subsequent observations, the most recent analysis concludes that there is no periodic structure in the redshift data (Schneider *et al.*, 2007). Therefore, the idea that time may be part of a three-dimensional continuum appears incorrect.

Tifft's model incorporates a serial time of the form that we are used to, since a worldline in any dimensional spacetime can be parametrized by a single real variable, which can be called a time.

We come now to a theoretical discussion by the theorist Tegmark of the mathematical consequences of having a genuine multi-dimensional form of time.

Tegmark analyses a flat spacetime with p time dimensions modelled by coordinates $\mathbf{t} \equiv \{t^1, t^2, \ldots, t^p\}$ and q spatial dimensions modelled by coordinates $\mathbf{x} \equiv \{x^1, x^2, \ldots, x^q\}$ (Tegmark, 1997). The important property here is the signature of the metric, denoted by (p, q). Drawing on experience in standard-signature $(1, 3)$ SR spacetime, or Minkowski spacetime, Tegmark's discussion focuses on a second-order partial differential wave equation for a spinless relativistic field φ of the form

$$\left\{ \sum_{i=1}^{p} \frac{\partial^2}{\partial t^i\, \partial t^i} - \sum_{j=1}^{q} \frac{\partial^2}{\partial x^j\, \partial x^j} \right\} \varphi(\mathbf{t}, \mathbf{x}) + V(\varphi(\mathbf{t}, \mathbf{x})) = 0, \qquad (1.1)$$

where V is some self-interaction term, such as a mass term, which does not depend on any derivatives of the field φ. The point here is the sign difference between the timelike coordinates $\{t^1, t^2, \ldots, t^p\}$ and the spacelike coordinates $\{x^1, x^2, \ldots, x^q\}$ in (1.1), which arises from an assumed line element of the form

$$ds^2 = (dt^1)^2 + \cdots + (dt^p)^2 - (dx^1)^2 - \cdots - (dx^q)^2. \tag{1.2}$$

The sign change in (1.2) is of the greatest importance to the modelling and interpretation of physics. In the case $p = 0$, referred to as the elliptic case, observers have no predictive power (Tegmark, 1997). There are no lightcones and no timelike worldlines in such spaces. This case corresponds to the imaginary-time scenario discussed by Minkowski in 1908 (Minkowski, 1908), which is frequently invoked in various branches of cosmology and particle physics in attempts to regularize mathematical divergences. There are numerous issues about this scenario that should cause concern (Jaroszkiewicz, 2002).

In the case of our Universe as we believe it to be, $p = 1$ and $q = 3$. Then the above differential equation is an example of a hyperbolic differential equation (Arfken, 1985). This case models the physically reasonable situation where an observer can use initial data over an initial spacelike hypersurface in relativistic spacetime to predict the final data over a final spacelike hypersurface. There are lightcones and timelike worldlines in such a spacetime. Tegmark concludes that the case $q < 3$ gives too simple a model and the case $q > 3$ leads to instability in the physics.

The remaining possibility, $p > 1$, is known as the ultrahyperbolic regime and leads to unpredictability.

Tegmark's analysis is based on what observers might see or be unable to see for various values of p and q, the value $p = 1$ being consistent with information flow in the form we are used to. In other words, observational criteria are used to decide what the spacetime architecture of the Universe might be.

There is no principle in GR that forbids a change of signature, apart from considerations such as those of Tegmark. The possibility that the signature changes dynamically has been considered. For instance, particle production from signature change from $(1, 1)$ to $(0, 2)$ was discussed by Dray *et al.* (1991).

1.2.5 Manifold time versus process time

The question raised earlier, namely that of whether time is an object or a process, leads to two mutually exclusive interpretations of time referred to as manifold time and process time, respectively (*Encyclopædia Britannica*, 2000). Manifold time regards time as a geometrical quantity, an objective thing having a single dimension and all the ordering properties of the real line. Manifold time represents an absolutist approach to time. An associated concept is the block universe (Price, 1997), which models spacetime as an object.

On the other hand, process time models time contextually: time is not anything that exists by itself but is an attribute of physical processes, a manifestation of change. Since change can be defined only relative to the memories of observers or their equivalent, process time implies the presence of observers and is compatible with and consistent with relational QM (Rovelli, 1996).

The difference between the manifold time and process time perspectives is important to us in this book. If time is indeed best described as part of a monolithic four-dimensional continuum, then discretizing time requires us to choose a preferred frame of reference.

On the other hand, if time is a manifestation or résumé of what a given observer experiences, then discretization of time need be considered only for *that* observer. An analogy can be drawn here with electron spin. In classical mechanics (CM), angular momentum is a continuous variable and a spinning particle can have any value of angular momentum. But any given observer detecting for quantized electron spin in a Stern–Gerlach experiment (Gerlach and Stern, 1922a, 1922b) can assign to it only one of two possible quantum spin values, a discretization of continuous angular momentum sometimes referred to as spatial quantization. The classical spatial continuum still exists in the formalism because the orientation parameters of the apparatus are not quantized, i.e., the direction in space of the main magnetic field axis is classical and can take on any value in QM.

The history of physics contains two important examples analogous to the manifold–process time debate: (i) classical thermodynamics treats temperature and entropy as classical attributes of continuous matter, whereas statistical physics interprets both of these concepts as statistical attributes of ensembles of systems in thermal equilibrium; and (ii) heat is interpreted as a substance in the theory of phlogiston, whereas the modern view is to regard it as part of a process.

An example of such an idealogical conflict in mathematics comes from probability theory, where the frequentist view of probability as an absolute quality of a random variable that can be measured approximately by sampling contrasts with the Bayesian view of probability as conditioned by prior information, i.e., a contextual approach to probability.

1.2.6 Multi-fingered time

Once we think of time as a manifestation of processes involving observers, we are naturally led to the idea that there may be as many times as there are observers. In SR, this is a well-understood feature of proper time: different observers following different worldlines experience time in a local, path-dependent way. This is the source of the so-called twin paradox, which is a paradox only if time is interpreted in the wrong way.

The multi-fingered time interpretation is compatible with relational QM (Rovelli, 1996). Moreover, time as it relates to the process of observation is naturally discrete: an observer prepares a quantum state by an initial time and registers outcome information at a subsequent final time. These times are reasonably well defined relative to that observer.

1.2.7 Temporal continuity

In addition to the above interpretational issues, there arises the question of which specific mathematical structures should be used in the modelling of time. Immediately we are faced with two choices: is time continuous or discrete? In CT, time is represented by a continuous parameter usually ranging over some interval $I_{\mathrm{if}} \equiv [t_{\mathrm{i}}, t_{\mathrm{f}}]$ of the real line, where t_{i} is the initial time and $t_{\mathrm{f}} > t_{\mathrm{i}}$ is the final time of some experiment. The universal convention in physics, which we follow in this book, is that, given two different values t_{i} and t_{f} of time, the larger value represents a physically later time in the laboratory, with 'later' being associated with the observed direction of the expansion and evolution of the Universe. On the other hand, DT is modelled as a sequence $\{t_n\}$ of real numbers, labelled by an integer n running from an initial value M to some final value $N > M$.

CT remains a powerful and popular model, which from the time of Newton onwards has been thoroughly explored and exploited. On the other hand, DT is still being developed. All the indications are that the importance of DT is growing, particularly on account of the impossibility of modelling CT exactly on a computer. Many CT models are approximated by appropriate discretizations of time so that they can be modelled on computers.

Our aim in this book is to discuss DT in those areas with which we are most familiar, but the importance of CT should not be overlooked. CT will be a central feature in much of our discussion and frequently used alongside DT as a parallel component of the discussion. It is possible to discuss the two views of time in the same context, provided that care is taken. To illustrate what we mean, consider a stone skipping over the surface of a pond. The pond's surface can be regarded as a continuum, but where the stone bounces off that surface is described as a discrete set.

Because CT is a central element of Newtonian CM, we may reasonably assume it is familiar to the reader. However, we shall review some of its basic features in order to highlight the differences between it and DT.

To understand CT we should understand the definition of a linear continuum. A continuum is a space, i.e. a set with certain properties. We need not concern ourselves with the nature of the points of the space.

Definition 1.2 A partially ordered set, or poset, S, is a set with a binary relation denoted by $\leqslant$, such that we have

(i) **reflexivity**: for every element x in S, $x \leqslant x$;
(ii) **antisymmetry**: if x and y are elements of S such that $x \leqslant y$ and $y \leqslant x$, then $x = y$; and

(iii) **transitivity**: if x, y and z are elements of S such that $x \leqslant y$ and $y \leqslant z$, then $x \leqslant z$.

Posets are important in SR because of the following exotic possibility: there may be elements u, v for which the binary relation $\leqslant$ is not defined. Minkowski spacetime $\mathcal{M}^4$, the four-dimensional spacetime of SR, has a Lorentzian metrical lightcone structure that creates this possibility. We can pick pairs of different events U, V in $\mathcal{M}^4$ such that, in some inertial frames, the times t_U, t_V assigned to them respectively satisfy $t_U \leqslant t_V$, and such that, in other inertial frames, we have $t'_V \leqslant t'_U$. Such pairs of events will be called relatively spacelike pairs.

We shall return to posets in Chapter 29, when we discuss causal sets.

To define a linear continuum, we need extra conditions. In particular, we need to eliminate the possibility of relatively spacelike relationships by introducing an extra condition, which turns a poset into a totally ordered set.

Definition 1.3 A **linearly ordered** or **totally ordered set** S is a poset with the additional property of

(iv) **totality**: for any two elements x, y of S, then $x \leqslant y$ or $y \leqslant x$.

The totality property of a totally ordered set essentially places a veto on finding a spacelike pair in S. An additional binary relation can be defined for each totally ordered set.

Definition 1.4 A **strict total order** on a totally ordered set is a binary relation $<$ such that $x < y$ if and only if $x \leqslant y$ and $x \neq y$.

Before we define a linear continuum, we need to define the concept of least upper bound.

Definition 1.5 Let S be a subset of a poset X. Then an element b of X is an **upper bound** for S if, for every element x of S, $x \leqslant b$.

Definition 1.6 Let S be a subset of a poset X. If there exists an element b_0 of X such that $b_0 \leqslant b$ for every upper bound of S, then b_0 is the **least upper bound** or **supremum** for S.

We note that the supremum, if it exists, is unique.

We now have the structures needed to define a linear continuum.

Definition 1.7 A *linear continuum* is a non-empty totally ordered set S such that

(1) S has the least upper bound property; and
(2) given any two different elements x and y of S such that $x < y$, there always exists another, distinct element z in S such that $x < z$ and $z < y$ (we write $x < z < y$).

Property (2) is at the heart of the difference between CT and DT. Suppose we have two values of time, t_1 and t_2, such that $t_1 < t_2$. If we know we are dealing with CT, then we can assume that we can find another time t such that $t_1 < t < t_2$. This need not be true in DT. It is easy to see from this that a linear continuum is necessarily of infinite cardinality, i.e., is a set with an infinite number of elements. Discrete sets can have either a finite cardinality or an infinite cardinality.[3]

The archetypical linear continuum is $\mathbb{R}$, the set of real numbers, which is generally used to model CT. To be more precise, intervals of time are modelled by interval subsets of $\mathbb{R}$.

In CT, and relative to a given observer's frame of reference, each observed event E is assigned a real number t_{E} called the assigned time of that event. Then, given two events E and F, we say that E is *later* than F if $t_{\mathrm{E}} > t_{\mathrm{F}}$, *simultaneous* with F if $t_{\mathrm{E}} = t_{\mathrm{F}}$, and *earlier* than F if $t_{\mathrm{E}} < t_{\mathrm{F}}$.

We will not discuss here how temporal assignment is carried out by any particular observer, as this would involve a lengthy discussion of clocks and measurement protocols. Suffice it to say that the advent of SR highlighted the importance of an observer's protocols in their definition of simultaneity.

We shall assume that an observer can always assign a time to an event that they observe. However, that is not the same thing as assuming that every point in a temporal continuum can be observed by a given observer. Any real measurement takes a non-zero amount of laboratory time to be completed, so there is in reality no possibility of proving that time is continuous.

Before the general acceptance of the atomic theory of matter, it was generally believed that matter is continuous. In the absence of any molecular or atomic scale, Fourier's principle of similitude was a formal statement of this belief: this principle asserts that the laws of physics are the same regardless of the scale of measurement. As we now know, it is an empirically false principle: we cannot cut material objects in half indefinitely without eventually running into molecular and atomic scales, below which matter behaves quite differently compared with its behaviour on larger scales. We should be prepared to expect the same with time.

The continuity of time has been essential to the development of CM, temporal derivatives (or fluxions in the language of Newton) being central elements in many of the laws of classical and QM. Temporal continuity is necessary to define the fundamental concepts of velocity and acceleration.

The assumed continuity of time and the modelling of it by the real numbers is so much part of theoretical science that it has been integrated into almost all theoretical frameworks as if it were some sort of geometrical dimension, analogous to the geometrical dimensions of physical space. This integration achieves its most complete form in the theories of special and general relativity, where the temporal and spatial dimensions are welded together into a single geometrical

[3] Surprisingly, a discrete set need not be countable. However, this is a relatively exotic mathematical concept, so we shall assume throughout this book that discrete means countable.

space known as spacetime. This was Minkowski's great leap in 1908 (Minkowski, 1908). Under the influence of novels such as *The Time Machine* by H. G. Wells (1895), the idea that time is the 'fourth dimension' is now an expression familiar even to non-scientists.

1.2.8 Temporal discreteness

Despite the ubiquity of the CT concept, there are important circumstances where time takes on a more discrete character. For instance, anything that is observed, i.e., any event, has a start and a finish as far as the observer is concerned, and both of these concepts are naturally associated with discreteness rather than continuity.

The question of the continuity or discreteness of time concerns the representation of time mathematically, but there are many other questions associated with time, namely the relationship between CM and QM, information acquisition and loss, the difference between observers and SUOs, decoherence versus state reduction, the loss of absolute simultaneity in relativity, and so on.

In this book it will be assumed mostly that we are dealing with exact ideas, not approximations to exact ideas. Certainly, we *could* imagine that DT is a useful approximation to CT in some situations, such as in computer programs. In any computer, the central processor has a characteristic fixed frequency and so in effect runs in terms of discrete clicks of its internal clock. Any dynamical system simulated on a computer will involve some version of DT mechanics. Numerical approximations in QM have been developed by Bender and others (Bender *et al.*, 1985b) via DT. We will not discuss such concepts in this book, our aim being to suppose that time is inherently discrete. This means we will try to deal with equations as if they were exact, rather than approximations.

An important question that we will address is the possible origin of temporal discreteness. We shall seek to understand where in physics temporal discreteness might originate. Some theorists, such as Hartland Snyder, have speculated that spacetime itself is fundamentally discrete. We shall discuss Snyder's model in Chapter 28. The common assumption that the Universe runs on CT principles may be a convenient simplification of a vastly more complex phenomenon, akin to the concept of temperature in statistical mechanics or centre of mass in CM. What really matters are observers, processes of observation and the information that can be gleaned from SUOs, and that always involves discreteness in one form or another. Everything else is metaphysics.

1.3 The chronon: historical perspectives

After three hundred years during which mathematical physics has been dominated by the continuous geometrical time of Galileo, Barrow, and Newton,

> *the idea that time is atomic, or not infinitely divisible, has only recently come to the fore as a daring and sophisticated hypothetical concomitant of recent investigations in the physics of atoms and elementary particles.*
>
> Whitrow, *The Natural Philosophy of Time*, p. 202 (Whitrow, 1980)

The notion that time may be discrete has a long history, which we discuss briefly in this section. In modern terminology, the proposed hypothetical 'quantum of time' is frequently referred to as a chronon, the earliest usage of this term known to us being in an article by Lévi in 1927 (Lévi, 1927).

The following is a list of thinkers who considered temporal discreteness in one form or another (Whitrow, 1980).

1.3.1 Zenocrates

Zenocrates was a pupil of Plato who made reference to possible discreteness of time, alluding to indivisible 'atoms' of time (Sambursky, 1959).

1.3.2 Aristotle

Aristotle gave an argument against the discreteness of time, which was based on the supposed continuity of matter, writing in the *Physica* (Aristotle, 1930)

> It is clear, then, that time is 'number of movement in respect of the before and after', and is continuous since it is an attribute of what is continuous. (Chapter 11)

and

> ... but in respect of size there is no minimum; for every line is divided *ad infinitum*. Hence it is so with time. (Chapter 12).

As we mentioned above in our reference to Fourier's principle of similitude, the notion that matter is a continuum is false. Therefore, the basis of Aristotle's view that time is continuous is empirically false.

1.3.3 The Sautrāntika

Time and the origin of the Universe was always of interest to ancient cultures and philosophers. The Sautrāntika, a Buddhist sect from the second and first centuries B.C., formulated a theory of reality that was based on atoms of time, in which everything exists for a chronon and then is replaced by a facsimile of itself.

1.3.4 Maimonides

Maimonides was a twelfth-century A.D. Jewish philosopher who wrote in his *Guide for the Perplexed* that

> Time is composed of time-atoms, i.e., of many parts, which on account of their short duration cannot be divided. This proposition also is a logical consequence of the first. The Mutakallemim[4] undoubtedly saw how Aristotle proved that time, space, and locomotion are of the same nature, that is to say, they can be divided into parts which stand in the same proportion to each other: if one of them is divided, the other is divided in the same proportion. They, therefore, knew that if time were continuous and divisible *ad infinitum*, their assumed atom of space would of necessity likewise be divisible. Similarly, if it were supposed that space is continuous, it would necessarily follow, that the time-element, which they considered to be indivisible, could also be divided. This has been shown by Aristotle in the treatise called *Acroasis*.[5] Hence they concluded that space was not continuous, but was composed of elements that could not be divided; and that time could likewise be reduced to time-elements, which were indivisible. An hour is, e.g., divided into sixty minutes, the minute into sixty seconds, the second into sixty parts, and so on; at last after ten or more successive divisions by sixty, time-elements are obtained, which are not subjected to division, and in fact are indivisible, just as is the case with space. Time would thus be an object of position and order.
>
> (Maimonides, 1190)

Three points to note are as follows.

1. Maimonides had no empirical evidence whatsoever for any of the above assertions.
2. According to his mathematics, an hour would be divided into at least 60^{10} time-atoms or chronons, so the chronon T_{M} according to Maimonides is no greater than approximately 10^{-9} seconds. This is far bigger, by a factor of about 10^{10}, than the shortest time intervals accessible to physicists in the laboratory at this time and bigger by a factor of 10^{35} than the Planck time postulated by some early-universe cosmologists to be the smallest possible interval of time.
3. The origin of Maimonides' chronon T_{M} is attributed by him to discreteness in space, the last sentence in the above quotation clearly supporting the view that he regarded time in terms of observation and observability. This is in tune with our own thoughts on the subject of DT in this book.

[4] Jewish philosophers of the style known as Kalam.
[5] Aristotle (1930).

1.3.5 Bartholomew the Englishman

In the ninth book of his popular encyclopaedia *De Proprietatibus Rerum*, which was written in 1239–1240 and translated by John Trevisa, Bartholomæus Anglicus (1240) asserted that an hour had 40 moments of time, each moment had 12 ounces of time and each ounce had 47 atoms of time. This means that one hour is equal to 22 560 atoms of time, so T_B, the Bartholomew chronon, is just 0.16 of a second, which is very long in comparison with other values for chronons.

1.3.6 Nicholas of Autrecourt

In the fourteenth century Nicholas of Autrecourt considered that matter, space, and time were all made up of indivisible atoms, points of space and instants of time, and that all generation and corruption took place by the rearrangement of material atoms. In 1347 he was forced by Church authorities to renounce his ideas and burn his writings.

1.3.7 Descartes

From cursory reading of his writings, there developed a tradition in philosophy that Descartes believed in DT. His ideas were quasi-religious in character, bringing together metaphysical notions involving existence, creation and suchlike, and elaborating on conjectures such as 'parts of time are mutually independent'. However, that view has been dissected and is now subject to reappraisal (Arthur, 1988; Secada, 1990), so that, currently, the evidence that he believed in DT is superficial.

1.4 The chronon: some modern perspectives

1.4.1 Pauli's theorem

Although we shall use the term 'chronon' in this book, we have some reservations about using it for the hypothetical shortest interval of DT, on the grounds that it may suggest without justification that there is a quantum origin for temporal discreteness. Not all discrete systems are quantum systems, a good example being coinage.

We should in general avoid adding the suffix 'on' to classical words and assuming that the result makes sense quantum mechanically. For example, there is currently no empirical evidence for the existence of gravitons. The concept of a quantum in the proper theoretical-physics sense of the word is meaningful only if there is a discrete subset in the spectrum of eigenvalues of a Hermitian operator representing some physical observable. It is not clear at this time, for example, what the observables are or should be in quantum gravity.

The notion of an operator for time has been mooted frequently throughout the history of QM and the subject continues to be discussed in significant detail (Muga and Egusquiza, 2008). It was pointed out by Pauli, however, that standard quantum physics does not readily support the notion of a time operator (Pauli, 1933). In Schrödinger wave mechanics, time is a parameter, not a dynamical variable, and is not associated with an observable in the way that position or momentum are.

Pauli's theorem, as it is conventionally called, asserts that there can be no consistent self-adjoint time operator $\hat{t}$ that is canonically conjugate to the energy operator (the Hamiltonian) $\hat{H}$. If there were, then we would expect the canonical commutation relation

$$[\hat{H}, \hat{t}] = i\hbar \tag{1.3}$$

to hold, just as for other canonically conjugate pairs of operators such as position and momentum, which satisfy the standard commutation relation $[\hat{p}, \hat{x}] = -i\hbar$.

A heuristic proof of Pauli's theorem goes as follows. In standard QM, conjugate pairs of operators can be interpreted as generators of canonically conjugate transformations. For example the position operator $\hat{x}$ generates momentum changes in momentum eigenstates $|p\rangle$ according to the rule

$$\exp(ik\hat{x}/\hbar)|p\rangle = |p + k\rangle \tag{1.4}$$

whilst the momentum operator $\hat{p}$ generates translations in space according to the rule

$$\exp(ia\hat{p}/\hbar)|x\rangle = |x - a\rangle. \tag{1.5}$$

Suppose now we assumed that there is a time operator $\hat{t}$ canonically conjugate to the Hamiltonian or energy operator $\hat{H}$ with the commutation rule (1.3). Then, given an energy eigenstate $|E\rangle$, we would expect the result

$$\exp(i\varepsilon\hat{t}/\hbar)|E\rangle = |E - \varepsilon\rangle, \tag{1.6}$$

which should hold for all real values of ε.

Now, if $|E\rangle$ is an energy eigenstate, then we can use the Baker–Hausdorff relation

$$e^{\hat{B}}\hat{A}e^{-\hat{B}} = \hat{A} + [\hat{B}, \hat{A}] + \frac{1}{2}[\hat{B}, [\hat{B}, \hat{A}]] + \cdots \tag{1.7}$$

for linear operators $\hat{A}$ and $\hat{B}$ and the commutator (1.3) to prove that $\exp(i\varepsilon\hat{t}/\hbar)|E\rangle$ is also an energy eigenstate. But if, as is commonly the case, the Hamiltonian is non-negative, then it can have no negative-energy eigenstates, and this is where the problem lies. Whatever the positive value of E, we could always choose a value for ε such that $E - \varepsilon$ is negative. This would then mean that we had generated an energy eigenstate with a negative energy, contrary to the given non-negativity of that Hamiltonian.

One way of understanding the significance of this argument is that translations in a single spatial dimension can be either to the left (say) or to the right without undermining any physical principles: in the absence of gravity, physical space does not seem to have a preferred direction. With time it is different. Relative to a given observer, the past is fixed whereas the future is not decided, according to the rules of QM. In essence, '*The past is a foreign country: they do things differently there.*' (Hartley, 1953). There is no forwards–backwards symmetry as far as physical time is concerned. The apparent temporal symmetry of unitary evolution in QM arises from neglecting the role of observers and the protocols of observation, which are not time-reversible.

The canonical commutation relation $[\hat{p}, \hat{x}] = -i\hbar$ leads to the Kennard–Heisenberg uncertainty relation $\Delta p \, \Delta x \geqslant \frac{1}{2}\hbar$ (Kennard, 1927; Heisenberg, 1927), where Δp is the uncertainty in momentum, etc. Any commutation relation such as (1.3) would suggest a corresponding time-energy uncertainty relation of the form $\Delta E \; \Delta t \geqslant \frac{1}{2}\hbar$. Although suggestive, such an idea requires care in its interpretation. Specifically, the observational architecture (i.e., the protocol of measurement) requires to be clarified (Taylor, 1987).

The subject of a time operator remains active, a recent analysis of the topic suggesting that loopholes exist, which may permit the notion of a time operator with discrete eigenvalues (Galapon, 2002). The fact is, though, that, throughout most of mathematical physics, time is regarded as a parameter labelling successive states in QM rather than as an eigenvalue of an observable. If we wanted to change this, we would have to dig deeper into the foundations of time and space, and that would probably require a better theory of what it means to be an observer. We shall return to this point when we come to discuss Snyder's quantized spacetime theory in Chapter 28.

1.4.2 Caldirola and the Italian School

The history of DT would not be complete without reference to the notable work of the Italian physicist Peiro Caldirola (1914–1984) (Caldirola, 1978) and successors such as Farias and Recami (Farias and Recami, 2010). Amongst many other interests in physics, Caldirola explored the DT structure of known elementary particles such as the electron and the muon, bringing general relativity into the picture. What is of interest to us is that his interest in DT sprang from the same source as ours: he and his successors were not interested in approximating CT per se but wanted to explore DT as a viable alternative model of reality.

There are two specific features of Caldirola's work that we comment on. First, his theories incorporate a unit of time that is enormously larger than the Planck time T_{P} which we shall discuss in the next chapter. We shall denote this relatively long timescale of the Caldirola chronon by T_{C}.

The Caldirola chronon is given in SI units by $T_{\text{C}} \equiv e^2/(6\pi\varepsilon_0 m_{\text{e}} c^3) \approx 6.2664 \times 10^{-24}$ s, where e is the electric charge of the electron, m_{e} is the mass of the electron, ε_0 is the permittivity of free space and c is the speed of light. There is no appearance here of either Planck's constant $\hbar$ or Newton's gravitational constant G, in contrast to the Planck time T_P, which is given by $T_P \equiv \sqrt{G\hbar/c^5} \approx 5.391 \times 10^{-44}$ s. The Caldirola chronon and the Planck time are not only very different in magnitude, by a factor of 10^{20}, but also their theoretical origins seem to be completely different.

There are various ways of coming across expressions for a Caldirola-type chronon, all based on some conjectured structure of the electron. What is common to these models is a 'worlds within world' ideology: in an attempt to explain fundamentals such as the finite mass of the electron or the electron-to-muon mass ratio, the electron is modelled to have a classical internal structure such as a rigid sphere, or a spinning disc, and treated as if conventional classical mechanical ideas can be applied.

In the early years of the twentieth century, there were many attempts along such lines to solve the notorious problem in classical electron theory that a true point-like electronic charge would carry with it divergent field energies. A review of Caldirola's early work on this aspect is given by Farias and Recami (2010). His work followed in the tradition of Dirac, who approached the electron divergence problem by modelling the worldline of a relativistic electron as a self-interacting tube of charge (Dirac, 1938).

The second notable feature of Caldirola's view of DT is his attempt to use the internal structure of the electron in conjunction with appropriate DT equations to solve the longstanding puzzle of the muon mass (Benza and Caldirola, 1981). The muon is very much like an unstable electron: it has the same electric charge and is involved in leptonic interactions, but has a much larger mass. Currently, there is no understanding of the ratio $r_{\mu e} \equiv m_\mu/m_e \approx 206.77$ of the mass of the muon to that of the electron.

Caldirola's attempt at an explanation starts with a proposed discretization of Schrödinger's wave equation (Farias and Recami, 2010), of the form

$$i\hbar\frac{\Psi(t+T_C,\mathbf{x})-\Psi(T-T_C,\mathbf{x})}{2T_C} = \overrightarrow{H}\Psi(t,\mathbf{x}). \tag{1.8}$$

Five points about this equation should be noted.

1. Time t is continuous, but the equation is a finite-difference equation in t.
2. Temporal discretization is symmetric in terms of past and future chronon contributions on the left-hand side of the equation.
3. The right-hand side does not involve the chronon.
4. The nature of the Hamiltonian operator $\overrightarrow{H}$ on the right-hand side is crucial to the modelling.
5. No attempt appears to be made to extend the model to quantum field theory.

Assuming a conventional energy eigensolution of the form $\overrightarrow{H}\Phi_E(\mathbf{x}) = E\Phi_E(\mathbf{x})$ and making the ansatz

$$\Psi(t, \mathbf{x}) = e^{-i\varepsilon t/\hbar}\Phi_E(\mathbf{x}), \tag{1.9}$$

equation (1.8) gives a relationship between the conventional energy E and the vibrational energy ε in the above ansatz, given by

$$\sin\left(\frac{\varepsilon T}{\hbar}\right) = \frac{TE}{\hbar}. \tag{1.10}$$

We encounter similar but not identical equations in our approach. For convenience, we refer to the dimensionless quantity $\varepsilon T/\hbar$ as a *vibrancy*, to distinguish it from E, which is an energy in the conventional sense of the word. On the basis of conditions such as (1.10), Caldirola arrived at a model explanation of the ratio $r_{\mu e}$. It has to be pointed out, however, that the assumptions made are based on a need to derive this result, and do not come from any other theoretical framework per se.

1.4.3 Finkelstein's space-time chronon

Many authors have discussed the notion that space as well as time may be discrete on certain scales. We shall discuss some concepts associated with this idea later. To illustrate some of the conclusions to which this idea leads us, we shall discuss briefly one particular line of discussion involving an algebraic view of things: Finkelstein's 'space-time code'. In a series of papers (Finkelstein, 1969, 1972a, 1972b, 1974; Finkelstein and Susskind, 1974), Finkelstein argued that the size of the chronon would be indicated by 'three conspicuous traces' (Finkelstein, 1969), as follows.

1. The spectrum of observed elementary particle masses.

   Given a mass m and the universal constants $\hbar$ and c (the speed of light), then the quantity $\tau_1 \equiv \hbar/(mc^2)$ has the physical dimensions of a time. Finkelstein suggested that the chronon, or shortest timescale relevant to the 'space-time code' discussed in his papers, would have a value dictated by a mass of the order of the muon mass. The muon is an unstable elementary particle approximately 200 times more massive than the electron. Inserting the known value of the muon mass into the above relation gives a value $\tau_1 \approx 6 \times 10^{-24}$ s. This is comparable to T_C, the chronon value calculated by Caldirola.
2. The size of atomic nuclei.

   If space is discrete then it would not be meaningful to discuss distances, areas and volumes below those on scales of a certain fundamental length denoted r_0. Finkelstein argued that there would be no way of experimentally pinpointing the spacetime coordinates of processes such as the emission of a photon from an atomic nucleus over a length scale comparable to its nuclear radius and over the transit time r_0/c, or the time that a photon would take to cross such a distance. If the relation between the chronon and the distance scale is assumed to be $r_0^4 \approx \tau_2^4 c^4$, then, taking r_0 to be the empirically determined charge radius of the proton (McAllister and Hofstadter, 1956), we find $\tau_2 \approx 2 \times 10^{-24}$ s.

3. High-energy particle scattering cross-sections.

   Finkelstein argues heuristically that, when two energy particles approach in a head-on collision in some laboratory with incident speeds v such that their Lorentz factors $\gamma \equiv 1/\sqrt{1 - v^2/c^2}$ are high, then, relative to the laboratory frame, each will appear to have a FitzGerald contraction (FitzGerald, 1889) along its direction of motion. Therefore, the effective time during which they could interact and scatter is reduced from what would be expected if the particles did not suffer FitzGerald contraction. From this, and assuming that there is a timescale below which meaningful interaction could not occur, Finkelstein concludes that experiments should show a preference for small-angle scattering, indicating that scattering was not really taking place, when $\gamma > \sigma/(\tau_3^2 c^2)$, where σ is the cross-section at high energies.

   The heuristic calculation runs as follows. Assume that a cross-section σ as measured in a laboratory is related to an effective radius a of each particle by $\sigma \approx a^2$ (in such heuristic calculations, factors of π, 2 and suchlike are overlooked: it is the orders of magnitude that are our concern). The volume V of each particle as seen in the laboratory is that of a sphere FitzGerald contracted in the direction of motion by a factor γ^{-1}, so the net interaction volume is approximately $V \approx a^3\gamma^{-1}$. The time T during which the particles overlap as seen in the laboratory frame is given approximately by $T \approx a\gamma^{-1}/c$. Hence the total spacetime four-volume of the interaction as seen in the laboratory is given by $VTc \approx a^4\gamma^{-2}$, where we multiply the interaction time T by the speed of light to convert it to a distance. Assuming that there is a chronon τ_3 associated with each of the fundamental time and space dimensions, we conclude that no meaningful discussion in terms of time and space can occur if

$$\tau_3^4 c^4 > VTc \approx a^4\gamma^{-2} \approx \sigma^2\gamma^{-2}, \tag{1.11}$$

   which agrees with the inequality stated in the previous paragraph. This argument is reminiscent of the 'old quantum mechanics' view of quantization, which was that there is a minimum 'volume' element in mechanical phase space given by Planck's constant h, below which it is not meaningful to discuss position and momentum with classical certainty.

   Looking at recent particle scattering data from the Large Hadron Collider (TOTEM Collaboration, 2012), the proton–proton total cross-section is approximately $\sigma \approx 98 \times 10^{-31}$ m^2 for a proton Lorentz factor of $\gamma \approx 7460$, so we find $\tau_3 \approx 1 \times 10^{-25}$ s.

The remarkable closeness of the three timescales τ_1, τ_2 and τ_3 should not be taken as a definitive assessment of the chronon. Much shorter theoretical timescales are now routinely discussed by grand unified theorists in early-universe cosmology, of the order of 10^{-36} s. On even smaller scales, quantum-gravity theorists assume that spacetime is discrete on the so-called Planck scale, which corresponds to a timescale of the order of 10^{-44} s. This timescale is far, far smaller than the temporal scales discussed by Finkelstein.

1.5 Plan of this book

The thirty chapters in this book fall naturally into six parts. These parts are successively more technical and abstract.

Part I, *Discrete time concepts*, consists of Chapters 1 to 7 and discusses DT concepts in general. We discuss the physics of time. Of importance here are the differences between the mathematics of the continuum and that of discrete sets. In particular, the relationship between infinitesimal calculus and its discrete analogue is discussed. We include a discussion of cellular automata, since these demonstrate in a clear way the flow of information throughout a discrete spacetime.

Part II, *Classical discrete time mechanics*, consists of Chapters 8 to 15 and begins a systematic formulation of the principles underpinning our approach to DT CM. Here we discuss the action sum, the Weiss action principle, virtual paths and conserved quantities.

Part III, *Discrete time quantum mechanics*, consists of Chapters 16 to 19 and discusses quantization principles relevant to DT mechanics.

Part IV, *Discrete time classical field theory*, consists of Chapters 20 to 25 and extends the quantization principles discussed in Part III to scalar, Dirac and Maxwell fields. Emphasis is placed on DT gauge invariance.

Part V, *Discrete time quantum field theory*, consists of Chapters 26 and 27 and deals with second quantization of the fields discussed in the previous part.

Part VI, *Further developments*, runs from Chapter 28 to Chapter 30. This part discusses DT concepts associated with spacetime and concludes the book.

Appendices are included in order to review some relevant topics such as coherent states, time-dependent oscillators and quantum registers.

2

The physics of discreteness

2.1 The natural occurrence of discreteness

Although the conventional experience of time gives a strong impression that time 'flows' in a continuous way, humans have always measured time in discrete units such as hours, days, months and years. In general, longer time periods require less sophisticated technology for their measurement. Stone Age observers could record the passage of days easily by the simple process of adding one more pebble to a pile of pebbles every time the Sun set. It would not take much data collection to create semi-permanent records of the passage of days, which could be analysed at leisure to reveal longer timescales such as the lunar month and the solar year. On the other hand, hours would require a theoretical division of a day into equal parts, which could in principle be recorded by counting turns of an hour-glass or burnt candles. But this requires relatively sophisticated technology and constant monitoring. One issue in antiquity was whether it was the period of daylight that was divided into twelve hours or whether it was the period from one midnight to the next midnight that was divided into twenty-four. The former definition of the hour had the unfortunate characteristic of depending on the season, being shorter in winter than during the summer.

Although mathematicians routinely deal with the continuum $\mathbb{R}$ of real numbers, the set of natural numbers $\mathbb{N} \equiv \{\text{one, two, three}, \ldots\}$ has long been regarded as special by mathematicians. Integers are important even to non-mathematicians, because counting objects such as cows in a field or coins in a purse is practical and useful.

In this book, integers will play an important role too, serving as labels for successive intervals of discrete time. If T is the value of the chronon, or smallest unit of time, then we will often refer to the time nT as n, if we know that T is independent of n.

2.2 Fourier-transform scales

In many branches of physics, experimentalists capture large amounts of data in the form of observed values $F(\tau)$ of some empirical variable F, at times parametrized in the laboratory by a single real parameter τ, which we can take to be lab time (time as measured in the laboratory). The value of F recorded at time τ will be denoted $F(\tau)$.

The architecture of an experiment is important. There are two commonly used architectures that are very different and knowing which is being used is crucial. In the first, which we shall refer to as single-run architecture, an SUO is monitored over a single long interval of lab time, which can usually be taken to be $(-\infty, +\infty)$. Discreteness comes into play at this point because data accumulated in any single run of an experiment whatsoever will always constitute a countable set $\{F(\tau_i) : i = 1, 2, 3, \dots, \tau_i < \tau_{i+1}\}$, simply because all apparatus is constructed of a finite number of atoms operating by the rules of quantum mechanics, and there is no way of registering any sort of final data in a continuous way. It is simply not true to say otherwise.

In the other architecture, a number N of repetitions or runs of a basic procedure are performed and a datum $F(\tau_i)$ is captured at the end of run i, for each $i = 1, 2, \dots, N$. The point of such an experiment is to compare the data from all the runs. To do this, the lab time τ_i at which the datum was captured is replaced by the run time $t_i \equiv \tau_i - \bar{\tau}_i$, where $\bar{\tau}_i$ is the lab time at the start of the ith run. Essentially, each run starts as if the lab time were reset to zero. This architecture will be referred to as ensemble-run architecture. A typical ensemble-experiment would be to measure the half life of an unstable particle.

Care should be taken when discussing concepts associated with ensemble-run experiments, because there may be factors precluding their validity. For example, if early-Universe dynamics is being discussed, there seems no clear empirical meaning to the notion of an ensemble of rapidly inflating universes being tested by an experimentalist. Ensemble-run experiments are generally assumed to be conducted over short timescales compared with the expansion scale of the Universe, and then they are generally in order. Surprisingly, this problem is an acute one for high-energy particle physics: the Large Hadron Collider is measurably affected by the position of the Moon on a daily basis and its disruptive effects must be corrected for during each run of the collider (Gagnon, 2012).

As a rule, some theory will be tested against the observations in the form of a comparison between the set of observations $\{F(\tau_i) : i = 1, 2, \dots, N\}$ and a theoretical prediction f. We shall concern ourselves here with the theoretical prediction, which will be a function of continuous time t.

If the independent variable coordinate t has an unbounded domain, then it is often convenient to consider a Fourier transform of the theoretical prediction with respect to that coordinate. This serves several purposes. First, derivatives of functions with respect to the original independent variable are replaced by algebraic multiplicative factors in the conjugate variable. Second, Fourier transforms collect information over the whole domain of the original coordinate and repackage it in terms of local functions of the conjugate variable, giving a novel perspective of the original function.

The theory of the Fourier transform is vast, so we give only some relevant properties. We shall give a physicist's definition, since our applications will be entirely within mathematical physics. Given a function $f(x)$ of some unbounded real parameter x, the Fourier transform $\tilde{f}(p)$ of f is defined here by

$$\tilde{f}(p) \equiv \int_{-\infty}^{\infty} dx\, e^{ipx/\hbar} f(x). \tag{2.1}$$

Here p is a real parameter and $\hbar$ is Planck's constant. This transform exists either as a regular function of p or in the sense of a distribution or generalized function (Gelfand and Shilov, 1964). Provided that integrability conditions are met, the inverse Fourier transform exists and we may write

$$f(x) = \int_{-\infty}^{\infty} \frac{dp}{2\pi\hbar} e^{-ipx/\hbar} \tilde{f}(p). \tag{2.2}$$

There are useful and powerful theorems associated with Fourier transforms. Of interest to us here is the Fourier-transform uncertainty principle.

Typically, functions f and $\tilde{f}$ that are related to physical observations will have a bell-shaped curve or some variant thereof. Therefore, we may assume that f is integrable and also square integrable. Without loss of generality we can rescale the function and assume that f is normalized to unity, which means that

$$\int_{-\infty}^{\infty} dx |f(x)|^2 = \int_{-\infty}^{\infty} \frac{dp}{2\pi\hbar} |\tilde{f}(p)|^2 = 1. \tag{2.3}$$

What will often be of interest to the experimentalist will be the approximate position of the peak of f and the width of f about this position. To characterize the data in these terms, we define the dispersions

$$D_{x_0}(f) \equiv \int_{-\infty}^{\infty} (x - x_0)^2 |f(x)|^2\, dx, \quad \tilde{D}_{p_0}(\tilde{f}) \equiv \int_{-\infty}^{\infty} \frac{dp}{2\pi\hbar} (p - p_0)^2 |\tilde{f}(p)|^2, \tag{2.4}$$

where x_0 and p_0 are arbitrary real numbers. Then it can be shown that, for normalized f (Stein and Shakarchi, 2003),

$$D_{x_0}(f) \tilde{D}_{p_0}(\tilde{f}) \geqslant \frac{\hbar^2}{4}. \tag{2.5}$$

This result translates to the Kennard–Heisenberg uncertainty principle on defining the uncertainty in position $\Delta_x(\Psi)$ relative to the wavefunction $\Psi(x)$ and the uncertainty in momentum $\tilde{\Delta}_p(\tilde{\Psi})$ by

$$\Delta_{\bar{x}}(\Psi) \equiv \sqrt{D_{\bar{x}}(\Psi)}, \quad \tilde{\Delta}_{\bar{p}}(\tilde{\Psi}) \equiv \sqrt{\tilde{D}_{\bar{p}}(\tilde{\Psi})}, \tag{2.6}$$

where $\bar{x}$ and $\bar{p}$ are defined by

$$\bar{x} \equiv \int_{-\infty}^{\infty} dx\, x|\Psi(x)|^2, \quad \bar{p} \equiv \int_{-\infty}^{\infty} \frac{dp}{2\pi\hbar} p|\tilde{\Psi}(p)|^2. \tag{2.7}$$

Then we have the famous uncertainty relation

$$\Delta_{\bar{x}}(\Psi)\tilde{\Delta}_{\bar{p}}(\tilde{\Psi}) \geqslant \frac{\hbar}{2}. \tag{2.8}$$

This result is fundamental to the measurement of time intervals on the shortest scales. We commented on Pauli's theorem in Chapter 1, but, if care is taken to define the protocols for the measurement of energy and intervals of time, then there is some validity to a time–energy uncertainty relation of the form

$$\Delta_t(\Psi)\tilde{\Delta}_E(\tilde{\Psi}) \gtrsim \frac{\hbar}{2}. \tag{2.9}$$

This should be interpreted as saying that, if we wish to measure smaller and smaller intervals of time, we will have to use greater and greater energies to do so.

Such an idea has immediate implications. Not only is it practically impossible to reach infinite energies (presumably by accelerating particles), but also there is no evidence that infinite energy exists in the Universe. Any attempt to accumulate a very large amount of energy in the form of mass would probably be undermined by GR effects such as the creation of a black hole in the apparatus used. This was a fear expressed at one time before the Large Hadron Collider started its operations. We conclude that there is no feasible way of measuring time accurately below a certain scale. It is believed by many physicists that such a scale is provided by the Planck time, discussed below, which is of the order of 10^{-44} s.

Lasers are devices that produce coherent pulses of light that can be made short enough in length to provide the shortest measurable timescales. The technology available manages to concentrate a lot of energy into narrow pulses, which translates into times on the scale of tens of attoseconds, viz., of the order of 10^{-16} s.

2.3 Atomic scales of time

The conventional definition of the second is one sixtieth of a minute, which is one sixtieth of an hour, which is one twenty-fourth of a day. As timescales became shorter and shorter and the need for accuracy increased, it was recognized by physicists that such a definition depended too much on the vagaries of the Earth's rotation. A more reliable definition based on atomic processes was found. These processes should be transferable around the solar system and beyond, since the laws of atomic physics presumably have universal validity.

In the standard international system of units (SI), the second is defined in terms of a mathematical count:

> The second is the duration of 9 192 631 770 periods of the radiation corresponding to the transition between the two hyperfine levels of the ground state of the caesium 133 atom.
>
> Bureau International des Poids et Mesures (2006)

Despite the apparently clean nature of this definition in terms of integers, which can be counted with absolute precision, there is an important caveat, viz.,

> This definition refers to a caesium atom at rest at a temperature of 0 K. This note was intended to make it clear that the definition of the SI second is based on a caesium atom unperturbed by black body radiation, that is, in an environment whose thermodynamic temperature is 0 K. The frequencies of all primary frequency standards should therefore be corrected for the shift due to ambient radiation, as stated at the meeting of the Consultative Committee for Time and Frequency in 1999.
>
> Bureau International des Poids et Mesures (2006)

This is a warning that the empirical determination of time at comparatively minute scales is fraught with technical difficulties. Suppose an error is made in such a count, over one second. The least significant is obviously plus or minus one, corresponding to a time of $1.087\ldots \times 10^{-10}$ s. This is far above the timescale expected of any chronon, or fundamental interval of time.

There is a difference between the accuracy of time measurement and the stability of clocks. A clock may operate with a measurable variability over a short period of time, but with an impressively low net loss in timekeeping over the long run. On this account manufacturers prefer to express the quality of their devices in terms of seconds lost per million years, rather than fractions of a second of uncertainty per day.

A different atomic scale of time can be found by considering the time light takes to cross a typical atom. The diameter of an atom is of the order of an *ångström unit*, or 10^{-10} m. Taking the speed of light to be close to 3×10^8 metres per second, this gives a time of the order 3×10^{-19} s. The attosecond, or 10^{-18} s, is the time light takes to cross three hydrogen atoms. A recent record for the shortest time measured, 12 *attoseconds*, or 12×10^{-18} s, was set in 2010 (Koke *et al.*, 2010).

2.4 De Broglie scales

In 1900, Planck devised the relation $E = h\nu$ in order to derive a theoretical fit to the observed black-body spectrum (Planck, 1900). Here E is the value of an energy quantum absorbed by an atomic oscillator vibrating with characteristic frequency ν and h is Planck's original constant. In 1905, Einstein went further and proposed that electromagnetic energy itself is quantized, with the energy E now being the energy of the particles of light (subsequently dubbed photons) and

ν the frequency of the light, which was assumed to be monochromatic (Einstein, 1905a).

In 1924, de Broglie proposed that the same relationship should apply in the case of massive particles, moving at sub-luminal speeds (de Broglie, 1924). Whilst the frequency of an electromagnetic wave was by then a well-established concept, de Broglie's proposal was a leap in the dark, because it mixed the discreteness of material particles with the continuity of waves in the same expression. De Broglie's idea was confirmed subsequently by Davisson and Germer in electron-diffraction experiments in 1927 (Davisson and Germer, 1927). Those experiments focussed on the wavelength of an electron wave, which is a spatial quantity. Our concern, on the other hand, is with frequency, which is a temporal quantity. A frequency ν is naturally associated with a time period T given by $T = 1/\nu$.

By 1924, the theory of SR was well established and well known to de Broglie. His ideas were based on SR, not Newtonian mechanics (de Broglie, 1924). According to SR, the energy $E_{\mathbf{p}}$ of a massive particle of rest mass m moving with three-momentum $\mathbf{p}$ relative to some inertial frame $\mathcal{F}$ is given by $E_{\mathbf{p}} = c\sqrt{\mathbf{p}\cdot\mathbf{p} + m^2c^2}$. Assuming that there is a monochromatic wave of frequency $\nu_{\mathbf{p}}$ satisfying the relation $E_{\mathbf{p}} = h\nu_{\mathbf{p}}$, we find $\nu_{\mathbf{p}} = ch^{-1}\sqrt{\mathbf{p}\cdot\mathbf{p} + m^2c^2}$. The characteristic period $T_{\mathbf{p}} \equiv 1/\nu_{\mathbf{p}}$ is therefore

$$T_{\mathbf{p}} = \frac{h}{c\sqrt{\mathbf{p}\cdot\mathbf{p} + m^2c^2}}. \tag{2.10}$$

In the rest frame of the particle, where $\mathbf{p} = \mathbf{0}$, this time is given by $T_0 = hm^{-1}c^{-2}$, which we shall call the de Broglie time. For an electron, we find $T_0 = 8.0935 \times 10^{-21}$ s, which is significantly smaller than the attosecond scale of 10^{-18} s encountered above.

Expressing $T_{\mathbf{p}}$ in terms of the speed v of the particle gives

$$T_{\mathbf{p}} = \frac{h}{mc^2}\sqrt{1 - \frac{v^2}{c^2}}. \tag{2.11}$$

We conclude from this that the de Broglie timescale is an extrinsic one, i.e., decreases as the speed of the particle increases relative to the observer. This suggests that a de Broglie wave should be interpreted not as an intrinsic property of an SUO but as a manifestation of the processes of observation.

To see this more clearly, consider the phenomenon of Fitzgerald contraction in SR. Classically, an SUO with a natural rest length scale L_0 will appear to have a length $L = L_0\sqrt{1 - v^2/c^2}$ when moving with speed v relative to some inertial frame. If now an observer at rest in that inertial frame attempted to analyse the shape of that SUO, using light for example, the wavelengths needed to retain accuracy would need to be correspondingly sharper. Our analysis of the Fourier transform in an earlier section supports this conclusion.

Now a reduction in the wavelength λ of observation means an increase in frequency ν, because monochromatic waves of light satisfy the relation $\lambda\nu = c$.

Therefore, any equipment used by an observer to measure smaller distances has to operate over shorter timescales, hence our result.

2.5 Hadronic scales

Most particles studied by high-energy physicists are unstable, with lifetimes ranging from just over 14 minutes for the free neutron to about 10^{-24} s in the case of the shortest-lived hadronic resonances. These resonances and their lifetimes are inferred from analyses of high-energy scattering cross-sections.

To see how this relates to heuristic expectations, consider a hadronic interaction involving a proton. The charge radius of the proton has been measured to be about 0.88×10^{-15} metres. Given that the speed of light c is about 3×10^8 metres per second, it takes light about 10^{-24} seconds to cross a distance comparable to the diameter of a proton. The heuristic argument here is that any short-lived hadronic resonance occurs as a result of structural changes in a proton or other baryon, and these cannot occur faster than such systems could stabilize over their structures. The fastest this could occur would be at the speed of light operating over their spatial extent.

Hadronic resonances are detected by peaks in differential cross-sections, and these have characteristic widths, denoted by Γ in units of energy. The lifetime τ of such a resonance is related to the width by $\Gamma\tau = \hbar$. For example, the Z boson, a spin-1 field involved in electroweak theory, has a reported full width of about 2.4 GeV. This gives a lifetime of about 2.5×10^{-25} seconds, which is one of the shortest timescales encountered in the laboratory, albeit indirectly.

2.6 Grand unified scales

For a long time, physicists believed that there were four independent interactions: electromagnetic, weak, strong and gravitational forces, each with very different individual characteristics. But, as high-energy physicists gradually extended the energy scales of their particle-scattering experiments, their data led them to suspect that, at certain energy scales, some or all of these forces would merge into a single, unified interaction. The great paradigm for the unification of disparate forces that stimulated the quest for unification was electromagnetism, the union of electricity and magnetism, which is commonly attributed to James Clerk Maxwell.

Eventually, an energy scale at which the electromagnetic and weak interactions were indeed successfully modelled by a single theory known as the electroweak theory was reached. Currently, physicists are exploring the unification of the electroweak interaction and the strong interaction. The relevant scale at which these hitherto disparate interactions become equal in strength is called the GUT (grand unified theory) scale. The current estimate of the energy of this scale is $E_{\text{GUT}} \approx 10^{16}$ GeV.

Experimentalists are very far from this scale currently: the Large Hadron Collider operates on a scale of about 10^4 GeV, i.e., a thousand billion times less than the GUT scale.

When translated into temporal terms by the formula $T_{\mathrm{GUT}} \approx \hbar/E_{\mathrm{GUT}}$, we find a timescale T_{GUT} of the order of 10^{-40} seconds. It is believed that in the very early Universe, on timescales comparable to this, the strong, weak and electromagnetic interactions had not separated out, so not even nuclei existed. Therefore, observers as we know them could not have existed then and so the interpretation of phenomena in this epoch requires some thought.

2.7 Planck scales

The Planck scales are currently believed to mark the absolute extremes of physics. On these scales all interactions are subsumed into one, which remains the focus of much speculation. Using the standard constants of physics, i.e., c the speed of light, $\hbar$ the reduced Planck constant and G the Newtonian constant of gravitation, we can use dimensional analysis to construct the Planck time T_{P}, the Planck length L_{P} and the Planck mass energy E_{P} in terms of c, G and $\hbar$. We find

$$\begin{aligned} T_{\mathrm{P}} &= \sqrt{\frac{G\hbar}{c^5}} \approx 5.4 \times 10^{-44} \text{ s}, \\ L_{\mathrm{P}} &= \sqrt{\frac{G\hbar}{c^3}} \approx 1.6 \times 10^{-35} \text{ m}, \\ E_{\mathrm{P}} &= \sqrt{\frac{c^5\hbar}{G}} \approx 2.0 \times 10^{9} \text{ J} = 1.3 \times 10^{19} \text{ GeV}. \end{aligned} \tag{2.12}$$

In such expressions, the absence of any numerical factors tells us that these scales are not derived from any theory. They are speculative, subject to debate and have been criticized as metaphysical (Meschini, 2006). Nevertheless, they serve as a signpost to the sort of timescale a DT theory might be based on. Certainly, any proposed chronon on the hadronic scale of 10^{-25} seconds stands a good change of being far too big to be compatible with early-Universe cosmology as it is currently understood.

3
The road to calculus

In this chapter we discuss the infinitesimal calculus and its discrete analogues.

3.1 The origins of calculus

Before the work of Newton and Leibniz established differential and integral calculus, the term calculus referred to any general body of mathematics. Afterwards, it became reserved almost exclusively for what should more properly be called the infinitesimal calculus, or the mathematics of infinitesimals. Infinitesimals are non-zero numbers that are smaller in magnitude than any finite number, the latter being numbers that can be assigned as magnitudes of physically measurable quantities.

Although all mathematical concepts are abstractions, infinitesimals are usually regarded as somehow 'more abstract' than finite numbers. The reason for this anomaly is that, whilst we use ordinary integers such as one, two, etc. for counting and fractions such as a half, one third, etc. for dividing up objects such as cakes, we do not encounter situations calling for explicit use of infinitesimals. In those circumstances where we encounter objects that appear very small on ordinary scales, such as atoms in chemistry and angles of optical resolution in astronomy, we use devices such as microscopes and telescopes to magnify them, so that they appear to be of finite size.

Infinitesimals are perhaps best regarded in terms of processes; that is, their value lies in what is done with them rather than with their individual qualities. In this respect, an infinitesimal is like the concept of a limit or a derivative. Indeed, calculus developed when, in separate countries, independently and almost simultaneously, Newton and Leibniz made the leap from thinking of infinitesimals in terms of very small numbers to thinking of them as a succession of ratios of smaller and smaller quantities. In other words, calculus emerged once a certain way of thinking about limiting processes had developed.

3.1.1 Limits

A fundamental concept that had to be grappled with before differential calculus could be rigorously established was that of a *limit*. The limit concept is somewhat counter-intuitive, being a discussion of what happens in the neighbourhood of a point in a continuum, **not** of what happens at that point.

The conventional modern definition of the limit of a real-valued function f of a single real variable x as x approaches some chosen value a is generally attributed to Weierstrass and goes as follows.

Definition 3.1 l is the limit of f as x tends to a if, given any positive number ε, there can be found a positive number δ_ε (which can depend on the choice ε) such that, for any x in the domain of f for which $|x - a| < \delta_\varepsilon$, we can be sure that $|f(x) - l| < \varepsilon$.

The conventional wisdom is that the above so-called 'epsilon–delta' approach to calculus that was developed in the nineteenth century overthrew the confusion of earlier workers about limits. However, there are arguments to support the view that the subtle difference between infinitesimals and finite numbers was appreciated by Newton and Leibniz (Blaszczyk *et al.*, 2013). For instance, in Section 1 of Book 1 of the *Principia*, Newton writes

> *And similarly the ultimate ratio of vanishing quantities is to be understood not as the ratio of quantities before they vanish or after they have vanished, but the ratio with which they vanish.*
>
> (Newton, 1687)

An in-depth analysis of the *Principia* has led to the suggestion that Newton was not thinking in terms of time discretized into small but still finite intervals when he discussed limits, but in terms of *infinitesimal* intervals, each of the same magnitude *dt* (Cohen, 1999). In other words, the suggestion is that Newton was thinking in terms of DT (discrete time) mechanics based on an infinitesimal notion of a chronon.

In recent years, mathematicians have attempted to revive the reputation of infinitesimals in a rigorous formalism known as non-standard analysis (NSA), which we shall discuss presently.

3.1.2 Zeno's paradoxes

Zeno's paradoxes have come down to us as evidence that the concept of the very small preoccupied thinkers in antiquity. Following his master Parmenides' thoughts about time and change, Zeno discussed more than forty situations involving time, space and motion where bizarre or paradoxical conclusions were reached via common-sense arguments. Of these paradoxes, nine are known in detail to modern scholars. Some of these nine are clearly flawed, in that the scenarios being discussed are treated in a limited, incomplete, or even clearly fallacious way. Others, however, retain the interest of modern thinkers because they touch upon the concepts of limits and continuity. We discuss two of the more famous paradoxes now.

The arrow

In the arrow paradox, Zeno argues that motion is impossible. His argument was paraphrased by Aristotle in the *Physica* (Aristotle, 1930):

> *If everything when it occupies an equal space is at rest, and if that which is in locomotion is always occupying such a space at any moment, the flying arrow is therefore motionless.*
>
> Aristotle, *Physics*, Book VI, Chapter 9

Aristotle disagreed with Zeno that there was a paradox, but his reasoning seems specious. Familiarity with Newtonian mechanics, which is based on second-order differential equations, points to an obvious problem with Zeno's thinking: instantaneous position alone is not a complete specification of the dynamical state of an object. What is missing is an additional statement about the object's instantaneous velocity.

In the arrow paradox (also known as the fletcher's paradox), Zeno states that, for motion to occur, an object must change the position which it occupies. He gives an example of an arrow in flight. He states that in any one (durationless) instant of time, the arrow is neither moving to where it is, nor to where it is not. It cannot move to where it is not, because no time elapses for it to move there; it cannot move to where it is, because it is already there. In other words, at every instant of time there is no motion occurring. If everything is motionless at every instant, and time is entirely composed of instants, then motion is impossible.

We shall touch upon this line of argument when we discretize time. The quantity of interest is the so-called system function, which depends on the instantaneous position $\mathbf{q}_n$ at DT nT and at the position $\mathbf{q}_{n+1}$ at the next instant of DT $(n+1)T$, where T is the chronon.

Achilles and the tortoise

In this paradox, intuitive arguments are used to 'prove' that the fast runner Achilles could never overtake the much slower tortoise in a race where the tortoise starts ahead of Achilles. The argument depends on a discretization of time into variable intervals and goes as follows.

Assume for simplicity that Achilles (A) can run twice as fast as the tortoise (T) and that A starts at position $x = 0$ whilst T starts at position $x = 1$. Suppose A can cover one unit (a metre) of x in one second and both A and T run in the positive-x direction at uniform speed. Let $x_{\mathrm{A}}(t)$ and $x_{\mathrm{T}}(t)$ be the positions of A and T, respectively, at time t.

At time zero, $x_{\mathrm{A}}(0) = 0$ and $x_{\mathrm{T}}(0) = 1$.

At the end of the first interval Δt_1, which has duration 1 second, A has reached $x_{\mathrm{A}}(1) = 1$, but now T has reached $x_{\mathrm{T}}(1) = 1 + 1/2 = 3/2$. Therefore, at the end of one second, T is ahead of A by half a metre.

At the end of the second interval Δt_2, which now has duration 1/2 second, A has reached $x_{\mathrm{A}}(1 + 1/2) = 3/2$, which is where T was at the end of the first interval, but T has been moving as well and has now reached position $x_{\mathrm{T}}(t_2) = 3/2 + (1/2 \times 1/2) = 7/4$. Hence A has not caught up with T by the end of the second interval of time.

The argument is continued indefinitely, with ever decreasing temporal intervals. By the end of the nth interval, the total time elapsed is $t_n = 1 + 1/2 + 1/4 + \cdots + 1/2^n = 2 - 1/2^n$. Therefore, at this time A is at position $x_{\mathrm{A}}(t_n) = 2 - 1/2^n$ whilst T is at position $x_{\mathrm{T}}(t_n) = 1 + t_n/2 = 2 - 1/2^{n-1}$, which is greater than $x_{\mathrm{A}}(t_n)$ for any finite value of positive n. Therefore, A has not caught up with T by the end of nth interval. Since the argument is valid for any positive integer n, Zeno's conclusion is that A *never* catches up with T.

This intuitive argument is appealing to the mathematically unsophisticated but is incorrect. Its resolution is elementary once the concept of a limit is understood. A sum of infinitely many positive numbers can converge to a finite number, as is the case here. The infinite sum $1 + 1/2 + 1/4 + \cdots$ converges to the limit 2, which is the time at which A would catch up with T.

We confirm this calculation with the following synthetic argument. Let t_{f} be the final time at which A overtakes T. Since their positions coincide at that time, we must have $x_{\mathrm{A}}(t_{\mathrm{f}}) = x_{\mathrm{T}}(t_{\mathrm{f}})$. But $x_{\mathrm{A}}(t_{\mathrm{f}}) = t_{\mathrm{f}}$ and $x_{\mathrm{T}}(t_{\mathrm{f}}) = 1 + t_{\mathrm{f}}/2$. Hence we must solve the equation $t_{\mathrm{f}} = 1 + t_{\mathrm{f}}/2$, which has the solution $t_{\mathrm{f}} = 2$ as stated.

3.1.3 Adequality

The term *adequality* refers to an approximate quantity or an equality occurring when terms proportional to squares of infinitesimals are neglected. It was used by the mathematician Fermat to find the local maxima of simple functions and for related problems, such as finding tangents to curves.

To illustrate the method, consider Fermat's example of how to find the local maximum of the function $f(x) = ax + bx^2$, where a and b are constants. To find a value x_0 of x where a local maximum or minimum of f may occur, we equate the value $f(x_0)$ of the function at x_0 to the value $f(x_0 + \varepsilon)$ of the function at the point $x_0 + \varepsilon$, where ε is infinitesimal. Hence

$$ax_0 + bx_0^2 = a(x_0 + \varepsilon) + b(x_0 + \varepsilon)^2. \tag{3.1}$$

Expanding on the right-hand side and setting terms of order ε^2 to zero gives the equation

$$0 = (a + 2bx_0)\varepsilon. \tag{3.2}$$

Infinitesimals are non-zero, so the only consistent conclusion is that $x_0 = -a/(2b)$. This is the result we would expect from standard calculus, which gives x_0 as a solution of the equation $f'(x) = 0$.

We can understand this method from the perspective of standard calculus. Suppose that, for an arbitrary differentiable function f, we make a Taylor expansion about a local stationary point x_0:

$$f(x_0 + \varepsilon) = f(x_0) + \frac{df}{dx}(x_0)\varepsilon + O(\varepsilon^2). \tag{3.3}$$

Then, upon equating the two sides of the equality and neglecting terms $O(\varepsilon^2)$, we arrive at the equation $df/dx(x_0) = 0$ for stationary points. This is equivalent to the adequality method.

The adequality method is related to the so-called *transcendental law of homogeneity*, a heuristic principle proposed by Leibniz, namely that, in any sum involving infinitesimals of different powers, only the lowest-power term should be retained and the remainder can be discarded. For instance, if a is a finite number and dx is an infinitesimal, then $a + dx = a$. Likewise, for finite a, b and infinitesimal dx, dy then $(a + dx)(b + dy) - ab = a\ dy + b\ dx$.

One of the distinguishing features of our approach to DT mechanics is that we do not neglect quadratic and higher powers of the chronon T in the equations of motion. In other words, our chronon is not regarded as infinitesimal. The significance of this is that it implies that the chronon might be measurable empirically.

3.1.4 Related principles

In the run-up to the development of the infinitesimal calculus, certain heuristic principles either involving infinitesimals or bypassing them were recognized and applied to various mathematical problems. *Cavalieri's principle*, also known as *the method of indivisibles*, asserts that two objects have equal volume if the areas of their corresponding cross-sections are in all cases equal. An application of this idea to plane geometry is the rule that triangles of equal base and equal height have the same areas.

The ancient Greeks developed *the method of exhaustion*, another technique for the calculation of various results, which did not rely on infinitesimals but did imply some form of limit process. For example, by approximating a circle by a polygon of n sides and taking the limit $n \to \infty$, Euclid was able to prove that the area of a circle is proportional to the square of its radius. Of course, as n gets larger, the length of each side of the approximating polygon tends to zero. However, for any finite n this length is still finite and therefore is not an infinitesimal. We can glimpse here the conceptual chasm between the concept of an infinitesimal, which is asserted to have a mathematical existence, and the process of a limit, where a leap of faith is made between finite quantities and the value of a limit.

3.2 The infinitesimal calculus and its variants

Long before DT mechanics was developed, mathematicians had been interested in alternatives to the infinitesimal calculus. One such alternative, an influential approach known as q-calculus, was developed principally by the Reverend F. H. Jackson (Jackson, 1910) in the first decade of the twentieth century, and on that account q-derivatives and q-integrals are often referred to as Jackson derivatives and Jackson integrals, respectively.

The novelties of q-calculus were rediscovered by mainstream mathematical physicists in the last quarter of the twentieth century and the subject remains at the centre of much current research into deformed mechanics both in its classical and in its quantum varieties. One reason for the popularity of q-calculus is that many theorists believe that our current views about space and time are too classical and that a 'deformed' theory of spacetime holds some promise. The subject is still under active development and may yet turn out to be of fundamental importance to science.

The inspiration for deformed mechanics comes from the history of QM (quantum mechanics). Until technology became good enough to detect departures from classical expectations, QM remained hidden from physicists. Such departures tend to be unobservable on conventional scales of measurement. For example, the typical quantum unit of angular momentum is Planck's constant h, which in SI units is $6.626\ldots \times 10^{-34}$ J s, an entirely negligible quantity in classical terms. Another example is that the classically minute size of atoms is determined by the value of Planck's constant.

As QM developed, the idea that QM could be regarded as a deformation, or slightly altered version, of CM (classical mechanics) emerged, the difference between the two being parametrized by Planck's constant. If this idea has any validity, then CM predictions should emerge in the limit $h \to 0$, provided that other quantities are scaled up in an appropriate fashion. For example, a particle of intrinsic spin j can be in a quantum state with z-component of angular momentum $m\hbar$, where $\hbar$ is the reduced Planck constant $h/(2\pi)$ and m can take values in the set $\{-j, -j+1, \ldots, j-1, j\}$. A classical picture of a particle with a non-zero, finite z-component of angular momentum can be obtained in the limit $h \to 0$ provided that j becomes large at the same time, such that the product mh remains finite in this limit. This idea was incorporated into Bohr's correspondence principle (Bohr, 1920).

We shall discuss q-calculus in the last section of this chapter, reviewing the essentials of Jackson derivatives and then Jackson integrals in preparation for our discussion in Chapter 14 of their generalization by Klimek, which may be used to give one particular approach to DT mechanics. In anticipation of the various similarities between continuous and DT formulations, we review here briefly two different approaches to integration.

3.2.1 The Riemann–Stieltjes integral

The Riemann–Stieltjes definition of an integral comes close to the sort of discrete sum we shall encounter later on in this book, the fundamental difference being that limits are taken in the case of the integral that are not available to us in the case of the sum.

In the following we shall use the notation $\mathbb{D}_n F_n \equiv F_{n+1} - F_n$ for any function or variable indexed by integer n.

Let f and g be bounded real-valued functions over some closed finite interval $I_{ab} \equiv [a, b]$ of the real line $\mathbb{R}$, with $a < b$. The aim is to construct a definition of the integral of the integrand f with respect to the *integrator* g. This is achieved via a limiting process.

First construct a partition $P_N \equiv \{t_0, t_1, \ldots, t_N\}$ of I_{ab}, a set of points x_i such that $a \equiv x_0 < x_1 < \ldots < x_{N-1} < x_N \equiv b$. The values $x_0, x_1, \ldots, x_N$ will be called nodes and the intervals $[x_i, x_{i+1}]$ will be called the links. The differences $\mathbb{D}_n x_n$ need not be all the same. The mesh or norm $|P_N|$ of the partition P_N is $\max\{\mathbb{D}_n x_n : 0 \leqslant n < N-1\}$.

Next, pick some point y_i in the ith link, i.e., y_i is in the interval $[x_i, x_{i+1}]$. Denote the set of y_i values by Y_{P_N}. Then define the well-defined finite sum

$$S_{P_N, Y_{P_N}}(f, g) \equiv \sum_{n=0}^{N-1} f(y_n) \mathbb{D}_n g(x_n). \tag{3.4}$$

The Riemann–Stieltjes integral A of f with respect to g over the interval I_{ab} exists if there is a number A such that, for any choice of number $\varepsilon > 0$, there exists a number $\delta > 0$ such that, for every partition P_N such that $|P_N| < \delta$ and for any possible Y_P,

$$|S_{P, Y_P}(f, g) - A| < \varepsilon. \tag{3.5}$$

We shall encounter expressions just like (3.4) in our development of DT mechanics. The big difference is that we shall not go to any limit.

Note that, since $|P_N| \equiv \max\{\mathbb{D}_n x_n : 0 \leqslant n < N-1\}$, the above definition of the Riemann–Stieltjes integral requires us to consider the limit of partitions such that $\lim_{N\to\infty} |P_N| \to 0$, i.e., ever finer and finer partitions of the fixed interval $[a, b]$.

3.2.2 The Lebesgue integral

The Lebesgue approach to integration was developed to deal with functions that could not be integrated in the Riemann–Stieltjes approach (Rudin, 1964). The method is to turn the problem around so that the summation, which is what an integral is in essence, is organized in a more useful way.

A useful analogy, attributed to Lebesgue, is with counting how much money there is in a bag of coins of different values, such as dollars, pounds and euros. The standard approach would be to take out one coin at a time and add its value to a running total. The final coin would add its value to the penultimate running total, giving the final monetary total.

The Lebesgue approach to integration reorganizes the adding up of the total monetary value differently. The bag is first emptied and the coins sorted into piles, each pile containing coins with the same face value: dollars go to the dollar pile, pounds to the pound pile and so on. Next, the number of coins in each pile is counted, giving a size or measure of that pile. Note that this measure has nothing to do with the individual coin values in any pile. After that, the monetary value of each pile is calculated by multiplying the measure of that pile (i.e., the number of coins in it) by the monetary value of any one coin in that pile. The result is a subtotal monetary value for that pile. Finally, all the monetary subtotals are added up to give the grand monetary total. The result should be the same as that calculated by the first method.

For a broad range of functions, the Lebesgue integral gives the same answer as the Riemann integral or the Riemann–Stieltjes integral and is given the notational representation

$$\int_a^b f(t)\mu(t), \tag{3.6}$$

where $\mu(t)$ is known as the (Lebesgue) measure.

The notation here serves three purposes. First, the $\int_a^b$ part tells us this is an integral over the interval $[a, b]$. Lebesgue integration can be extended, however, to cover much more general sets than intervals of real numbers, something which other forms of integration would not do easily. Next, the $f(t)$ tells us that we are integrating a specific function f of a real variable, denoted by t, over the interval $[a, b]$. Finally, and this is the crucial part, the $\mu(t)$ tells us what measurement value we should place on any given subset of the interval of integration.

For certain classes of function, the Riemann and Riemann–Stieltjes integrals may fail to give an answer whereas the Lebesgue integral may give an answer. An example of such a function is the following: let f be the real-valued function over the real interval $[0, 1]$ such that $f(x) = 0$ for x rational and $f(x) = 1$ for x irrational. What is the integral of f over the given interval, assuming it exists?

Any standard approach based on partitioning the interval and calculating upper and lower sums will fail, because the upper and lower bounds will always turn out to be one and zero, respectively, for this particular function.

On the other hand, the bag analogy given above comes in useful here. The 'bag' is now the interval $[0, 1]$ and there are only two coin values: zero and one. On sorting out all the zero-value coins, we see that they occur on the pile of rationals in that interval, whilst the value-one coins occur on the pile of irrationals. Hence the integral according to Lebesgue's strategy is given by

$$\begin{aligned}\int_0^1 f(t)\mu(t) &= 0 \times \{\text{measure of set of rationals over } [0, 1]\} \\ &\quad + 1 \times \{\text{measure of irrationals over } [0, 1]\}.\end{aligned} \tag{3.7}$$

We state without proof that the set of rationals in $[0,1]$ has measure zero whilst the set of irrationals in $[0,1]$ has measure one. Hence we conclude

$$\int_0^1 f(t)\mu(t) = (0 \times 0) + (1 \times 1) = 1. \tag{3.8}$$

Two comments are in order: (i) infinitesimals do not appear to have played a role in this form of integration; and (ii) we have discussed Lebesgue integration because the concept of measure will be invoked when we come to discuss the question of q-integrals (or Jackson integrals) in the last section of this chapter, and its generalization, phi-integrals, in Chapter 14.

In the next section, we go to the other extreme and discuss a formalism that takes infinitesimals very seriously.

3.3 Non-standard analysis

We mentioned earlier that Newton and Leibniz had an intuitive notion of infinitesimals, but the development of epsilon–delta methods in defining limits, continuity and differentiability appeared to kill off that line of thought.

However, the situation has changed remarkably over the last few decades. Mathematicians have developed rigorous extensions to the real numbers, the so-called hyperreals, which incorporate the notion of infinitesimals in a rigorous way. The resultant theory is called non-standard analysis (NSA) (Robinson, 1966). The essential feature of NSA is that every ordinary or finite real number x has associated with it a set of hyperreals, which are regarded as infinitely close to x. An infinitesimal is then a hyperreal that is infinitesimally close to the real number zero. Typically, we denote such a hyperreal by $x' \equiv x + dx$, where x' is a hyperreal, x is a finite real and dx is to all intents and purposes an infinitesimal hyperreal.

The calculus of hyperreals bears striking similarity to Fermat's adequality methods, an excellent account of the application of NSA to differential and integral calculus being given by Keisler (2012).

At this time, we have not investigated the construction of a viable DT mechanics based on NSA. In that context, if the chronon were regarded as an infinitesimal in the sense of NSA and not just a very small time interval, as discussed in Chapter 2, the advantage would be that there would be no question of finding its value. By definition, infinitesimals have no finite scale.

It is in a sense gratifying that the intellectual giants Newton and Leibniz have been vindicated by NSA in their intuition about infinitesimals. It has to be said, however, that there has been fierce criticism of NSA from constructivists (Bishop, 1977), on what appear to be subjective grounds. It is our intuition that NSA may yet turn out to have a lot to offer to DT mechanics, providing a model for the chronon that does not require a scale. We saw in the previous chapter how much the size of the chronon changed in the minds of theorists as higher and higher energy scales were encountered by physicists. The chronon viewed as an infinitesimal avoids the problem of scale altogether. The infinitesimals and their reciprocals may yet provide the link between the safe finiteness of observation and the dangerous infinities of the continuum.

3.4 q-Calculus

q-Calculus is a modification of standard infinitesimal calculus that, rather like NSA, may be thought of as a half-way house between the continuous and the discrete, but without the infinitesimals. A parameter q is introduced, which takes the independent real variable t from value t_0 to a new value, qt_0.This allows us to form the equivalent of the differential quotient in standard infinitesimal calculus, but without the taking of the limit $qt_0 \to t_0$. First we discuss the q-derivative, the q-analogue of the standard derivative, and then we discuss q-integration.

3.4.1 The q-derivative

In standard infinitesimal calculus, the derivative df/dt of a function f of a real variable t, evaluated at $t = t_0$, is defined in terms of a limit applied to the differential quotient, viz.,

$$\frac{d}{dt}f(t_0) \equiv \lim_{h \to 0} \frac{f(t_0+h) - f(t_0)}{(t_0+h) - t_0}, \tag{3.9}$$

assuming this limit exists. The q-derivative (also known as the Jackson derivative) evaluated at $t = t_0$ is defined as the quotient

$$\frac{d_q}{d_q t}f(t_0) \equiv \frac{f(qt_0) - f(t_0)}{qt_0 - t_0}, \quad q \neq 1. \tag{3.10}$$

This q-quotient is perfectly well defined for all non-zero values of t_0 and all values of q not equal to unity. The above q-derivative is often written $D_q f(t)$, which we shall use.

At this point we need to sound a note of caution. Suppose we are given a real- or complex-valued function f of a real variable t over a finite interval $[a, b]$, with $a < b$. We shall assume that f has no singularities in this interval. Now consider the q-derivative defined as in (3.10). We are strictly allowed to construct it in the interval $[a, b]$ only for values of t such that $a \leqslant t \leqslant b$ (of course) and such that $a \leqslant qt \leqslant b$. If either or both of these conditions does not hold then we would be attempting to evaluate a function outside of its domain of definition, which is not permitted.

This issue has an impact on the notion of q-integration. We shall comment on this again presently.

The relationship between the q-derivative and the standard derivative df/dt becomes clear if we introduce a new parameter T in place of q, such that

$$q = 1 + \frac{T}{t}, \quad t \neq 0,\ T \neq 0. \tag{3.11}$$

This gives

$$D_q f(t) \equiv \frac{f(t+T) - f(t)}{T}, \tag{3.12}$$

from which we deduce

$$\lim_{q \to 1} D_q f(t) = \lim_{T \to 0} \frac{f(t+T) - f(t)}{T} = \frac{df}{dt}(t). \tag{3.13}$$

The q-derivative has the following properties.

1. **Linearity**. For any suitable functions f and g, and constants a and b, we have

$$D_q\{af(t) + bg(t)\} = a\ D_q f(t) + b\ D_q g(t). \tag{3.14}$$

2. **Product rule**. There are two equivalent forms:

$$\begin{aligned} D_q\{f(t)g(t)\} &= g(t)D_q f(t) + f(qt)D_q g(t) \\ &= g(qt)D_q f(t) + f(t)D_q g(t). \end{aligned} \tag{3.15}$$

3.4.2 The q-integral

In standard calculus, integration is frequently discussed as the process of reversing differentiation. Given a function f of t, a primitive, or antiderivative, for f is defined as any function F of t such that

$$\frac{d}{dt}F(t) = f(t). \tag{3.16}$$

A common notation for the antiderivative is

$$F(t) = \int f(t)dt. \tag{3.17}$$

A standard theorem in calculus states that two antiderivatives of the same function differ at most by a constant.

An analogous theory can be constructed for q-calculus. A function $F(t)$ is a q-antiderivative of $f(t)$ if

$$D_q F(t) = f(t). \tag{3.18}$$

By analogy with (3.17) we write

$$F(t) = \int f(t)d_q t, \tag{3.19}$$

where $d_q t$ refers to a 'q-measure', left unspecified in practice.

We can readily find a formal solution to the problem of finding a q-antiderivative as follows. First we introduce a *q-scaling* operator M_q, which has the property

$$M_q f(t) = f(qt). \tag{3.20}$$

Then (3.18) can be written in the form

$$(1 - M_q)F(t) = (1 - q)tf(t), \tag{3.21}$$

which has the formal solution

$$\begin{aligned} F(t) &= (1 - M_q)^{-1}\{(1-q)tf(t)\} \\ &= \sum_{n=0}^{\infty} M_q^n \{(1-q)tf(t)\} \\ &= \sum_{n=0}^{\infty} \{(1-q)(q^n t)f(q^n t)\}. \end{aligned} \tag{3.22}$$

Hence a formal definition of the q-antiderivative, or Jackson integral, is

$$F(t) = (1-q)t\sum_{n=0}^{\infty} q^n f(q^n t). \tag{3.23}$$

The sum does not always converge, even if an antiderivative exists. With convergence in mind, the uniqueness of the q-antiderivative is addressed by the following theorem (Oney, 2007).

Theorem 3.2 *If $0 < q < 1$, then, up to an additive constant, any function $f(t)$ has at most one q-antiderivative that is continuous at $t = 0$.*

3.4.3 The definite Jackson integral

In standard calculus, integration is more than inverse differentiation. In particular, the concept of a definite integral has great applicability. In q-calculus, the same remark applies, leading to the definition of the definite q-integral, also referred to as the definite Jackson integral: For $0 < b < \infty$ we define the definite q-integral to be given by

$$\int_0^b f(t)d_q t \equiv (1-q)b\sum_{j=0}^{\infty} q^j f(q^j b), \tag{3.24}$$

assuming the series converges.

This opens the door to a definition of the q-integral over the interval $[a, b]$, where $0 < a < b$:

$$\int_a^b f(t)d_q t \equiv \int_0^b f(t)d_q t - \int_0^a f(t)d_q t \tag{3.25}$$

$$= (1-q)\sum_{j=0}^{\infty} q^j \left[f(q^j b)b - f(q^j a)a\right]. \tag{3.26}$$

Assuming that we can rearrange summation terms without problems, we readily find the q-analogue of the fundamental theorem of integration:

$$\int_a^b D_q f(t) d_q t \equiv f(b) - f(a). \tag{3.27}$$

An important remaining problem to address is the failure of (3.26) to converge in the limit $b \to \infty$. A formal resolution is given by Oney (2007). For $q < 1$ the improper integral on $[0, \infty)$ is redefined by

$$\int_0^{\infty} f(t) d_q t \equiv \sum_{j=-\infty}^{\infty} \int_{q^{j+1}}^{q^j} f(t) d_q t, \tag{3.28}$$

which leads to the formal definition, for $q \neq 1$ (Oney, 2007),

$$\int_0^{\infty} f(t) d_q t \equiv |1 - q| \sum_{j=-\infty}^{\infty} q^j f(q^j). \tag{3.29}$$

3.4.4 Discussion

We return to the cautionary remark we made above about the domain of definition of the q-derivative. Suppose f is defined only over the interval $[0, b]$ and we are attempting to integrate $D_q f$ from 0 to b, as in (3.24). Then if $q > 1$ we would be attempting to sum over terms containing undefined expressions of the form $f(q^n b)$. On the other hand, assuming $0 < q < 1$ and $b > 0$, then each term in (3.24) is well defined.

This is not the same issue as the question of whether the sum in (3.24) exists. The point is one of the domain of definition. Therefore, we shall assume in common with most authors that $0 < q < 1$. The same issue of the domain of definition arises with the phi-integral discussed in Chapter 14.

We can make explicit contact with DT under special circumstances, as follows. The definition of the bounded q-integral (3.26) given above involves an infinite summation on the right-hand side for almost all values of a and b. Here we assume f is defined over $(0, b]$, so each term in the summation is well defined. Nevertheless, for almost all values of a such that $0 < a < b$, we still have to consider the convergence of the infinite sum in (3.26).

However, consider the special case $a = q^N b$, which we shall refer to as a *q-resonance*. Then, assuming convergence whenever required in rearranged summations, we readily find

$$\begin{aligned} \int_{q^N b}^b f(t) d_q t &\equiv (1 - q) \sum_{j=0}^{\infty} q^j \left[f(q^j b) b - f(q^{j+N} b) q^N b \right] \\ &= (1 - q) b \sum_{j=0}^{N-1} q^j f(q^j b). \end{aligned} \tag{3.30}$$

Now we define $t_n \equiv q^{N-n}b$, so $t_0 = a$ and $t_N = b$, and then we find

$$\int_{q^N b}^{b} f(t)d_q t = \sum_{n=0}^{N-1} f(t_{n+1})T_n, \tag{3.31}$$

where $T_n \equiv t_{n+1} - t_n = q^{N-n-1}(1-q)b$ is a variable 'chronon', i.e., a time-step of increasing size. We note that $T_{n+1} = q^{-1}T_n > T_n$ and

$$\sum_{n=0}^{N-1} T_n = q^{N-n-1}(1-q)b = (1-q^N)b = b-a, \tag{3.32}$$

as expected.

An important point to note is the forwards bias on the right-hand side of (3.31), by which we mean that the function f is evaluated at the forwards end of each chronon interval. This can be traced back to the definition (3.10) of the q-derivative, which by inspection is reverse biased for $q < 1$.

The position of the function in a link interval is a critical issue here and in other fields, such as stochastic integration. In the theory of stochastic processes, there are two prominent definitions of stochastic integral, known as the Itō integral and the Stratonovich integral. The Itō integral uses the process at one end of a link, the equivalent of (3.31), whilst the Stratonovich integral takes the average of the stochastic processes at the ends of a link.

4
Temporal discretization

4.1 Why discretize time?

There are several reasons why we might want to discretize time, the following being a list of those that readily come to mind.

1. **We might believe as a matter of principle that time really is discrete, rather than continuous or part of a continuum**, as will be discussed in Section 28.1.

   Our motivation here might be a belief in some deeper and more comprehensive view of space and time, such as Snyder's quantized spacetime algebra (Snyder, 1947a, 1947b).
2. **We want just to explore the advantages and disadvantages of an alternative model of time, with no prejudice one way or another.**

   This is the point of view that we find most appealing in this list. There is no proof that time is continuous or discrete. It is conceivable that DT (discrete time) mechanics might lead to a testable prediction that could not be formulated in CT (continuous time).
3. **We might be sampling some quantity at regular or irregular intervals of time.**

   This is in fact how all experiments are done, because matter is discrete, not continuous. All information is acquired discretely and there really are no truly continuous processes of information acquisition, if we believe in quanta and the atomic theory of matter.
4. **We want to approximate a continuous function as a discrete function in order to make a numerical estimate of some integral or the solution to a differential equation.**

   This was the motivation for the DT approach to quantum mechanics taken by Bender and collaborators (Bender *et al.*, 1985a, 1985b, 1993). Their work influenced us when we started our numerical simulations of soliton scattering.

5. **We are writing a computer game or performing a computer simulation.**

   In such situations we have no choice but to discretize equations of motion, because all computers process information in discrete ticks of their internal processor clocks. This was the scenario which triggered our own interest in developing exact principles of DT mechanics. During our computer simulations of scattering collisions of non-topological solitons (Jaroszkiewicz, 1994b), we encountered a problem that is frequently observed in such situations: naive temporal discretization was generating numerical approximations to nominally exactly conserved quantities such as energy that were *not* being conserved over a given run of the computer simulation. This led us to look for systematic methods of constructing dynamical invariants that would be precisely conserved modulo the discretized equations of motion, rather than merely approximated to. Inevitably, this led to a search for the principles underlying DT mechanics as an exact form of mechanics, rather than thinking of DT mechanics as no more than an approximation to CT mechanics.

4.2 Notation

In this section we introduce the operator notation used throughout this book. This notation is motivated by, and to some extent legitimized by, Taylor's theorem in real analysis. This theorem can be regarded as a way of discussing discrete changes in a continuous parameter.

We restrict our attention here to real-valued functions of a real variable x. Taylor's theorem deals with a subset of such functions known as real analytic functions and states the following. Consider a real-valued smooth function f of a real variable x with the value of the function at x denoted by $f(x)$. Since the function is smooth, we can also determine its nth derivative $f^{(n)}(x)$ at x, where n is any positive integer. Assuming that the function f has a Taylor-series expansion about x and that h is of lesser magnitude than some critical number known as the radius of convergence of that Taylor series at x, then we can write

$$f(x+h) = f(x) + hf'(a) + \cdots = \exp\left\{h\frac{d}{dx}\right\} f(x), \tag{4.1}$$

the exponential in (4.1) being defined in terms of its standard power-series expansion.

We can improve the notation by defining the displacement operator $\overrightarrow{U}_h$:

$$\overrightarrow{U}_{x,h} \equiv \exp\left\{h\frac{d}{dx}\right\}, \tag{4.2}$$

and then (4.1) can be written as

$$f(x+h) = \overrightarrow{U}_{x,h} f(x). \tag{4.3}$$

There is no need to restrict the displacement h to positive values, the only criterion being that we must not attempt to displace outside the radius of convergence.

The set of real-valued functions over an interval is a commutative ring, which means that, given any two such functions f and g and any two real numbers λ and μ, then $\lambda f + \mu g$ and λfg are also real-valued functions over that interval. Then, provided that Taylor's theorem applies to each function, we find that the displacement operator is linear relative to linear combinations of functions, i.e.,

$$\overrightarrow{U}_{x,h}\{\lambda f + \mu g\}(x) = \lambda f(x+h) + \mu g(x+h). \tag{4.4}$$

On the other hand, it is non-linear, or invasive, relative to products of functions, i.e.,

$$\overrightarrow{U}_{x,h}\{f.g\}(x) = f(x+h).g(x+h), \tag{4.5}$$

and to compositions of functions, i.e.,

$$\overrightarrow{U}_{x,h}(f \circ g)(x) = f \circ g(x+h). \tag{4.6}$$

Example 4.1

$$\overrightarrow{U}_{x,h}(h)\cos(\sin\ e^{x} + x^2) = \cos(\sin\ e^{x+h} + (x+h)^2). \tag{4.7}$$

We shall incorporate the linearity and invasiveness properties of the displacement operator in the DT situation. In general, we shall use open-face symbols such as $\mathbb{U}_n$ for DT operators. Consider a finite sequence $\{f_n : M \leq n \leq N\}$ of values of some dynamical variable f, where, in this context, the integer n plays the role of a discrete parameter index. By analogy with the continuum case discussed just above, we introduce the **discrete displacement operator** $\mathbb{U}_n$ acting on the symbol n such that for any expression f_n involving n we define

$$\mathbb{U}_n f_n = f_{n+1}, \tag{4.8}$$

valid for $M \leq n \leq N-1$. It will be inconvenient to keep restating the domain of validity of this operator so we shall henceforth assume that the above operation is well defined in practice, i.e., is never applied outside of its domain of applicability $[M, N-1]$ in the index n.

As with the continuous displacement operator $\overrightarrow{U}_h$, the discrete displacement operator $\mathbb{U}_n$ acts invasively on arbitrarily complicated functions of n, as the following examples demonstrate.

Example 4.2

$$\begin{aligned}
\mathbb{U}_n\{n^2\} &= (n+1)^2,\\
\mathbb{U}_n\left\{\frac{y_n}{x_n}\right\} &= \frac{y_{n+1}}{x_{n+1}},\\
\mathbb{U}_n\{n^n\} &= (n+1)^{n+1},\\
\mathbb{U}_n\{\cos^n(\sin n)\} &= \cos^{n+1}(\sin(n+1)),
\end{aligned} \tag{4.9}$$

and so on.

Two successive applications $\mathbb{U}_n\{\mathbb{U}_n f_n\}$ of the displacement operator are denoted by $\mathbb{U}_n^2$. This readily extends to any positive integer p, so p applications of $\mathbb{U}_n$ are denoted by $\mathbb{U}_n^p$, again assuming that we are within the domain of applicability of the combined operator. Given this, we readily see that for positive or negative integers p, q we have the properties

$$\begin{aligned} &\text{(i)} \qquad \mathbb{U}_n^p f_n = f_{n+p}, \\ &\text{(ii)} \qquad \mathbb{U}_n^p \mathbb{U}_n^q = U_n^{p+q}. \end{aligned} \tag{4.10}$$

In this notation, the symbol n is more than just an integer. It has the status of an *indexed name* or a *symbol* in computer algebra software packages such as Maple. An explicit or evaluated integer is unaffected by the temporal displacement operator, so that for example $\mathbb{U}_n 3 = 3$.

A symbol that is used in the above way will be called an *active* symbol or index. If we have two or more active symbols, such as m and n, then the action of their respective displacement operators leaves any other active symbol alone, i.e.,

$$\begin{aligned} \mathbb{U}_n f_n &= f_{n+1}, & \mathbb{U}_m g_m &= g_{m+1}, \\ \mathbb{U}_n g_m &= g_m, & \mathbb{U}_m f_n &= f_n. \end{aligned} \tag{4.11}$$

As with the differential operator $\overrightarrow{U}_{x,h}$, we shall generally discuss the action of $\mathbb{U}_n$ on n without any explicit reference to existence, provided that we are not evaluating the result at a given value of n, in which case we must check whether we are in the domain of validity of the notation. For instance, it is legitimate to write

$$\mathbb{U}_n \left\{ \frac{1}{n} \right\} = \frac{1}{n+1} \tag{4.12}$$

as a general formula, but it would not be correct to then evaluate the result at $n = -1$.

Throughout this book we shall represent various operators acting on a DT index with open-face symbols, such as the displacement operator $\mathbb{U}_n$. A number of useful operators of this kind can be constructed using $\mathbb{U}_n$ as follows:

(1) the **inverse displacement operator** $\bar{\mathbb{U}}_n \equiv \mathbb{U}_n^{-1}$, defined by

$$\bar{\mathbb{U}}_n f_n \equiv f_{n-1}; \tag{4.13}$$

(2) the **forwards difference operator** $\mathbb{D}_n$, defined by

$$\mathbb{D}_n \equiv \mathbb{U}_n - 1; \tag{4.14}$$

(3) the **backwards difference operator** $\bar{\mathbb{D}}_n$, defined by

$$\bar{\mathbb{D}}_n \equiv 1 - \bar{\mathbb{U}}_n; \tag{4.15}$$

(4) the **forwards averaging operator** $\mathbb{A}_n$, defined by

$$\mathbb{A}_n \equiv \frac{1}{2}\{\mathbb{U}_n + 1\}; \tag{4.16}$$

(5) the **backwards averaging operator** $\bar{\mathbb{A}}_n$, defined by

$$\bar{\mathbb{A}}_n \equiv \frac{1}{2}\{1 + \bar{\mathbb{U}}_n\}; \tag{4.17}$$

(6) the **non-locality operator** $\mathbb{S}_n$, defined by

$$\mathbb{S}_n \equiv \frac{\mathbb{U}_n + 4 + \bar{\mathbb{U}}_n}{6}; \tag{4.18}$$

(7) the **forwards discrete fluxion (time derivative) operator** $\mathbb{T}_n$, defined by

$$\mathbb{T}_n \equiv T_n^{-1}\,\mathbb{D}_n; \tag{4.19}$$

where T_n is the time-step $\mathbb{D}_n t_n \equiv t_{n+1} - t_n$; and

(8) the **backwards discrete fluxion operator** $\bar{\mathbb{T}}_n$, defined by

$$\bar{\mathbb{T}}_n \equiv T_{n-1}^{-1}\,\bar{\mathbb{D}}_n. \tag{4.20}$$

Because every one of the above constructions is defined in terms of the single operator $\mathbb{U}_n$, they all commute.

4.3 Some useful results

The notation introduced in the previous section can be used to obtain some useful results.

4.3.1 The summation theorem

Given a differentiable real-valued function f of some real variable u, the second fundamental theorem of calculus tells us that

$$\int_a^b \frac{d}{du} f(u)du = f(b) - f(a). \tag{4.21}$$

The summation theorem is the discrete analogue of this result.

Theorem 4.3 *Let $\{f_n : M \leq n \leq N\}$ be a real-valued finite sequence, with $M < N$. Then*

$$\sum_{n=M}^{N-1} \mathbb{D}_n f_n = f_N - f_M. \tag{4.22}$$

4.3.2 Summation by parts

Given two differentiable real-valued functions f and g of some real variable u, standard integration by parts gives us the result

$$\int_a^b \frac{d}{du} f(u)g(u)du = f(b)g(b) - f(a)g(b) - \int_a^b f(u)\frac{d}{du}g(u)du. \tag{4.23}$$

The discrete analogue of this result is the following theorem.

Theorem 4.4

$$\sum_{n=M}^{N-1} \mathbb{D}_n f_n.g_n = f_N.g_{N-1} - f_M.g_M - \sum_{n=M+1}^{N-1} f_n.\bar{\mathbb{D}}_n g_n. \tag{4.24}$$

When $g_n = 1$ for all n, we recover Theorem 4.3.

4.3.3 Mean-value summation

We may use the above theorems to sum some important classical sums. The method is based on the mean-value theorem in standard real analysis. This theorem asserts that, if $f : [a,b] \to \mathbb{R}$ is a continuous function on the closed interval $[a,b]$ and differentiable on the open interval (a,b), where $a < b$, then there exists a c in (a,b) such that

$$f'(c) = \frac{f(b) - f(a)}{b - a}.$$

The method starts by dividing the interval $[a,b]$ into a partition $\{a \equiv x_M < x_{M+1} < \ldots < x_{N-1} < x_N \equiv b\}$ and applying the above theorem to each subinterval $[x_n, x_{n+1}]$. The theorem tells us that we can find a value c_n such that $x_n < c_n < x_{n+1}$ and

$$f'(c_n) = \frac{f(x_{n+1}) - f(x_n)}{x_{n+1} - x_n} = \frac{\mathbb{D}_n f(x_n)}{\mathbb{D}_n x_n}. \tag{4.25}$$

Hence summing over the partition and applying theorem (4.3) we find

$$\sum_{n=M}^{N-1} \mathbb{D}_n x_n.f'(c_n) = f(x_N) - f(x_M). \tag{4.26}$$

This leads to the method of summation, as the following examples show.

Example 4.5 Take $f(x) = x$. Then $f'(x) = 1$. Next, take $x_n = n$ and $M = 0$, which from (4.26) gives

$$\sum_{n=0}^{N-1} 1.(n+1-n) = N - 0, \tag{4.27}$$

or

$$\sum_{n=0}^{N-1} 1 = N, \tag{4.28}$$

as expected.

Example 4.6 Take $f(x) = x^2$. Then $f'(x) = 2x$. Taking $x_n = n$, we find

$$2c_n = \frac{(n+1)^2 - n^2}{n+1-n} = 2n + 1. \tag{4.29}$$

Hence

$$\sum_{n=0}^{N-1} (2n+1) = N^2, \tag{4.30}$$

or

$$\sum_{n=0}^{N-1} 2n + \sum_{n=0}^{N-1} 1 = N^2, \tag{4.31}$$

which, using the previous result, gives the standard result

$$\sum_{n=0}^{N-1} n = \frac{N(N-1)}{2}. \tag{4.32}$$

Example 4.7 Taking $f(x) = x^3$ and the same partition as before, we find

$$f'(c_n) = 3n^2 + 3n + 1. \tag{4.33}$$

Using the previous two examples, we find the result

$$\sum_{n=0}^{N-1} n^2 = \frac{N(N-1)(2N-1)}{6}. \tag{4.34}$$

4.4 Discrete analogues of some generalized functions

Generalized functions such as the Heaviside theta Θ and the Dirac delta δ were used heuristically by physicists long before they were given a rigorous basis by mathematicians. Mathematicians initially viewed these concepts with suspicion, on account of the non-rigorous ways physicists used to deal with them. They are not smooth functions in the usual sense: the Heaviside theta models discontinuous (on/off) switches whilst the delta models point-like densities. Such generalized functions have applicability in many branches of physics and engineering. Eventually, rigorous mathematical theories of generalized functions, or distributions, as they are also called, were developed (Gelfand and Shilov, 1964). In one real dimension x, these generalized functions have the properties

$$\Theta(x) = \begin{cases} 0, & x < 0, \\ 1, & x > 0, \end{cases} \tag{4.35}$$

$$\int_{-\varepsilon}^{\varepsilon} \delta(x)dx = 1, \quad \forall \varepsilon > 0.$$

The 'values' $\Theta(0)$ and $\delta(0)$ are usually undefined, but Fourier-series analysis typically returns a value $\Theta(0) \mapsto \frac{1}{2}$, which is the average of the left and right limits of Θ as x tends to zero. In our DT analogue Θ_n for the Heaviside theta, we shall choose $\Theta_0 = 0$.

The above generalized functions are not differentiable everywhere in the conventional sense because they are not continuous everywhere. However, an important formal relationship between Θ and δ is

$$\frac{d}{dx}\Theta(x) = \delta(x). \tag{4.36}$$

This relationship is meaningful in the right context, which is inside an integral performed on any element of a class of smooth functions known as *test functions* (Streater and Wightman, 1964).

We do not encounter any such issues when we discuss the DT analogues of $\Theta(x)$ and $\delta(x)$, denoted by Θ_n and δ_n, respectively. We define, for integer n,

$$\Theta_n = \begin{cases} 0, & n \leqslant 0, \\ 1, & n > 0, \end{cases} \tag{4.37}$$

$$\delta_n = \begin{cases} 0, & n \neq 0, \\ 1, & n = 0. \end{cases}$$

From these definitions we can read off a number of useful results:

$$\begin{aligned} 1 &= \Theta_n + \delta_n + \Theta_{-n}, \\ \Theta_{n+1} &= \Theta_n + \delta_n, \\ \Theta_{n-1} &= \Theta_n - \delta_{n-1}, \\ \Theta_{1-n} &= \Theta_{-n} + \delta_n, \\ \Theta_{-1-n} &= \Theta_{-n} - \delta_{n+1}. \end{aligned} \tag{4.38}$$

We deduce

$$\mathbb{D}_n\Theta_n = \delta_n \quad \text{and} \quad \mathbb{D}_n\Theta_{-n} = -\delta_{n+1}, \tag{4.39}$$

the discrete analogues of the formal property (4.36).

4.5 Discrete first derivatives

In the following, we assume that n, M and N are integers, with $M < N$.

Given a discrete set of independent parameter values $\{x_n : M \leq n \leq N\}$, with an associated set of dependent variable values $\{y_n : M \leq n \leq N\}$, the question of the meaning, if any, of the concept of a derivative of the y dependent variables with respect to the parameter values x arises.

To answer this question, consider the construction of a first derivative in standard calculus. Given a function $f(t)$ of a continuous parameter t, the first step is to construct the differential quotient

$$\Phi(f; t_0, h) \equiv \frac{f(t_0 + h) - f(t_0)}{h} \tag{4.40}$$

for $h \neq 0$ and f well defined over some interval containing t_0 and $t_0 \pm h$. In standard calculus we then take the limit h tends to zero for fixed t_0. If the differential quotient tends to a finite limit then we call that limit the derivative $\dot{f}(t_0)$ of the function f at t_0.

To explore possible discretizations of the first derivative, we consider Taylor's theorem for a real-valued function f on some interval $[a, b]$ (Rudin, 1964). If the second derivative $\ddot{f}$ is continuous on $[a, b]$ and the third derivative exists on (a, b), then for $a \leq t_0 - T < t_0 < t_0 + T \leq b$ we may write

$$f(t_0 + T) = f(t_0) + \dot{f}(t_0)T + \frac{1}{2}\ddot{f}(t_0)T^2 + O_1(T^3), \tag{4.41}$$

$$f(t_0 - T) = f(t_0) - \dot{f}(t_0)T + \frac{1}{2}\ddot{f}(t_0)T^2 + O_2(T^3), \tag{4.42}$$

in standard error notation.

There are three useful approximations for the first derivative that can be derived from (4.41) and (4.42).

1. The forwards derivative, obtained by rearranging (4.41):

$$\dot{f}(t_0) = \frac{f(t_0 + T) - f(t_0)}{T} + O_3(T). \tag{4.43}$$

2. The backwards derivative, obtained by rearranging (4.42):

$$\dot{f}(t_0) = \frac{f(t_0) - f(t_0 - T)}{T} + O_4(T). \tag{4.44}$$

3. The symmetric derivative, obtained by subtracting (4.42) from (4.41):

$$\dot{f}(t_0) = \frac{f(t_0 + T) - f(t_0 - T)}{2T} + O_5(T^2). \tag{4.45}$$

An obvious approach to temporal discretization is simply to ignore the error term in the forwards derivative, i.e., make the replacement

$$\dot{f}(t_0) \to \frac{f(t_0 + T) - f(t_0)}{T} \tag{4.46}$$

everywhere in the CT Lagrange formalism. So, for example, we construct the discretized 'Lagrangian' L_{D} from the CT Lagrangian $L(\dot{x}, x, t)$ by defining

$$L_{\mathrm{D}}(x_{n+1}, x_n, n) \equiv L\left(\frac{x_{n+1} - x_n}{T}, x_n, nT\right), \tag{4.47}$$

where $x_n \equiv nT$.

This approach has some problems.

1. The use of the forwards time derivative in place of $\dot{x}$ introduces a bias, which may mean that the equations of motion are not time-reversal-invariant in some sense, whereas Lagrangian-based equations of motion are generally of forms that are invariant with respect to time reversal.

2. Quantities such as energy, which may be conserved in CT, might not be formally conserved in DT.
3. Such naive temporal discretizations are inadequate for dealing with gauge covariant derivatives.

Higher-order discrete derivatives are constructed along the same lines as the above.

4.6 Difference equations

Difference equations are the DT analogues of differential equations in conventional mechanics. There are certain concepts in the study of differential equations that carry over into difference equations with some amendments, which we need to discuss.

We shall assume in this section that we are discussing a system characterized by a single dependent degree of freedom denoted by x_n, where n is an integer representing the independent DT parameter n labelling successive states of the system. The time parameter is assumed to start at integer M and end at integer $N > M$. The CT analogue of such a system is characterized by a differentiable function $x(t)$ of a real variable t, which is assumed to run over some interval.

In each form of mechanics, the objective is to determine the final state of the system, given the difference or differential equation relevant to the mechanics used, from known initial conditions.

Differential equations are generally classified by a number of properties, some of which make the equations insolvable, whilst others allow complete solutions. An important property of any differential equation is its *order*, which is the highest derivative in that differential equation. Typically, physicists deal with first-order and second-order differential equations. The order of a differential equation determines how many initial conditions are required in order to find a complete solution.

The analogue of order occurs in difference equations, but there are some pitfalls. Depending on how we have discretized a differential equation, the order of the difference equation may appear to be different from the order of the differential equation from which that difference equation was derived. The following worked example demonstrates some important points relevant to later chapters.

Example 4.8 We will investigate the discretization of the first-order differential equation

$$\frac{dx(t)}{dt} = x(t), \qquad x(0) = 1, \tag{4.48}$$

defined over the interval $[0, 1]$, the objective being to use the equation and the initial condition to determine the value of $x(1)$.

This differential equation has the exact solution $x(t) = e^t$, so we know that $x(1) = e$ is the solution in the continuum case.

Our task is to replace (4.48) with a discretized version. We shall discretize the left-hand side only, leaving the right-hand side alone. We shall use a regular temporal spacing, defining $t_n \equiv nT$, where n is an integer and T is a positive time step referred to as the chronon, and defining $x_n \equiv x(t_n)$.

The value of the chronon is not fixed but a choice. We shall choose T to be the reciprocal of some positive integer N, so that $NT = 1$. This means that we are dividing the CT interval $[0, 1]$ over which our differential equation is defined into a lattice of $N + 1$ equally spaced points. At the end of the day we shall be interested in taking the continuum limit $N \to \infty$ in order to compare with the continuum solution.

Following on from our discussion about the discretization of derivatives in the previous section we will investigate three possibilities.

Forwards differencing. On replacing the time derivative in (4.48) by the forwards difference, we arrive at the difference equation

$$\mathbb{D}_n x_n = T x_n, \qquad x_0 = 1, \tag{4.49}$$

for $0 \leq n < N$. This can be rearranged in the form

$$x_{n+1} = (1 + T)x_n = (1 + T)^{n+1} x_0, \tag{4.50}$$

which immediately leads to the formal solution

$$x_n = \left(1 + \frac{1}{N}\right)^N. \tag{4.51}$$

In the limit $N \to \infty$, we have

$$x_\infty \equiv \lim_{N\to\infty} \left(1 + \frac{1}{N}\right)^N = e, \tag{4.52}$$

as expected.

Backwards differencing. Now consider the difference equation

$$\bar{\mathbb{D}}_n x_n = T x_n, \qquad 0 < n \leqslant N, \qquad x_0 = 1. \tag{4.53}$$

We find the formal solution

$$x_n = \frac{1}{(1 - T)^n}, \qquad T \neq 1. \tag{4.54}$$

Hence, in the continuum limit,

$$x_\infty \equiv \lim_{N\to\infty} \frac{1}{\left(1 - \dfrac{1}{N}\right)^N} = \frac{1}{e^{-1}} = e, \tag{4.55}$$

as before.

Symmetric differencing. The previous discretizations have a natural bias, either towards the future in the case of forwards differencing or back to the past, in the case of backwards differencing. A third alternative is to symmetrize the differencing. Consider the discretization scheme

$$\bar{\mathbb{A}}_n \mathbb{D}_n x_n = T x_n, \tag{4.56}$$

which is equivalent to

$$x_{n+1} - x_{n-1} = 2T x_n. \tag{4.57}$$

As a check, we write

$$x_n = x(t_n), \qquad t_n \equiv nT, \tag{4.58}$$

and then, assuming that Taylor expansion is valid in powers of the chronon T, we have

$$\begin{aligned} x_{n+1} &\approx x_n + T\dot{x}_n + \frac{1}{2}T^2\ddot{x}_n + O(T^3), \\ x_{n-1} &\approx x_n - T\dot{x}_n + \frac{1}{2}T^2\ddot{x}_n + O(T^3). \end{aligned} \tag{4.59}$$

Inserting these expressions into (4.57) and taking the continuum limit gives consistency with (4.48).

To solve (4.57) we assume a power-series ansatz: $x_n = C^n$, where C is to be determined. This leads to the quadratic equation

$$C^2 - 2TC - 1 = 0, \tag{4.60}$$

which has two solutions, given by

$$C_1 = T + \sqrt{T^2 + 1}, \quad C_2 = T - \sqrt{T^2 + 1}. \tag{4.61}$$

We see here the appearance of a second solution, C_2, which has unusual behaviour. Equation (4.57) is a *second-order* difference equation and, like a second-order differential equation, has two solutions. Then, because the difference equation (4.57) is linear homogeneous, the general solution is given by the linear combination $x_n = AC_1^n + BC_2^n$, where A and B are arbitrary. In our case, we have only one piece of initial information, which is that $x_0 = 1$, so the best we can do at this point is to assert that

$$x_n = AC_1^n + (1 - A)C_2^n, \tag{4.62}$$

with A arbitrary. This solution therefore is not unique.

It is of course disconcerting that a discretization of a first-order differential equation, which has only one solution, should generate two solutions. The question is, how can we relate these solutions to the continuum case?

To answer this question and fix a sensible value of A we need to understand the nature of the second solution C_2. To this end, consider a Taylor expansion of C_1 and C_2 in powers of T:

$$C_1 \equiv T + \sqrt{T^2+1} \approx 1 + T + \frac{1}{2}T^2 - \frac{1}{8}T^4 + O(T^5),$$
$$C_2 \equiv T - \sqrt{T^2+1} \approx -1 + T - \frac{1}{2}T^2 + \frac{1}{8}T^4 + O(T^5). \quad (4.63)$$

In the large-N limit, the C_1^N solution is well behaved and tends towards the exponential limit found with the previous discretizations, i.e.,

$$\lim_{N\to\infty} C_1^N = \lim_{N\to\infty}\left(1 + \frac{1}{N} + \frac{1}{2}\frac{1}{N^2} - \frac{1}{8}\frac{1}{N^4} + O(N^{-5})\right)^N = e. \quad (4.64)$$

On the other hand, the other solution, C_2^N, does **not** have a limit as $N \to \infty$, essentially because it changes sign (more or less) every time N increases by one.

We shall refer to solutions that behave like the C_1^n solution as *ordinary* solutions, whereas we describe solutions to difference equations that oscillate with increasing n in the manner of the C_2^n solution as **oscillon** solutions.

We can remove the oscillon solution to equation (4.57) by imposing the continuity requirement that the large-N limit must exist for all acceptable solutions. Imposing this condition on (4.62) then requires us to take $A = 1$ and so the oscillon solution is discarded.

It is important to note that the C_2^n solution does **not** vanish as n tends to infinity. It oscillates without diverging or vanishing. Therefore, there is no sense in which we can imagine that the oscillons go away in a natural way.

This brings us to a matter of principle that is central to this book. **IF** we believe that time really is continuous, then equation (4.48) is our starting point and also our end point: discretization is just a method of solving a differential equation in a certain way. However, the aim of this book is not to develop approximations to CT mechanics, but to explore the notion that time really is discrete at a fundamental level. Therefore, we should in principle *start* with a difference equation and make no insistence about the large-n behaviour. The reason is quite simple: DT mechanics is richer than CT mechanics because the assumption of differentiability is quite a strong imposition on the behaviour of functions. The oscillon solution is not an *ad hoc* approximation to the difference equation (4.57) but one of the two exact solutions to it. Therefore, if the limit $T \to 0$ is not taken, we cannot avoid the oscillons.

We shall investigate oscillon-type solutions in greater detail in later chapters.

4.7 Discrete Wronskians

In the theory of differential equations, suppose that f and g are two non-zero complex-valued solutions to the same second-order differential equation. The question of whether or not these solutions are linearly independent often arises. A powerful technique for investigating this issue is to consider the consequences of the assumption that they are in fact linearly dependent over some interval $[a, b]$ of the independent variable x, which is assumed real. If this be the case, then linear dependence means that there exist non-zero complex numbers α and β such that

$$\left.\begin{aligned} \alpha f(x) + \beta g(x) &= 0 \\ \alpha f'(x) + \beta g'(x) &= 0 \end{aligned}\right\} \quad x \in [a, b]. \tag{4.65}$$

If it can be shown that either α or β is zero, or that both are, then the functions are declared to be linearly independent. The second of the conditions above follows from the first condition and the known differentiability of the solutions over the given interval. Writing (4.65) in matrix form gives

$$\begin{bmatrix} f(x) & g(x) \\ f'(x) & g'(x) \end{bmatrix} \begin{bmatrix} \alpha \\ \beta \end{bmatrix} = \begin{bmatrix} 0 \\ 0 \end{bmatrix}. \tag{4.66}$$

If the determinant $W(x) \equiv f(x)g'(x) - f'(x)g(x)$ of the matrix on the left-hand side does not vanish over the required interval then the matrix inverse exists over that interval and then $\alpha = \beta = 0$. $W(x)$ is known as the Wronskian (Arfken, 1985). The conclusion is that, if the Wronskian does not vanish over the given interval, then the functions are linearly independent. It should be noted that the vanishing of the Wronskian over an interval does not imply the linear dependence of the two functions.

Consider now the generic second-order, linear homogeneous difference equation

$$x_{n+1} + a x_n + b x_{n-1} = 0, \tag{4.67}$$

where a and b are constants. We shall encounter such equations frequently throughout this book.

Suppose that $\{x_n\}$ and $\{y_n\}$ are two solutions, and suppose we wish to investigate whether these two solutions are linearly independent. Following the Wronskian method above, suppose that these two solutions are indeed linearly dependent. This means that there exist non-zero constants α and β such that $\alpha x_n + \beta y_n = 0$ for all values of n in some discrete interval $[M, N]$. Hence we can write

$$\begin{aligned} \alpha x_n + \beta y_n &= 0, \\ \alpha x_m + \beta y_m &= 0, \end{aligned} \tag{4.68}$$

for any two different values n and m in the interval $[M, N]$. These equations can be rearranged into the matrix form

$$\begin{bmatrix} x_n & y_n \\ x_m & y_m \end{bmatrix} \begin{bmatrix} \alpha \\ \beta \end{bmatrix} = \begin{bmatrix} 0 \\ 0 \end{bmatrix}. \tag{4.69}$$

Then we conclude that $\alpha = \beta = 0$ and the solutions are linearly independent if the discrete Wronskian W_{nm}, the determinant of the 2×2 matrix on the left-hand side, is non-zero, i.e., we require

$$W_{nm} \equiv \begin{vmatrix} x_n & y_n \\ x_m & y_m \end{vmatrix} = x_n y_m - x_m y_n \neq 0. \tag{4.70}$$

As an example, consider the difference equation (4.57). Consider the two solutions

$$x_n \equiv (T + \sqrt{T^2 + 1})^n, \quad y_n \equiv (T - \sqrt{T^2 + 1})^n \tag{4.71}$$

for $n = m + k$, where $k = 1, 2, 3, \ldots$. Then $W_{nm} = 0$ gives

$$(T + \sqrt{T^2 + 1})^k - (T - \sqrt{T^2 + 1})^k = 0, \quad k = 1, 2, \ldots, \tag{4.72}$$

which cannot be satisfied for any T. Hence the solutions are linearly independent.

5

Discrete time dynamics architecture

5.1 Mappings, functions

In this book, the term architecture refers to a non-mathematical description of the sequence of events which collectively describes an experiment.

Continuous time (CT) classical mechanics (CM) is based on some explicit and implicit assumptions that generate architectures radically different from those describing quantum processes. In the following list, which is not complete by any means, most of these assumptions may appear obvious and overstated, but one or two are subtle and should be brought to light, since they have an enormous impact on the theories which are based on them. Our comments are in square brackets at the end of each item in the list.

1. At any given time, a system under observation (SUO) exists in a unique physical state with absolute properties independent of any observer [a standard classical metaphysical assumption].
2. The physical state of an SUO at any given time t may be represented by a single element $\Psi(t)$ called a state of some fixed state space U (or universe) [a reasonable mathematical assumption].
3. The state $\Psi(t')$ of an SUO at any time t' later than t is also an element of U, not necessarily the same state $\Psi(t)$ at the earlier time [a reasonable classical mathematical assumption].
4. Observers are exophysical, meaning that they stand outside of U [a nearly universal metaphysical assumption; observers are not discussed in detail in CM or QM simply because there is no comprehensive theory of what constitutes an observer].
5. Information from an SUO can be absorbed in principle by an observer with one hundred percent efficiency and without affecting any of the properties of the SUO [a reasonable mathematical assumption analogous to neglecting friction in mechanics].

6. Observers are not necessary for the existence of SUOs [a nearly universal metaphysical assumption].
7. There is a general sense of time, which affects both observers and SUOs in the same way: both age in the same direction of time as marked by the expansion of the wider universe [a reasonable physical assumption].

Figure 5.1(a) illustrates the basic architecture of CT CM, Newtonian mechanics based on Newton's three laws of motion being the archetypical implementation of this architecture. In the *Principia* (Newton, 1687), Newton outlined what he meant by the time parameter, which he called absolute time. He also referred to the physical space in which SUOs move about as absolute space.

Discrete time has one advantage over CT: DT can readily support changes in architecture that would be problematical for CT. Consider the conventional architecture shown in Figure 5.1, where a state of an SUO moves around a fixed background universe denoted by U. Continuous time evolution of a state of an SUO is represented in Figure 5.1(a). Discrete time can use either the same architecture, represented by Figure 5.1(b), or replace it by the architecture shown in Figure 5.2. In the architecture of Figure 5.2, the background universe changes at each time-step. In this architecture, the background universe U_n at time n is mapped into the next background universe U_{n+1} at time $n+1$. The dynamics of SUO states is now given by maps of elements of U_n into U_{n+1}.

Changes in the background universe U_n might be caused by the observer bringing new detectors on line or decommissioning old detectors, for example.

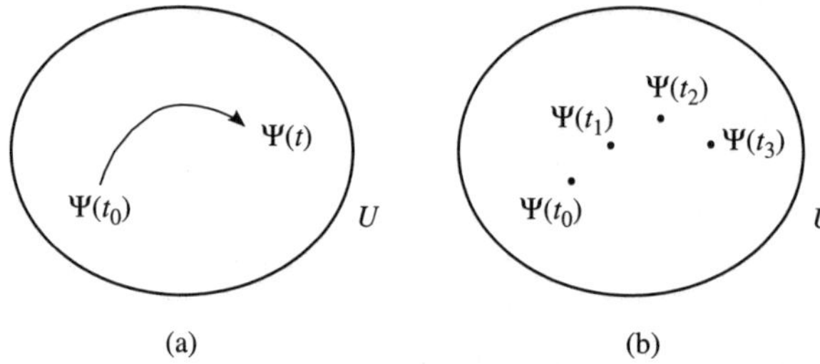

Figure 5.1. The architecture of (a) an SUO state evolving in continuous time and (b) an SUO state as it evolves in discrete time, in a fixed background universe U.

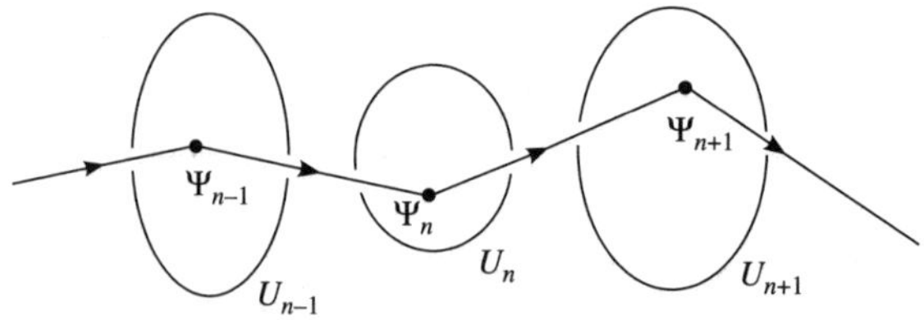

Figure 5.2. The architecture of state evolution in an evolving background universe model of dynamics.

The point is that U_{n+1} might be very different from U_n. Suppose U_n and U_{n+1} were finite vector spaces with different dimensions. Since such dimensions are integers, it is difficult to see how such a mapping of U_n to U_{n+1} could be described naturally in CT.

We shall refer to the first view of mechanics shown in Figure 5.1 as the single-universe paradigm and to the later architecture shown in Figure 5.2 as the multi-stage-universe paradigm.

Discrete time mechanics with the multi-stage architecture has been discussed by theorists such as Ikeda and Maeda (1978). These authors exploit the mathematical definition of a function to map states of an SUO from one background universe to the next. Functions play a crucial role in our formulation of DT mechanics also.

Using functions to map from one set to another permits us to discuss several varieties of DT mechanical model, each of which is characterized by how information from present and past states of an SUO is collected and processed in order to work out future states of that SUO. A given architecture for such a process will be called a dynamical architecture. Most of traditional DT CM runs on one of a relatively small number of dynamical architectures, but it is possible to invent quite bizarre forms of mechanics by devising novel architectural plans, as we shall see in the chapter on cellular automata, Chapter 7.

Functions are defined in terms of sets (Howson, 1972).

Definition 5.1 A *function* f is an ordered triple $f \equiv (F, D, R)$, where F, D and R are sets that satisfy the following rules:

(i) F is contained in the Cartesian product $D \times R$; and
(ii) for each element x in D, there is exactly **one** element y in R such that the ordered pair (x, y) is in F.

The set D is the domain of the function and R is the range of the function. Points to note are as follows.

1. None of the three sets F, D and R need be sets of numbers such as the real or the complex numbers.
2. Every element x of the domain D has a unique partner y in the range R called the image of x under the function f or the value of the function at x, denoted by $y = f(x)$.
3. The definition does not exclude the possibility that two different elements x_1 and x_2 of D have the same image in R, i.e., it is possible to have $f(x_1) = f(x_2)$ with $x_1 \neq x_2$. If this happens to be the case then the inverse function does not exist. In DT mechanics, we are invariably interested only in moving forwards in time, so the issue of whether the function mapping states of an SUO from one local universe to the next can be undone is not a physical one. So-called time-reversal experiments are in reality always carried out in the same temporal direction as that of the wider Universe beyond the laboratory.

In such experiments, time reversal reduces to the question whether successive states have properties consistent with hypothetical time-reversal (viz., those properties seen in time-reversed videos), such as reversed velocities.

4. The range R of a function is distinguished by some authors from the image set $f(D)$, which is the set of all values of the function. Other authors identify R with $f(D)$ and reserve the term co-domain for a set that contains the image set $f(D)$. It is an advantage in physical applications to make a distinction between the image set and the co-domain, since significant issues of physics may be involved. For example, in special relativity, the speed of a point particle is changed under a Lorentz boost, but, whatever the final value is, it is predicted to be bounded above by the speed of light. This has stimulated experimentalists to search for tachyons, hypothetical particles moving faster than light (Kowalczyński, 2000) and to investigate the possibility of physical information being transmitted at superluminal speeds (Scarani *et al.*, 2000).

The above definition of a function contains an explicit directionality: the function takes us *from* an element of D *to* an element in R. Figure 5.3 illustrates this ordering, which is suggestive of an arrow of time. This 'mathematical arrow of time' has the structure of the logical implication **if** P **then** Q'', which also has this sense of time. An alternative and useful perspective is to think of a function as a form of classical computation or processing of information: x is the input information, f represents the software program processing this information and $y \equiv f(x)$ is the output. Arrows indicate the direction of the mathematical arrow of time, which we also refer to as the direction of causal implication. A simplified diagrammatical representation is shown in Figure 5.4.

The link between functions, classical and quantum causality, Bayesian networks and the Arrow of Time is a rapidly developing field with a potentially vast applicability beyond that of mechanics. In our discussions we take a rather deterministic view of causality. By this we mean that if variable Y is 'caused' by variable X then there is a functional relationship of the form $Y = f(X)$. Such deterministic relationships are found even in quantum mechanics. For example, the Schrödinger equation is a fully deterministic equation: if we specify the

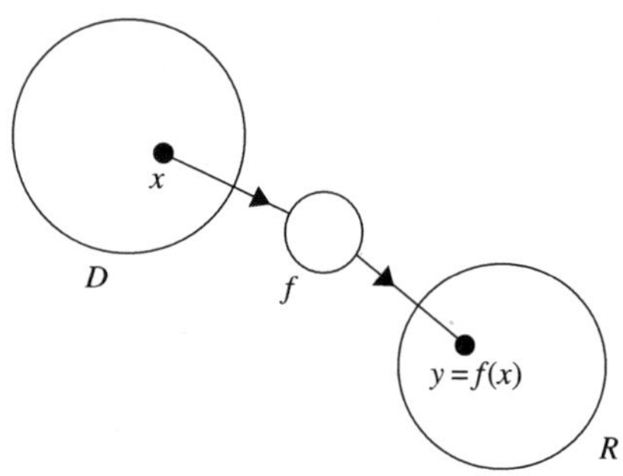

Figure 5.3. Graphical representation of a function.

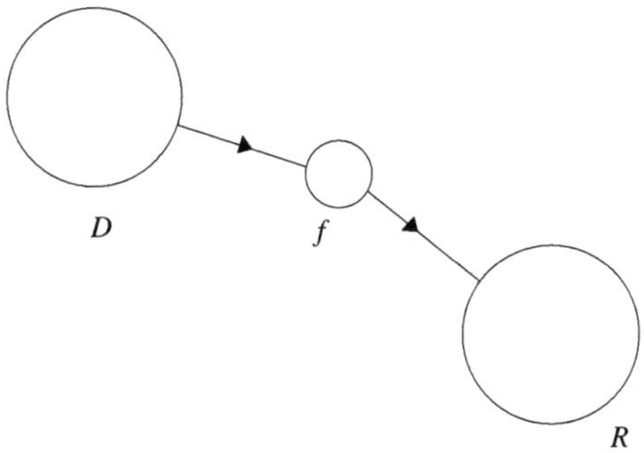

Figure 5.4. A function as a causal relationship.

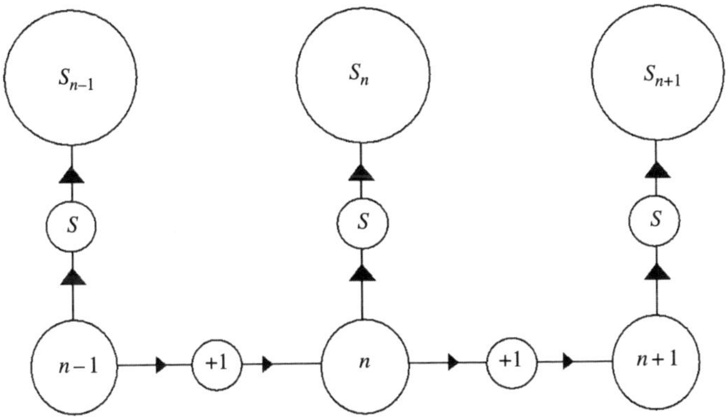

Figure 5.5. The topology of a generalized sequence.

initial wavefunction at time t_0, then in principle the wavefunction is uniquely determined for all subsequent times.

Of course, the real world contains noise, the effects of which on causal relationships can radically alter the functional dependences. However, quantum mechanics has shown in numerous scenarios that, with care, the radical effects of noise can be much reduced if not entirely eliminated. We have in mind here experiments where quantum interference is observed.

5.2 Generalized sequences

We define a generalized sequence (s_n) as a function from the natural integers $\mathbb{N}$ into some set $\mathcal{S} : n \to s_n,\ n = 0, 1, 2, \ldots$. Diagrammatically, we represent it as in Figure 5.5.

Arguably the most general discussion of DT dynamics would be in terms of generalized sequences. The elements s_n of such a sequence will be called *states*, n will be called the (*discrete*) *time* and S is the *state space*.

In this architecture, we imagine that an SUO starts in some state s_M at initial time M and then jumps from s_M to some other state s_{M+1} by time $M+1$, with s_{M+1} also in S. This process is repeated until some final time N.

In the most general scenario, acausal dynamics, successive elements $s_n, s_{n+1}, \ldots$ would not have any logical or computable connection, apart from the fact that they are related by an increasing time label. An example might be s_n as the nth digit in the decimal representation of π (it seems a reasonable conjecture to assert that there is no closed formula for s_n). Alternatively, we could take s_n to be an outcome of a random variable.

However, we live in a physical universe with causal principles (or so we believe) and time is not just a label for a sequence of events. We believe that successive states have something to do with each other. This means we believe in *causality*, the notion that some events influence other events.

Before we get down to discussing the specific nature of the states, there are two fundamentally different perspectives of time and dynamics that involve our view of reality and the Universe. If we believe that the past, present, and future 'exist' in some sense independently of how we exist or interact with the Universe, then the index n in the sequence $\{s_n\}$ represents time in its usual sense and the collection of past, present and future states lumped into one big set is called the *block universe* perspective. Classical mechanics in its usual formulation fits into this pattern. This includes the Newtonian paradigm of space and time and the relativistic paradigm of spacetime.

The alternative vision of dynamics is in terms of *process time* (Encyclopædia Britannica, 1993), which is the idea that all that 'exists' is the present, with the future being undetermined. This is a contextual view of dynamics, since we need to specify what the 'present' is.

On a final point, we may ask about where continuous time comes in. There is no problem here really. All we need to do is introduce some fundamental time unit T, and consider sufficiently large discrete times n in the limit of large n and vanishing T such that $nT \equiv t$ is fixed. Then we define

$$s(t) \equiv \lim_{\substack{n\rightarrow\infty, \\ T\rightarrow 0, \\ nT \text{ fixed}}} s_n, \qquad (5.1)$$

assuming this makes sense. This doesn't always happen in DT mechanics, as our example of oscillon solutions encountered in the previous chapter shows (Norton and Jaroszkiewicz, 1998b).

5.3 Causality

Simply viewing time and dynamics as in Figure 5.5 gives no insight into how the Universe evolves. To go further it is necessary to introduce some element of *causality*, which for our purposes is the notion that knowledge of the past and present plays a role in determining the future. The dynamical scheme in

Figure 5.5 will be called a *zeroth-order* dynamical system. In such a system only knowledge of n *and* (what is very important) the *rules* $\mathcal{R}_n$ of the dynamics is required.

5.4 Discrete time

In CT, dynamical variables $\mathbf{q} \equiv (q^1, q^2, \ldots, q^r)$ are continuous functions of time, denoted by physicists by $\mathbf{q}(t)$. In DT, $\mathbf{q}$ is a function of discrete time and written $\mathbf{q}_n \equiv \mathbf{q}(T_n)$.

First-order DT mechanics occurs when there are functional relationships between successive states of the form

$$\Phi(\mathbf{q}_n, \mathbf{q}_{n+1}, n) \underset{c}{=} 0, \quad M \leqslant n < N, \tag{5.2}$$

where M and N are the initial and final times, respectively. It is possible that the function Φ depends on n, but we ignore this possibility at this stage since it is easily incorporated into the discussion if required. In general, given $\mathbf{q}_n$, there need not be a unique solution for $\mathbf{q}_{n+1}$. Diagrammatically, we have the situation shown in Figure 5.6.

Here large bubbles represent initial and final sets whilst smaller bubbles represent the mappings or functional relationships between them. Lines without arrows (as in Figure 5.6) represent *implicit* relationships whilst lines with arrows (as in Figure 5.7) represent explicit mappings, with the direction of the arrow denoting the direction of the causal implication involved.

Example 5.2

$$\mathbf{q}_n \cdot \mathbf{q}_n - \mathbf{q}_{n+1} \cdot \mathbf{q}_{n+1} \underset{c}{=} 0. \tag{5.3}$$

In this example, there is in general no unique solution for $\mathbf{q}_{n+1}$ in terms of $\mathbf{q}_n$

If equations (5.2) can be solved for the $\mathbf{q}_{n+1}$ then knowledge of $\mathbf{q}_n$ alone is sufficient to give the dynamical evolution in the form

$$\mathbf{q}_{n+1} \underset{c}{=} \mathbf{s}(\mathbf{q}_n), \quad M \leqslant n < N, \tag{5.4}$$

where $\mathbf{s}$ is some solution function.

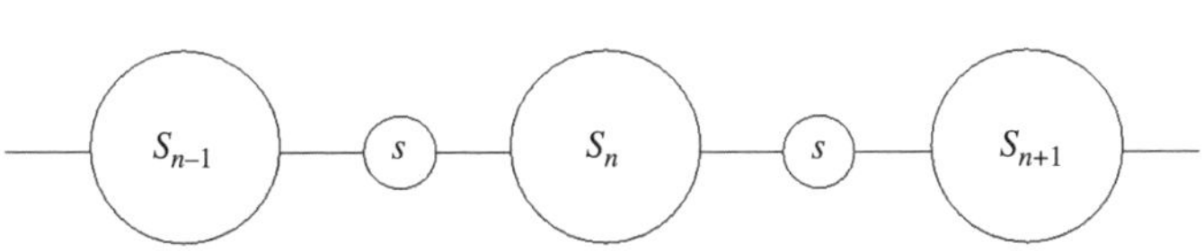

Figure 5.6. First-order discrete time mechanics architecture.

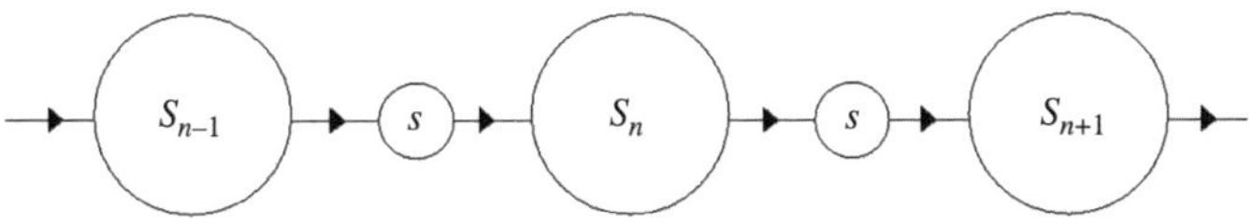

Figure 5.7. First-order causal implication.

When this occurs, we have the process of *causal implication*, shown schematically in Figure 5.7.

Second-order discrete time mechanics is when there are functional relationships between successive states of the form

$$\Phi_n(\mathbf{q}_{n-1}, \mathbf{q}_n, \mathbf{q}_{n+1}, n) \underset{c}{=} 0, \quad M < n < N, \tag{5.5}$$

where M and N are the initial and final times, respectively. Such a scheme is represented diagrammatically in Figure 5.8.

If equations (5.5) can be solved for $\mathbf{q}_{n+1}$ explicitly then we may write

$$\mathbf{q}_{n+1} \underset{c}{=} \mathbf{s}(\mathbf{q}_n, \mathbf{q}_{n-1}, n), \tag{5.6}$$

and then the diagram reduces to Figure 5.9.

Example 5.3 The discrete time harmonic oscillator. Assume each event (large circles) is a copy of $\mathbb{R}$, the reals. Suppose the system evolves such that the state at time n is some real number x_n. The discrete time oscillator is defined by the explicit equations

$$x_{n+1} \underset{c}{=} 2\eta x_n - x_{n-1}, \tag{5.7}$$

where η is real.

The discrete time harmonic oscillator equation of motion (5.7) is time-reversal-invariant, i.e., it is invariant with respect to the interchange

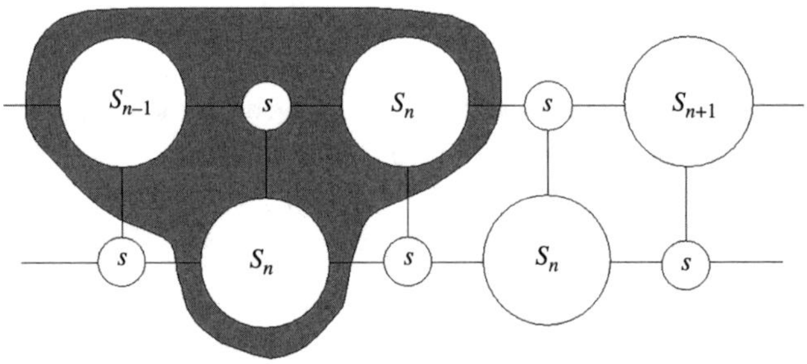

Figure 5.8. Second-order discrete time.

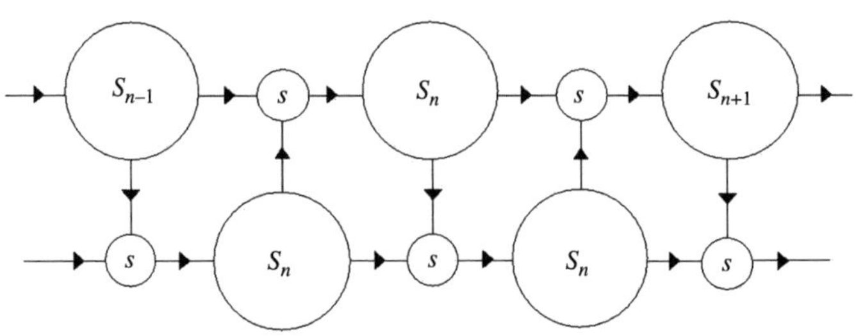

Figure 5.9. Causal implications in second-order discrete time.

$$x_{n+1} \leftrightarrow x_{n-1}. \tag{5.8}$$

In order to obtain a unique trajectory, we may specify x_M and x_{M+1}. This corresponds to *Newtonian evolution*, which generally arises from some *second-order* differential equation with boundary conditions equivalent to *initial position* and *initial velocity*.

There is an issue here concerning the topology of the second-order dynamical diagrams. Information about a quantum state can be acquired only once, but the *discrete time* oscillator discrete topology shows that information is being acquired *twice* from each event. The resolution of this lies in the recognition that, in the quantum theory of the oscillator, the causal topology of Figure 5.9 relates to the *operators* of position rather than to the states, which involve first-order equations. This is just like the Heisenberg operator equations of motion, which often obey second-order-in-time differential equations, whilst the Schrödinger equation is of first order in time.

5.5 Second-order architectures

Temporal discretization of Newton's equations of motion becomes a necessity when computer simulations are developed, such as in computer games. Consider a dynamical system in continuous time with one degree of freedom $q(t)$ subject to the equation of motion

$$\ddot{q}(t) \underset{c}{=} f(q(t)), \tag{5.9}$$

where a unit mass is assumed and f is the force on the particle. The first step is to move away from configuration space, which involves the single second-order differential equation (5.9), towards phase space, which involves position, momentum p and two first-order equations of motion:

$$\begin{aligned} \dot{q} &\underset{c}{=} p, \\ \dot{p} &\underset{c}{=} f(q). \end{aligned} \tag{5.10}$$

The next step is to divide the temporal interval over which the motion is being discussed into a number of equal segments of duration T, referred to as the chronon. At each time $t = nT$ we write $q_n \equiv q(nT)$ and $p_n \equiv p(nT)$. Finally, some step-by-step algorithm is chosen to simulate the motion in discrete time. We now discuss two discretization architectures used in such simulations. The advantages and disadvantages of each are discussed by Stern and Desbrun (2006).

5.5.1 *The explicit Euler architecture*

In this scheme, we use the forwards differencing scheme

$$\mathbb{D}_n q_n \underset{c}{=} T p_n, \quad \mathbb{D}_n p_n \underset{c}{=} T F(q_n), \tag{5.11}$$

which is equivalent to

$$q_{n+1} \underset{c}{=} q_n + T p_n, \quad p_{n+1} \underset{c}{=} p_n + T f(q_n). \tag{5.12}$$

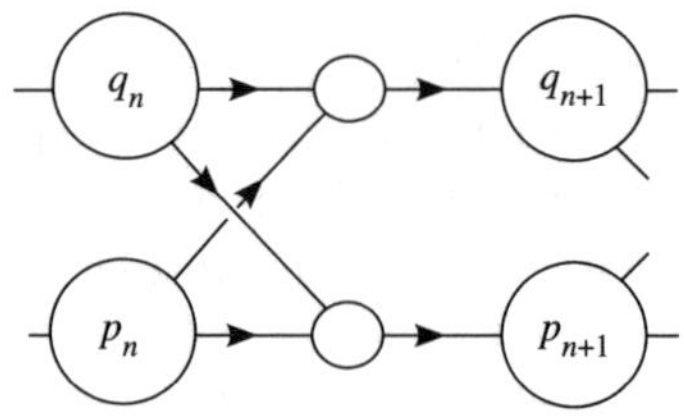

Figure 5.10. Information flow in the explicit Euler differencing scheme.

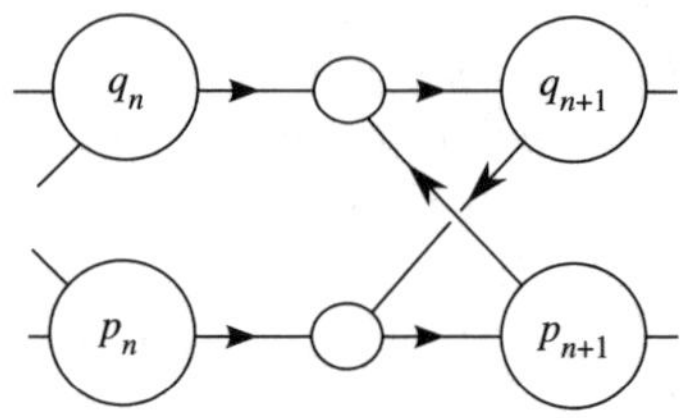

Figure 5.11. Information flow in the implicit Euler differencing scheme.

The information-flow architecture is shown in Figure 5.10, demonstrating non-planarity in the flow, but causality is explicitly maintained. Non-planarity means in this context that we cannot replace, say, q_n with q_{n+1} first and then replace p_n with p_{n+1}: we have to perform the calculations in some other register, or repository of information, and only then replace q_n and p_n at the same time, before the next time-step is taken.

5.5.2 The implicit Euler architecture

In this scheme, we use the backwards differencing scheme

$$\bar{\mathbb{D}}_n q_n \underset{c}{=} T p_n, \quad \bar{\mathbb{D}}_n p_n \underset{c}{=} T F(q_n), \tag{5.13}$$

which is equivalent to

$$q_{n+1} \underset{c}{=} q_n + T p_{n+1}, \quad p_{n+1} \underset{c}{=} p_n + T f(q_{n+1}). \tag{5.14}$$

This apparently minor modification of the explicit Euler scheme discussed above has severe consequences, as shown in Figure 5.11. Now not only is non-planarity unavoidable but also causality is clearly undermined: in order to calculate q_{n+1} we need to know p_{n+1}, but we cannot determine p_{n+1} without first knowing q_{n+1}. Therefore, the flow of information occurs in a temporal loop for this scheme, which means that it violates causality as we know it.

We should ask, given this violation of causal implication, how could we ever use the implicit Euler scheme in a practical calculation? In practice, what is done is to seek solutions consistent with the implicit relations, using some reasonable iteration scheme.

6

Some models

In this chapter we discuss some discrete time (DT) equations to illustrate various novelties, postponing the presentation of a systematic approach to DT mechanics to later chapters.

6.1 Reverse engineering solutions

In Newtonian mechanics, Newton's second law of motion,

$$m\frac{d^2}{dt^2}\mathbf{r} = \mathbf{f}, \tag{6.1}$$

for a particle moving under the action of an applied force $\mathbf{f}$ can be used in two ways. We can either integrate (6.1) for a given force and initial conditions in order to determine the trajectory $\mathbf{r}(t)$ of the particle, or we can do the opposite and determine from (6.1) what the force $\mathbf{f}$ has to be in order to 'cause' a given trajectory. We shall refer to this second usage of the force law as reverse engineering. A similar process of reverse engineering is encountered in general relativity (GR). We can use Einstein's equation (Rindler, 1969)

$$R_{\mu\nu} - \frac{1}{2}Rg_{\mu\nu} = \kappa T_{\mu\nu} \tag{6.2}$$

in two ways. On the one hand, if we are given the energy-momentum tensor $T_{\mu\nu}$ then (6.2) is a second-order partial differential equation for the metric $g_{\mu\nu}$, which can be solved in principle. An example is the energy-momentum tensor for a point particle, which leads to the Schwarzschild (black-hole) metric. On the other hand, given the line element $ds^2 = dx^\mu\, g_{\mu\nu}\, dx^\nu$ relative to some coordinate patch, we can read off the components $g_{\mu\nu}$ of the metric tensor, work out all terms on the left-hand side of (6.2) and hence deduce the components $T_{\mu\nu}$. This second approach was used by Gödel to find the energy-momentum tensor over a spacetime containing closed timelike curves (Gödel, 1949).

In this section we shall perform reverse engineering on a number of DT systems. Instead of trying to solve a difference equation, we shall use reverse engineering to derive the DT equations of motion from a given trajectory.

6.1.1 First-order linear solutions

Suppose we are given a solution to a DT equation for a one-dimensional system of the form

$$x_n = a + bn, \qquad n = 0, 1, 2, \ldots, \tag{6.3}$$

where a and b are constants. The highest power of n is one, so we shall refer to this equation as being of first order. By inspection, we see that $a = x_0$.

Replacing n by $n+1$ gives

$$x_{n+1} = a + b(n+1). \tag{6.4}$$

Subtracting (6.3) from (6.4) then gives the first-order difference equation

$$x_{n+1} = x_n + b, \tag{6.5}$$

which is our reverse-engineered equation of motion.

We check this reverse-engineered equation as follows. First, rewrite (6.5) as the difference equation

$$\mathbb{D}_n x_n = b. \tag{6.6}$$

Now apply the summation theorem (4.22), summing from integer $n = M$ to $N-1$. This gives

$$\sum_{n=M}^{N-1} \mathbb{D}_n x_n = \sum_{n=M}^{N-1} b, \tag{6.7}$$

or

$$x_N = x_M + (N - M)b. \tag{6.8}$$

Setting $N = n$, $M = 0$ then gives (6.3) as required.

6.1.2 Quadratic solutions

Suppose now that we are given a second-order trajectory of the form

$$x_n = a + bn + cn^2, \tag{6.9}$$

where a, b and c are constants. We reverse engineer it as follows, noting that $a = x_0$ as before. The procedure here is the same as for the linear case: eliminate n in favour of special values of the dynamical variable. In this case, we write

$$\begin{aligned} x_{n-1} = {} & a + b(n-1) + c(n-1)^2 = x_n - b - 2cn + c, \\ x_{n+1} = {} & a + b(n+1) + c(n+1)^2 = x_n + b + 2cn + c. \end{aligned} \tag{6.10}$$

Adding these two equations then eliminates the term linear in n, giving the required equation of motion

$$x_{n+1} = 2x_n - x_{n-1} + 2c. \tag{6.11}$$

Before we verify that solutions of this equation are of the form (6.9), we make the following observations.

1. Both (6.3) and (6.9) are well defined for negative values of the temporal index n. In fact, there is no asymmetry in principle in favour of one direction of time over the other. This is reflected in the reverse-engineered equations of motion (6.5) and (6.11). The parameter b in (6.5) is analogous to a velocity, which should change sign under time reversal, whilst the parameter c in (6.11) is analogous to an acceleration, which does not change sign under time reversal.
2. Equations (6.5) and (6.11) are equivalent to temporal discretization of the differential equations
$$\frac{d}{dt}x(t) = v, \quad \frac{d^2}{dt^2}x(t) = g,$$
respectively, where v is a constant velocity and g is a constant acceleration. These equations are discretized according to the rules
$$\frac{d}{dt}x(t) \to \frac{\mathbb{D}_n x_n}{T}, \quad \frac{d^2}{dt^2}x(t) \to \frac{\mathbb{D}_n \bar{\mathbb{D}}_n x_n}{T^2}. \tag{6.12}$$
3. The generic solution
$$x_n = a_0 + a_1 n + a_2 n^2 + \cdots + a_r n^r \tag{6.13}$$
should lead to an rth order difference equation that is a temporal discretization of the rth order differential equation
$$\frac{d^r}{dt^r}x(t) = k, \tag{6.14}$$
where k is a constant.

As a useful check, we formally solve (6.11) by first writing it in the form
$$\mathbb{D}_n \bar{\mathbb{D}}_n x_n = 2c \tag{6.15}$$
and then applying the summation theorem (4.22) from $n = M$ to $n = N - 1$:
$$\sum_{n=M}^{N-1} \mathbb{D}_n \bar{\mathbb{D}}_n x_n = \sum_{n=M}^{N-1} 2c, \tag{6.16}$$
which gives
$$\bar{\mathbb{D}}_N x_N = \bar{\mathbb{D}}_M x_M + (N - M)2c. \tag{6.17}$$
Now setting $M = 0$, replacing N with n and operating on both sides with $\mathbb{U}_n$ gives
$$\mathbb{D}_n x_n = x_0 - x_{-1} + 2c + 2cn. \tag{6.18}$$
Now applying the summation theorem (4.22) again from $n = 0$ to $n = N - 1$ gives a solution equivalent in form to the original solution (6.5),
$$x_n = x_0 + (x_0 - x_{-1} + c)n + cn^2, \tag{6.19}$$
if we make the identification $a \equiv x_0$ and $b \equiv x_0 - x_{-1} + c$.

We note that the solution (6.19) is not unique, since we can add to x_n any solution to the homogeneous equation $x_{n+1} = 2x_n - x_{n-1}$.

6.1.3 Reciprocal motion

Consider reverse engineering the trajectory

$$x_n = \frac{1}{a+bn}, \tag{6.20}$$

where a and b are incommensurate. The first step is to define $y_n \equiv x_n^{-1}$. Then

$$y_n = a + bn, \tag{6.21}$$

which is identified as the first equation (6.3) we reverse engineered above. Hence the equation of motion for y_n is

$$y_{n+1} = y_n + b. \tag{6.22}$$

Taking reciprocals gives the desired equation of motion

$$x_{n+1} = \frac{x_n}{1+bx_n}. \tag{6.23}$$

6.2 Reverse engineering constants of the motion

In this section we consider constructing DT equations of motion not from given solutions $\{x_n\}$ but from a weaker condition, involving constants of the motion. Since our main interest in this book is second-order difference equations, it turns out that first-order invariants of the motion are involved as a rule.

Given a dynamical sequence $\{x_n\}$, a first-order invariant C^n is a function of the form $C^n \equiv C(x_{n+1}, x_n)$ that satisfies the second-order equation

$$C^n \underset{c}{=} C^{n-1}. \tag{6.24}$$

The reverse engineering question now is as follows: given a first-order function C, what second-order difference equation is compatible with (6.24)?

Consider three successive elements of a dynamical sequence, x_{n-1}, x_n and x_{n+1} such that $x_{n+1} \underset{c}{=} \Phi(x_n, x_{n-1})$ for some function then (6.24) translates into the statement

$$C(x_n, \Phi(x_n, x_{n-1})) \underset{c}{=} C(x_{n-1}, x_n). \tag{6.25}$$

Since the equation of motion is of second order in DT, there will be two arbitrary constants associated with the general solution. Since (6.25) has to hold for general solutions, we deduce that the reverse-engineered problem is solved if we can find a function $\Phi(x,y)$ of two independent variables such that, for a given function $C(x,y)$ of two variables, we have the relation

$$C(x, \Phi(x,y)) = C(y,x) \tag{6.26}$$

for all values of x and y independently.

6.2.1 The anharmonic oscillator

For an arbitrary given C, it may be impossible to find a Φ that satisfies (6.26). It is relatively simple to deal with invariants that are quadratic functions of their arguments, so we shall discuss a class of invariant that contains quadratics as a special case: these are equivalent to quartic functions in their arguments and this leads to the DT anharmonic oscillator. Specifically, consider the time-reversal-invariant constant of the motion

$$C(x,\Phi) = ax^2 + a\Phi^2 + bx\Phi + cx^2\Phi^2. \tag{6.27}$$

Here a, b and c are constants and time-reversal invariance refers to the fact that $C(x,\Phi) = C(\Phi,x)$. This condition is imposed because, in all our applications, this symmetry emerges from the fact that we derive our DT equations of motion by discretizing canonical Lagrangians.

Condition (6.26) then gives

$$ax^2 + a\Phi^2 + bx\Phi + cx^2\Phi^2 = ay^2 + ax^2 + byx + cy^2x^2. \tag{6.28}$$

This gives a quadratic in Φ with two solutions:

$$\Phi_1 = y, \quad \Phi_2 = -\frac{bx}{a+cx^2} - y. \tag{6.29}$$

The first solution, Φ_1, is trivial, but the second, Φ_2, will be shown to be a discretization of the anharmonic oscillator, which will be discussed later in the book. Specifically, the reverse-engineered equation of motion that has the invariant $C^n \equiv ax_{n+1}^2 + ax_n^2 + bx_nx_{n+1} + cx_n^2x_{n+1}^2$ is

$$x_{n+1} \underset{c}{=} -\frac{bx_n}{a+cx_n^2} - x_{n-1}. \tag{6.30}$$

6.3 First-order discrete time causality

In this section we investigate a DT system described by a sequence $\{x_n : n = 0,1,2,\ldots\}$ of real or complex coordinate values satisfying the first-order linear inhomogeneous equation

$$x_{n+1} = ax_n + j_n, \quad n = 0,1,2,\ldots, \tag{6.31}$$

where we are given the initial datum x_0, a is a constant and there is an externally applied source interaction given by the sequence $\{j_n : n = 0,1,2,\ldots\}$. Our objective is to find a closed expression for x_n, $n > 0$, in terms of the given parameters and interactions.

By rewriting (6.31) in the form

$$(\mathbb{U}_n - a)x_n = j_n, \quad n = 0,1,2,\ldots, \tag{6.32}$$

where $\mathbb{U}_n$ is the classical time-step operator defined in Chapter 4, we see that solutions consist of two parts: a complementary solution and a particular solution, corresponding to the complementary functions and particular integrals encountered in the theory of linear inhomogeneous ordinary differential equations.

6.3.1 Complementary solution

A complementary solution y_n satisfies the homogeneous equation

$$y_{n+1} = a y_n, \quad n = 0, 1, 2, \ldots. \tag{6.33}$$

This is solved by noting that

$$y_n = a y_{n-1} = a^2 y_{n-2} = \ldots = a^k y_{n-k}, \tag{6.34}$$

so we deduce the solution

$$y_n = a^n y_0, \qquad n = 0, 1, 2, \ldots. \tag{6.35}$$

6.3.2 A particular solution

A particular solution z_n satisfies the inhomogeneous equation

$$(\mathbb{U}_n - a) z_n = j_n, \quad n = 0, 1, 2, \ldots. \tag{6.36}$$

Now suppose there is a DT Green function $G_{n,m}$ such that

$$(\mathbb{U}_n - a) G_{n,m} = \delta_{n,m}, \tag{6.37}$$

where $\delta_{n,m}$ is the Kronecker delta. Then a particular solution is

$$z_n = \sum_{m=0}^{\infty} G_{n,m} j_m, \quad n = 1, 2, \ldots, \tag{6.38}$$

assuming that $j_m = 0$ for $m < 0$.

The assumed vanishing of the source for $n < 0$ is an important input into this problem, reflecting the situation commonly encountered in physics, namely that SUOs do not exist before the initial time $n = 0$.

6.3.3 The Green function

The final step is to find the Green function or propagator, $G_{n,m}$. There are two important solutions, referred to as the causal propagator and the acausal propagator, respectively.

The causal propagator: $G^{(+)}_{n,m}$

From (6.37) we assume that the inverse operator $(\mathbb{U}_n - a)^{-1}$ exists, giving

$$\begin{aligned} G^{(+)}_{n,m} &= (\mathbb{U}_n - a)^{-1} \delta_{n,m} \\ &= \mathbb{U}_n^{-1} (1 - a \mathbb{U}_n^{-1})^{-1} \delta_{n,m} \\ &= \mathbb{U}_n^{-1} \sum_{p=0}^{\infty} a^p \mathbb{U}_n^{-p} \delta_{n,m}, \end{aligned} \tag{6.39}$$

leading to the formal solution

$$G^{(+)}_{n,m} = \sum_{p=0}^{\infty} a^p \delta_{n-p-1,m}. \tag{6.40}$$

The reason for why (6.40) is a causal propagator is that, in (6.39), the expansion is in terms of $\mathbb{U}_n^{-1}$, which means that information from the past is used to construct the solution in the present.

The acausal propagator: $G^{(-)}_{n,m}$

The acausal propagator requires expanding the inverse $(\mathbb{U}_n - a)^{-1}$ using $\mathbb{U}_n$, which means that information from the future is used to calculate the solution in the present, as follows:

$$\begin{aligned} G^{(-)}_{n,m} &= (\mathbb{U}_n - a)^{-1}\delta_{n,m} \\ &= -a^{-1}(1 - a^{-1}\,\mathbb{U}_n)^{-1}\delta_{n,m} \\ &= -a^{-1}\sum_{p=0}^{\infty} a^{-p}\,\mathbb{U}_n^p\delta_{n,m}, \end{aligned} \tag{6.41}$$

leading to the formal solution

$$G^{(-)}_{n,m} = -\sum_{p=0}^{\infty} a^{-1-p}\delta_{n+p,m}. \tag{6.42}$$

6.3.4 The retarded particular solution

Using the causal propagator (6.40), the retarded particular solution $z_n^{(+)}$ is given by

$$z_n^{(+)} = \sum_{m=0}^{\infty} G^{(+)}_{n,m} j_m, = \sum_{m=0}^{n-1} a^{n-m-1} j_m, \qquad n = 1, 2, \ldots. \tag{6.43}$$

6.3.5 The advanced particular solution

On the other hand, the acausal propagator (6.42) gives the advanced particular solution

$$z_n^{(-)} = \sum_{m=0}^{\infty} G^{(-)}_{n,m} j_m = -\sum_{m=n}^{\infty} a^{n-1-m} j_m, \quad n = 1, 2, \ldots. \tag{6.44}$$

6.3.6 Causality

It is easy to verify that the causal solution

$$x_n^{(+)} = a^n x_0 + \sum_{m=0}^{n-1} a^{n-m-1} j_m \tag{6.45}$$

satisfies the equation of motion (6.31) directly, whereas the acausal solution

$$x_n^{(-)} = a^n x_0 - \sum_{m=n}^{\infty} a^{n-1-m} j_m \tag{6.46}$$

satisfies the equation

$$x_n^{(-)} = a^{-1} x_{n+1}^{(-)} - a^{-1} j_n, \tag{6.47}$$

but this solution is equivalent to (6.31).

The fundamental difference between the causal and acausal solutions is that the former requires knowledge of the past in order to predict the next position in the future, whilst the latter requires knowledge of the future to retrodict information about the past. From a physical observer's process time point of view, only the causal solution is relevant: memory does *not* hold information from the future.

However, there are situations where retrodiction may be required, such as in early-Universe cosmology and archaeology. In such cases, information is collected over some finite interval, such as $\{M, M+1, \ldots, N\}$ in the case of DT, and then that information is used to retrodict the past at times before M. However, such retrodiction cannot be undertaken until the time $N+1$, because that is the earliest time at which the required boundary information is available.

6.4 The Laplace-transform method

The Laplace transform is a powerful method of solving certain differential equations. It is related to the Fourier transform but is well adapted to problems involving causal evolution from time zero to time $+\infty$. It can be readily adapted to the solution of linear inhomogeneous DT equations (Spiegel, 1965) and is related in that respect to the Z-transform.

We shall apply the method to the DT equation

$$x_{n+1} = a x_n + b, \quad n = 0, 1, 2, \ldots, \tag{6.48}$$

with a and b constants. The case $a = 1$ is solved by inspection: we find $x_n = x_0 + nb$, $n = 0, 1, 2, \ldots$.

The case $a \neq 1$ is more interesting. In the Laplace-transform method we write (6.48) as a difference equation for a function of a continuous real variable t, perform a Laplace transform of this equation and extract the solution using tables.

Equation (6.48) is rewritten as the difference equation

$$x(t) = a x(t-1) + b, \quad t > -1. \tag{6.49}$$

The study of such equations is a branch of the field of difference-differential equations (Pinney, 1958). Such equations occur naturally whenever retardation effects are taken into account in mechanical systems. For example, the stability of ships is affected not only by instantaneous forces, such as gravity and wind direction, but also by the time lag of water swirling around in the bilges. Such a system would be modelled by a difference-differential equation.

We shall impose the initial condition that $x(t) = 0$ for $-1 < t \leqslant 0$, since we are interested in the solution to (6.49) only for $t > 0$. Also, we need to relate $x(t)$ to x_n. With the given initial condition in mind, we define $x_n \equiv \lim_{\varepsilon \to 0^+} x(n - \varepsilon)$ for $n = 0, 1, 2, \ldots$, so we will find a solution to the DT problem with initial condition $x_0 = 0$.

The Laplace transform $\tilde{x}(s)$ is defined by

$$\tilde{x}(s) \equiv \int_0^\infty e^{-st} x(t) dt, \tag{6.50}$$

where s is the transform parameter. Taking the Laplace transform of (6.49) of both sides gives

$$\tilde{x}(s) = ae^{-s}\tilde{x}(s) + \frac{b}{s}, \tag{6.51}$$

using $x(t) = 0$ for $t < 0$. Hence

$$\tilde{x}(s) = \frac{b}{s(1 - ae^{-s})}. \tag{6.52}$$

The inverse Laplace transform is readily found by noting that for the Heaviside generalized function $\Theta(t - \tau)$ the Laplace transform is easily found for $\tau > 0$. Performing a power-series expansion of the denominator in (6.52) and inverting term by term (Spiegel, 1965) gives the solution

$$x_n \equiv \lim_{\varepsilon \to 0^+} x(n - \varepsilon) = b \sum_{k=0}^{n-1} a^k, \quad n = 1, 2, 3 \ldots, \tag{6.53}$$

with the initial condition $x_0 = 0$. The general solution to (6.48) would be given by the addition to (6.53) of an arbitrary solution to the homogeneous equation $x_{n+1} = ax_n$.

As confirmation of these methods, setting $j_n = b$, $n = 1, 2, 3, \ldots$, in the retarded solution (6.43) obtained by the Green-function method gives the same result as (6.53).

7

Classical cellular automata

7.1 Classical cellular automata

In the previous chapter we gave some examples of discrete time (DT) processes involving a small number of dynamical variables. However, the world around us is a complicated dynamical system with a truly vast number of degrees of freedom. Therefore, we should be prepared to discuss DT models with arbitrary numbers of degrees of freedom. In particular, we should be interested in cellular automata (CAs), models representing dynamical variables filling all of physical space.

Physical space is the three-dimensional arena of position that we see all around us and in which particle systems under observation (SUOs) appear to move. We shall not debate the question of whether physical space is real, a relational manifestation or a mental construct. For all practical purposes, CA theory requires us to think of it as real, particularly in computer simulations, where specific cells in the CA are given individual array labels that persist throughout a simulation. We shall see this explicitly when we come to discuss spreadsheet mechanics later on in this chapter.

In the applications that we investigate in this chapter, the architecture is that of DT evolution in a fixed background universe, shown in Figure 5.1(b). In Chapter 29 we discuss the scenario of a CA with an evolving background universe as shown in Figure 5.2.

It is important to distinguish physical space from configuration space, the mathematical space representing all the spatial dynamical degrees of freedom used to describe a many-particle SUO. For instance, if we had an SUO consisting of one hundred point particles each sitting at its own place in three-dimensional physical space, then configuration space would have three hundred dimensions. In the CA approach, the hundred particles would occupy one hundred cells of the CA, the other cells registering no content.

Another point to consider is the relationship with phase space. Modelling the analogue of velocity or momentum in a CA can be done in two ways: (i) a memory is kept of the state of the SUO at time n and at time $n-1$, or (ii) the CA array space is doubled so as to carry momentum information as well as positional information. The latter approach is the CA analogue of the cotangent-bundle $T^*\mathcal{C}$ techniques (phase space) used in advanced continuous time (CT) classical mechanics (CM), where $\mathcal{C}$ is a configuration-space manifold, whilst the former approach is the CA analogue of the direct-product $\mathcal{C} \times \mathcal{C}$ formalism used in temporally discretized versions of CT CM, as discussed by Marsden and West (2001) and in Chapter 15.

The focus of a CA is not on particles per se but on physical space. To this end, each point or cell of the CA is assigned one or more dynamical variables, each of which has an empirical meaning over and above that of position. For instance, we may assign a dichotomic variable to each cell, with the two possible cell values denoted 0 or 1. Typically the value 0 would represent the absence of a particle in that cell whilst the value 1 would represent the presence of a particle in that cell.

There is a point here to do with the realistic modelling of observation. The conventional approach is to assume that a cell value always exists and can be known unambiguously. However, a more realistic approach would be to take into account the possibility that (i) there is no apparatus associated with the cell, or (ii) the detector is faulty and the observer would read unreliable values in a cell. We have formulated an approach to this issue that takes into account **four** possibilities, the elements of the power set $\mathcal{P} \equiv \{\emptyset, \{0\}, \{1\}, \{0,1\}\}$ of the bit value set $\{0,1\}$ (Jaroszkiewicz, 2010). The empty-set element $\emptyset$ in $\mathcal{P}$ represents a state of detector non-existence because the observer has not yet constructed any apparatus in the cell, the elements $\{0\}$ and $\{1\}$ represent the two normal value states of a bit cell given that the apparatus has been constructed and is working properly, and the element $\{0,1\}$ represents the possibility that the apparatus exists but is either faulty (so that we cannot be sure whether the value is 0 or 1) or has been deliberately decommissioned by the observer. Each of these four possibilities is physically significant and represents in a simplified form precisely what happens in the real world of experiment. We discuss this approach to CA dynamics in Chapter 29.

There is no current evidence for temporal or spatial discreteness, so conventional physics goes to the continuum limit of CAs and models physical space by a continuum of cells, requiring a continuum of coordinates to label separate cells. This leads to field theory, in which dynamical degrees of freedom are assigned to each point of continuous space. A discussion of field theory as the limit of a CA is given by Eakins and Jaroszkiewicz (2005).

Although the principles of CAs appear different from those of standard CM, the two approaches can be related to each other if the latter is interpreted in a certain way. By this we mean the following. Consider the CM statement that a point particle is at a position P with Cartesian coordinates (x_P, y_P, z_P) at time t. That is an extremely efficient statement of the particle's position if we consider what such a statement actually implies. Specifically, **if** the particle is known to be at position (x_P, y_P, z_P), then **it is not at any other position in physical space**. In other words, the truth of one statement implies the truth of an infinite number of other statements. That is an impressive efficiency. Moreover, none of the implied information is irrelevant: knowing where a particle is **not** is arguably as significant a piece of information as knowing where it is. That logic is the basis for the optimum method of searching a house for a missing bunch of keys: we look everywhere once until we find the keys.

How do we account for the efficiency of CM in giving us this information? The answer is: context. **If** we know for sure that we have a one-particle system, then only one point in physical space could be occupied by that particle at any given time. It is this knowledge that would justify the conclusion that all the other points in physical space had to be unoccupied.

This is not a trivial argument, as can be seen by considering an N-particle system where $N > 1$. Then knowing where just one of the particles is is not enough information to generate implications about any of the other cells. A field theory is rather like the limiting case $N \to \infty$, where we have to say something at each point in physical space and the economy of point-particle mechanics has been totally lost. A CA is much more like a field theory in this respect than a particle theory, because we have to specify the state of each cell at each time.

7.2 One-dimensional cellular automata

In this section we discuss CAs based on a one-dimensional physical space. Such a model consists of an infinite number of cells, each labelled by a single integer i running from $-\infty$ to $+\infty$. Typically, the ith cell C^i is considered to be a nearest neighbour of the $(i-1)$th cell $\mathcal{C}^{i-1}$ and of the $(i+1)$th cell $\mathcal{C}^{i+1}$.

The indexing of the cells has two roles. First, it gives an identity to a cell over time. This underlines an aspect of the physics which it is easy to overlook, namely, the natural tendency of humans to objectivize those observed phenomena which persist over sufficiently long timescales as to warrant attention: individual sea waves are not named because they generally disappear relatively quickly,[1] whereas mountains persist on longer timescales (Hildebrandt, 2012). This applies to the space concept as well. According to Mach (1912) and Rovelli (1996), space is not a thing per se but a relationship between things. Nevertheless, there is a lot of advantage in giving space its own identity and indeed, in the case of electromagnetism, general relativity and quantum field theory, its own physical properties. Since there are many points of physical space, we give each one of them its own label, which in the case under discussion is the superscript i.

[1] Relative to human timesscales such as a minute, day, or hour. Frozen water can persist sufficiently long to be objectivized, as in the case of glaciers.

It is possible to discuss CAs where cell identity is not permanent. We shall discuss such a possibility towards the end of this book. The phenomenon of persistence, i.e., the apparent endurance of objects in time, is contextual, in that a process may be transient according to one observer's scale of time and enduring according to another observer's scale of time. What we can say with some confidence is that no object known to us has existed from the infinite past and no object will endure forever. So it is a matter of timescale how long a cell's identity can be meaningful relative to a given observer. In the real world that we actually experience, SUOs appear to come and go, being created and annihilated according to the shifting patterns of energy and matter in the Universe. Observers themselves do not endure and can be just as transient as SUOs. We can only speculate on the eventual development of a more comprehensive theory of observation than the ones we have now, which will take into account all the timescales of persistence which we see all around us and the physical laws which differentiate SUOs and observers (Jaroszkiewicz, 2010).

The second role of the index is to provide a measure of a distance relationship. By convention, cells $\mathcal{C}^i$ and $\mathcal{C}^{i+1}$ would be regarded as nearest neighbours but cells $\mathcal{C}^i$ and $\mathcal{C}^{i+1000}$ would not. This distance relationship and the physics underpinning a CA will determine the encoding of the dynamics. For instance, a CA model of Newtonian gravitation might have long-range interaction between relatively distant cells, whereas a simulation of GR would be restricted to nearest-neighbour interactions. We shall see when we come to discuss spreadsheet mechanics that the restriction to nearest-neighbour interactions induces the emergence in the course of the evolution of a causal structure analogous to the lightcone structure of relativistic spacetime. Such a structure disappears if interaction by 'action at a distance' between cells that are not nearest neighbours is permitted.

There is a distinct relationship between the existence of lightcone-like structure in physics and the fact that the laws of physics tend to be second-order hyperbolic differential equations. The reason is two-fold: (i) a derivative is the limiting case of a comparison of cellular information contained in nearest-neighbouring cells; and (ii) differential equations of motion dictate how dynamical information is transmitted from each cell to its neighbours in the course of time. We recall at this point our discussion in Section 1.2 of Tegmark's analysis of the dimensions of time.

Experience of CAs has led some theorists to the conjecture that the lightcone structure of relativistic spacetime may be an emergent consequence of the laws of CA physics (Minsky, 1982; Wolfram, 1986; Fredkin, 1990), rather than a preordained structure that the laws of physics have to respect. Whatever the truth, it is the case that in CA theory the specification of interactions between cells is the single factor that generates either causal structure or apparent chaos. We shall see this directly when we come to discuss spreadsheet mechanics below.

We shall discuss the simplest sort of model, wherein, at any given DT t_n, the ith cell C^i contains an observable value v_n^i of some measurable quantity. We shall not discuss who or what the observer is. We shall assume that cell values can be automatically transmitted to other cells during the course of the dynamics and contribute to the calculation of the values in those cells at subsequent times.

Discrete time is explicitly assumed in this formulation. It is a straightforward matter to use the information in the cells at time t_n, and possibly at earlier times, to determine the information in the cells at time $n+1$. The set $S_n \equiv \{\ldots, v_n^{-2}, v_n^{-1}, v_n^0, v_n^1, v_n^2, \ldots\}$ of all cell values at time n represents the state of the system at time n. In the models we are concerned with, the basic aim is to determine S_{n+1} given complete knowledge of S_n and as many of the earlier states S_{n-1}, $S_{n-2}, \ldots$ as the dynamical rules require.

Even in the case of the simplest CA, namely the one-dimensional system, there are vastly many forms of dynamics we could discuss and some sort of classification of architecture, or type of interaction, would be helpful.

7.2.1 Classical versus quantum

Generally, CAs are discussed either as fully classical or else with some degree of QM incorporated. All the CAs in this chapter are classical. We discuss an approach to QM CAs in Chapter 29.

7.2.2 Implicit versus explicit

In the theory of classical CA, dynamical relationships are inevitably chosen to be explicit. By this we mean that, given initial data $\{S_0, S_{-1}, \ldots\}$, we can work out S_1 from knowledge of the dynamical rules. Then, with that knowledge about S_1, we can work out S_2, and so on. We refer to this as causal implication and write $\{S_0, S_{-1}, \ldots\} \to S_1$, $\{S_1, S_0, \ldots\} \to S_2$, etc. Note that this process need not be reversible, i.e., knowledge of the data set $\{S_1, S_2, S_3, \ldots\}$ might not be enough to let us work out S_0.

Example 7.1 Consider a CA with two cells, such that, at time n, the state $\mathbf{v}_n$ of the automaton is given by the column array

$$\mathbf{v}_n \equiv \begin{bmatrix} v_n^1 \\ v_n^2 \end{bmatrix}, \tag{7.1}$$

where v_n^1 and v_n^2 are real, and suppose that the state $\mathbf{v}_{n+1}$ is given by

$$\mathbf{v}_{n+1} \equiv \mathsf{M}\mathbf{v}_n, \quad n = 0, 1, 2, 3, \ldots, \tag{7.2}$$

where M is a real 2×2 matrix. Then we can always work out $\mathbf{v}_1$, $\mathbf{v}_2$ and so on, if we are given $\mathbf{v}_0$. But, if we are given $\mathbf{v}_1$ and M happens to be singular, then we can never work out $\mathbf{v}_0$.

The problem with implicit dynamical relations, which are typically equations of the form $\Phi(S_1, S_0, S_{-1}, \ldots) = 0$, is that there might not be any unique solution for S_1, they might be unphysical, or, even more undesirably, the solution set for S_1 might be empty. We encounter the second possibility, unphysical solutions, in Lee's approach to DT mechanics, a form of DT mechanics where the time intervals themselves become dynamical variables. The possibility arises there that a nominally forwards time interval $t_{n+1} - t_n$ is actually negative.

7.2.3 First-order dynamics

The simplest CA architecture corresponds to the first-order dynamics shown in Figure 5.7. In that figure, the presence of arrows indicates that S_{n+1} is explicitly related to S_n, i.e., determined directly from knowledge of S_n, rather than being in some form of implicit relationship with it. A dynamics based on implicit relationships is inevitably a weaker form of mechanics than one based on explicit relationships, because the latter need not have a unique solution in terms of the required variable. In the explicit form of first-order dynamics, S_{n+1} is determined completely and uniquely from knowledge of S_n alone, the generic first-order rule being

$$v^i_{n+1} = \Phi^i(S_n), \qquad i = \ldots, -1, 0, 1, \ldots, \tag{7.3}$$

where Φ^i is a function of the cell values at time n.

7.2.4 Second-order dynamics

In this form of CA, S_{n+1} is determined from knowledge of S_n and S_{n-1}, and an example of the architecture is shown in Figure 5.9. In this mechanics, we need a memory of two configurations of the CA in order to evolve forwards in DT, so this is the analogue of the $\mathcal{C} \times \mathcal{C}$ approach to DT CM discussed by Marsden and West (2001) and in Chapter 15.

7.3 Spreadsheet mechanics

A spreadsheet such as Excel may be used to demonstrate some of the features of the CA referred to above. We now discuss a simulation of a one-dimensional first-order CA, with the spatial index running down the screen and with successive values of DT running from left to right on the screen. A typical spreadsheet such as Excel has columns labelled A, B, C, etc. from left to right and rows labelled 1, 2, 3, etc. running from top to bottom. These column and row labels are not cells of the CA itself but reference labels running on the top and left borders of the spreadsheet. They are fixed reference points in the program screen display.

The CA is set up and run with the following sequence of steps, which should be followed as given here.

1. Format all cells of the spreadsheet to show output results as shown in Figure 7.1.
2. In column A, enter the number zero (0) starting from row 1 and going down several hundred rows. This column represents all cells at DT $t = 0$ and the zeros indicate the quiescent state, akin to empty space.
3. In cell B2, which means column B, row 2, enter the formula

$$= (A1 + A2 + A3)/3, \tag{7.4}$$

   where A1 refers to the cell in column A, row 1, and so on.
4. Copy formula (7.4) from B2 into every cell below B2 in column B. Note that the spreadsheet automatically changes the formula in each cell to take into account the location of that cell. For example, cell B5 contains the formula

40	A	B	C	D	E	F	G	H
41	0	0	0	0	0	0	0	0
42	0	0	0	0	0	0	0	0
43	0	0	0	0	0	0	0	457
44	0	0	0	0	0	0	1372	3201
45	0	0	0	0	0	4115	8230	12803
46	0	0	0	0	12346	20576	28807	35208
47	0	0	0	37037	49383	61728	68587	73617
48	0	0	111111	111111	123457	123457	123457	121628
49	0	333333	222222	222222	197531	185185	172840	163237
50	1000000	333333	333333	259259	234568	209877	193416	179698
51	0	333333	222222	222222	197531	185185	172840	163237
52	0	0	111111	111111	123457	123457	123457	121628
53	0	0	0	37037	49383	61728	68587	73617
54	0	0	0	0	12346	20576	28807	35208
55	0	0	0	0	0	4115	8230	12803
56	0	0	0	0	0	0	1372	3201
57	0	0	0	0	0	0	0	457
58	0	0	0	0	0	0	0	0

Figure 7.1. A one-dimensional cellular automaton: time runs left–right, space runs top–bottom. Cell 50 starts with value 1000000 at time $n = 0$, creating a lightcone-like wave of causal information.

'= (A4 + A5 + A6)/3' whereas cell B16 contains the formula '= (A15 + A16 + A17)/3'.

5. Now copy the whole of column B into every column on its right, i.e., into column C, D, etc. Again, the spreadsheet automatically adjusts the formula for each cell, so, for example, N35 contains the formula '= (M34+M35+M36)/3'.
6. At this point, every cell to the right of column A should have the number zero (0) in it, representing the evolution of the quiescent state from column A to the other columns in the simulation. The formula works out the explicit value to be all zero for each cell in all the columns to the right of column A.
7. Now pick some cell in column A, such as A50, as shown in Figure 7.1. This is to ensure that we start well away from the actual edges of the spreadsheet, such as cell A1, which are not part of the simulation and affect the dynamics if approached too closely. Type into the chosen cell the number 1000000 (one million). This represents a point source at time $n = 0$, perturbing the quiescent state at that time.
8. Immediately, the spreadsheet changes and a 'lightcone'-like set of non-zero numbers appears to the right of column A, with its vertex at cell A50 and spreading out with increasing DT.

The following comments are relevant.

1. What this example shows is that *local* rules of interaction, or, more correctly, local rules for the transmission of information, can generate causal structures such as lightcones. From this perspective, lightcones do not occur because of the properties of space (the spreadsheet) per se but because of the specific way in which information is distributed between nearest neighbours.

   If the interaction between nearest neighbours is altered, then the lightcone may be altered. It is possible to have a directional bias to the lightcone, analogous to curvature of space, if there is no directional (i.e., up–down) symmetry in the nearest-neighbour interaction.
2. The reader is encouraged to investigate possible modifications to the rules including the following.

   Second-nearest, third-nearest, etc. interactions.

   Second-order interactions, such as $v_n^i = (v_{n-1}^i + v_{n-2}^{i+1} + v_{n-2}^{i-1})/3$, and so on. The possibilities are endless.

   Attempted time travel: given the output shown in Figure 7.1, change the formula in cell G50 to '= (F49 + P50 + F51)/3'. The result may be a surprise to the reader, but suggests that Microsoft (creators of Excel) know all about the pitfalls of time travel, particularly the Grandfather paradox. This is related to the choice of the retarded Green function rather than the advanced Green function discussed in the previous chapter.

Experience with spreadsheet dynamics provides an economical, immediate and instructive approach to understanding differential equations, cellular automata and DT mechanics and is highly recommended.

7.4 The Game of Life

Conway's Game of Life is a two-dimensional CA, played over a set of square cells, which carry only one of two possible values at any given stage (Gardner, 1970). One of these values is zero, interpreted as 'dead', whilst the other value is one, interpreted as 'alive'. The two dimensions are spatial, with time being represented by successive images of the cells.

This CA is particularly fascinating to run as a computer simulation and has stimulated much research into cellular automata. Free versions can be readily found on the Internet. In such simulations, each cell is coloured either black or white, depending on whether it is occupied or unoccupied at a given time. Typical simulations can involve a grid of many thousands of such cells. In Figure 7.2, which shows part of such a grid at two successive times, the configuration on the left changes to the configuration on the right according to specific rules. The original game involves an infinite two-dimensional orthogonal grid of square cells. Every cell has eight immediate neighbours and all nine cells are taken into account in determining what happens to the central cell.

At each time step the following possible transitions occur.

1. Any alive cell with fewer than two live neighbours dies.
2. Any alive cell with two or three alive neighbours remains alive.
3. Any alive cell with more than three alive neighbours dies.
4. Any dead cell with exactly three alive neighbours becomes alive.

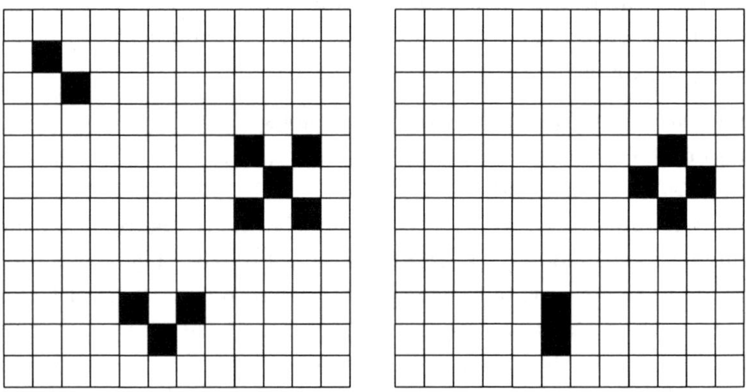

Figure 7.2. Successive configurations in the Game of Life.

The game starts with some initial configuration and then runs, in principle, forever. Several points are of interest.

1. There is an inherent irreversibility in this game, since two different configurations can step into the same one. This means that time cannot run 'backwards' in a deterministic way.
2. It is possible to have initial states that could not themselves have been the end products of any other state. Such initial states are known as 'Garden of Eden' states.
3. The rules can be generalized. The original version as defined by Conway is given the nomenclature 23/3, which is interpreted as follows. The numbers 2 and 3 on the left of the / indicate the number of alive cells surrounding an alive cell which would be needed in order to guarantee its survival. The number 3 on the right of the / shows how many alive cells surrounding a dead cell are required in order to turn it into an alive cell. In principle, games could range from 0/0 to 012345678/012345678. It seems that the 23/3 Conway game is borderline in some sense. Various patterns survive and/or move around, but many initial configurations eventually dissipate and fizzle out.

7.5 Cellular time dilation

In this section we use spreadsheet mechanics to investigate time dilation in CA. Time dilation is generally associated with special relativity (SR), where it manifests itself as an observed decrease in the rate of change of a moving clock relative to the clocks fixed in an observer's reference frame. Time dilation is never an increase in the rate of a moving clock.

In SR, the effect is well understood as a subtle interplay between the metrical structure of spacetime and the protocols of observation. By our use of this term, we mean that the observation of a moving clock is not a simple matter. Care must be taken to specify what is meant by an observed time interval: observers measuring the passage of time as registered by a moving clock necessarily interact with it at different places in their rest frame (otherwise the clock would not be moving). When observational protocol is carefully observed, SR predicts that, if a standard clock moves with speed v relative to an inertial frame $\mathcal{F}$, then observers at rest in $\mathcal{F}$ will measure that clock as going slower by a factor of $\sqrt{1 - v^2/c^2}$ relative to their own standard clocks, which are at rest in $\mathcal{F}$. Here c is the speed of light. Relativistic time dilation has been confirmed empirically by measurement of the increased lifetime of muons as they pass through the Earth's atmosphere (Li and Richardson, 2009).

Our interest in this topic was triggered by Marvin Minsky's paper on the simulation of physics with cellular automata (Minsky, 1982). Minsky was influenced by Fredkin (Fredkin and Toffoli, 1982) and Feynman, pioneers of cellular automata and quantum computation, respectively (Feynman, 1982). According to Minsky,[2] Feynman did not accept that a finite volume of physical space could contain an infinite number of dynamical degrees of freedom. The logical inference, then, is that space is discretized in some way, and from that it is a short leap to think in terms of DT. A modern echo of Feynman's view is the holographic principle ('t Hooft, 1993; Susskind, 1995), which asserts that the dynamical state of gravitational degrees of freedom inside a volume of space can be encoded in a finite way over the surface of that volume.

In an attempt to simulate physics, Minsky considered how the particle concept might emerge in a CA universe, focussing on particle wave packets and force concepts rather than the cellular field concept discussed above (Minsky, 1982). Of particular interest to us is his discussion of the 'slowing down of internal clocks' (Minsky, 1982), by which he meant relativistic time dilation. By considering the detailed computational processes evolving a particle-like wave packet, Minsky concluded that '*The faster a nonphoton moves, the slower must proceed its internal computations!*'.

Minsky noted certain problems with the detailed calculations in his simulations: (i) the slowing down of internal clocks went as the square of the SR prediction; (ii) there was no symmetry between different frames of reference, a problem typical of models where space and time are discretized in absolute terms; and (iii) length contraction was not accounted for.

We shall refer to Minsky's time dilation as computational time dilation. Such a phenomenon is to be expected in any process of non-local information exchange that takes place in DT steps. We have in mind here computers based on parallel processors exchanging information.

Suppose we have a complex of processors situated at various locations in space and exchanging data between themselves. Suppose further that processor A needs information from processor B to complete A's part of the overall calculation. Then A cannot complete its task until B has sent that information and that information has arrived at A. If, for some reason, the information from B is delayed, then A will inevitably be delayed in its computation. This is computational time dilation: the effective slowing down of a computer due to the delays in arrival of information throughout that computer.

Our approach is similar to that of Minsky, being based on a CA architecture. Time occurs in two radically different forms in our simulation (Jaroszkiewicz, 1999), both of them discrete. One form, which we shall refer to as lab time, is the time of events as registered by standard clocks fixed in cells distributed spatially in a one-spatial-dimensional inertial laboratory, referred to as the array, or the lab. We shall represent discrete lab time and cell position by a two-dimensional array, just like the spreadsheet automaton discussed above.

Lab time is the time in a block universe (Price, 1997), with the integrability property common to all coordinate times: any two events in the laboratory have

[2] Private communication.

unique values of lab time that are not contextual, i.e., do not depend on any paths between those events. The lab time of an event will be denoted by a small Latin letter.

The other form of time, referred to here as processor time, is quite different in origin and functionality from lab time. Imagine that at this point we adopt Fredkin's view of the Universe, namely that physics is synonymous with information (Fredkin, 2001) and that our laboratory *is* actually a spreadsheet. Processor time is the time of the computer chip running the simulation, instants of processor time being labelled by capital Latin letters. The central processor evolving the CA (the laboratory) plays the role of an exophysical observer standing outside of that universe or laboratory, overseeing its dynamical evolution.

7.5.1 Simulation architecture

The architecture of the simulation is as follows. Processor time runs from $T = 0$ to some final positive integer $T = N$ that, when taken sufficiently large enough, may be regarded as infinity. At each instant T of process time, the laboratory is in a state $\Lambda(T)$ of the two-dimensional spreadsheet array, which includes past, present and future laboratory times. As T changes from, say, $T = M$ to $T = M + 1$, the state $\Lambda(T)$ changes according to an algorithm $\mathcal{A}$ defined below. Between time M and time $M + 1$, the central processor (the external observer) scans each cell in the laboratory, both spatially and over all values of lab time, and makes all changes possible in each cell as stipulated by the algorithm $\mathcal{A}$, **at that value of processor time**. A fundamental constraint on this architecture is that, during the update from time T to time $T+1$, each cell in the laboratory is looked at once and once only, and how it changes is dictated by the information held in state $\Lambda(T)$, and is not affected by changes made after time T. This architecture means that there is a separate block-universe spreadsheet image for each state $\Lambda(T)$, $T = M, M + 1, \ldots, N$.

Successive lab images are not unrelated because the algorithm $\mathcal{A}$ essentially provides the 'laws of physics' relating $\Lambda(T + 1)$ to $\Lambda(T)$.

7.5.2 Time dilation

Time dilation in SR concerns the behaviour of clocks and relative rates of processes, not the actual processes themselves. It is generally accepted in SR that time dilation affects all processes in the same way, including atomic and biological processes. Likewise, what is of interest to us here is not the specific information being transmitted from cell to cell by the algorithm $\mathcal{A}$, but the value of processor time T when this is done. In Figure 7.3 the integers represent not the data being processed but the values of processor time T at which the calculation was completed for that cell. The algorithm $\mathcal{A}$ therefore has irreversibility built into it from the perspective of processor time but not of lab time: if the contents of

–8	–7	–6	–5	–4	–3	–2	–1	0	1	2	3	4	5	6	7	8	9	10	11	12	13	14	15	16	17
17												0	0	1	2	3	7	8	12	13	17	18	22	23	**S**
16			B									0	0	1	2	6	7	11	12	16	17	21	22	26	27
15		A	C	E								0	0	1	5	6	10	11	15	16	20	21	25	26	30
14			D							0	0	0	0	4	5	9	10	14	15	19	20	24	25	29	30
13									0	0	1	2	3	4	8	9	13	14	18	19	23	24	28	29	33
12									0	0	1	2	3	7	8	12	13	17	18	22	23	27	28	32	33
11									0	0	1	2	**Q**	7	11	12	16	17	21	22	26	27	31	32	36
10									0	0	1	5	6	10	11	15	16	20	21	25	26	30	31	35	36
9							0	0	0	0	4	5	9	10	14	15	19	20	24	25	29	30	34	35	39
8						0	0	1	2	3	4	8	9	13	14	18	19	23	24	28	29	33	34	38	39
7						0	0	1	2	3	7	8	12	13	17	18	22	23	27	28	32	33	37	38	42
6						0	0	1	2	6	7	11	12	16	17	21	22	26	27	31	32	36	37	41	42
5						0	0	1	5	6	10	11	15	16	20	21	25	26	30	31	35	36	40	41	45
4				0	0	0	0	4	5	9	10	14	15	19	20	24	25	29	30	34	35	39	40	44	45
3			0	0	1	2	3	4	8	9	13	14	18	19	23	24	28	29	33	34	38	39	43	44	48
2			0	0	1	2	3	7	8	12	13	17	18	22	23	27	28	32	33	37	38	42	43	47	48
1			0	0	1	2	**P**	7	11	12	16	17	21	22	26	27	31	32	36	37	41	42	46	47	51
0			0	0	1	5	6	10	11	15	16	20	21	25	26	30	31	35	36	40	41	45	46	50	51
–1	0	0	0	0	4	5	9	10	14	15	19	20	24	25	29	30	34	35	39	40	44	45	49	50	54
–2	0	1	2	3	4	8	9	13	14	18	19	23	24	28	29	33	34	38	39	43	44	48	49	53	54
–3	0	1	2	3	7	8	12	13	17	18	22	23	27	28	32	33	37	38	42	43	47	48	52	53	57
–4	0	1	2	6	7	11	12	16	17	21	22	26	27	31	32	36	37	41	42	46	47	51	52	56	57
–5	0	1	5	6	10	11	15	16	20	21	25	26	**R**	31	35	36	40	41	45	46	50	51	55	56	60

Figure 7.3. Time runs from left to right. The frame velocity is chosen such that the ratio of the spatial step to temporal steps is 3/5. Solid skew lines represent lines of constant t' and x'. Numbers in bold indicate some of the cells which have integer coordinates in the primed frame and the dotted lines show a typical forwards lightcone.

a cell are altered by the processor at processor time T_0, then that cell remains unchanged for all subsequent (i.e. greater) values of T.

The algorithm utilizes the local discrete topology of cellular automata, i.e., the relationship between adjacent cells, which makes it independent of the details of the dynamical rules relating state values in adjacent cells. Therefore, the results are generic and should find their analogues in a wide variety of cellular automata. The conclusion is that time dilation is not specific to Minkowski spacetime but occurs also as a generic feature of any non-localized system such as a parallel-processing computer where information is processed by a central processing unit governed by an internal clock. The essential reason is that some or all component processors may have to wait various clock cycles before certain local computations can be completed, simply because necessary information is not yet available to them, the net result being equivalent to a slowing down of clocks in the system, i.e., time dilation.

7.5.3 The algorithm

The objective of the simulation is to construct a block-universe image $\Lambda(T)$ of the spreadsheet for each value of processor time T, for $T = 0, 1, 2, \ldots$. Each cell $C_n^i(T)$ in $\Lambda(T)$ has a spatial position i, a lab time n, and a display that is either blank or an integer referred to as the completion time. Figure 7.3 shows the completion time in each cell for remote future processor time, i.e., essentially for $\Lambda(\infty)$. Time dilation is embedded in the set of completion times shown in $\Lambda(\infty)$, but this has to be analyzed carefully to reveal the patterns expected from SR.

The completion time in $C_n^i(\infty)$ is the value of processor time at which the dynamical calculation was completed for that cell. The completion time either has no value or else it is an integer. The image $\Lambda(\infty)$ shows display times for all the cells. Those with null display times are shown blank on screen whilst those cells with a non-empty value show an integer.

The actual datum in each cell that would be used in a calculation is not shown in Figure 7.3, but its existence is implied. If the display time is null, then the item of physical data is also null, i.e., does not exist. If a display time exists, then the physical datum exists by implication and would have been deposited in that cell by the processor at processor time equal to the display time.

The specific equation solved in the simulation has the architecture shown by the cells labelled A, B, C, D and E in Figure 7.3. It is derived from a simple discretization of the massless wave equation

$$\frac{\partial^2}{\partial t^2}\varphi(t,x) - \frac{\partial^2}{\partial x^2}\varphi(t,x) \underset{c}{=} 0, \tag{7.5}$$

namely

$$\frac{\varphi_{n+1}^i - 2\varphi_n^i + \varphi_{n-1}^i}{c^2\,\Delta T^2} - \frac{\varphi_n^{i+1} + 2\varphi_n^i - \varphi_n^{i-1}}{\Delta X^2} \underset{c}{=} 0, \tag{7.6}$$

where c is the speed of light and ΔT and ΔX are the step intervals for time and space, respectively. Equation (7.6) can be solved explicitly for φ_{n+1}^i, which means that cell E can be computed if data from cells A, B, C and D are available to the processor. If any of this information is absent, E cannot be calculated.

Processor time starts at $T = 0$. At this point $\Lambda(0)$ consists of some of the cells in spacetime (the two-dimensional array) displaying the number *zero*, implying that they contain non-null physical data, whilst all the other cells display a blank, meaning that they have null physical data. The state of a blank cell corresponds to a void state $|\emptyset)$, as discussed above. The sequence of processor steps is as follows.

1. Before any initial data are distributed into $\Lambda(0)$, each cell $C_n^i(0)$ contains the following spreadsheet formula:

$$\begin{array}{l}=\mathrm{IF}(\mathrm{AND}(\mathrm{C}_{n-2}^{i}!=\text{``''},\mathrm{C}_{n-1}^{i-1}!=\text{``''},\mathrm{C}_{n-1}^{i}!\text{-}\text{``''},\mathrm{C}_{n-1}^{i+1}!=\text{``''}),\\ \mathrm{MAX}(\mathrm{C}_{n-2}^{i},\mathrm{C}_{n-1}^{i-1},\mathrm{C}_{n-1}^{i},\mathrm{C}_{n-1}^{i+1})+1,\text{``''})\end{array}$$

2. A zero ('0') is inserted into each cell C_n^i in $\Lambda(0)$ for which initial data are to be set up. The display changes from blank to the number zero for each such cell. Note that the initial data need not be derived from the algorithm $\mathcal{A}$ using data from times earlier than $T = 0$. It is quite possible to have a 'Garden of Eden' initial state, reflecting the fact that what happens before an experiment starts need have nothing to do with the experiment per se, apart from providing a starting point. This could include constructing the apparatus (Jaroszkiewicz, 2010) and then initializing it.
3. During the interval of processor time between $T = N$ and $T = N + 1$, the processor examines the contents (not the formula) in each cell in $\Lambda(N)$. If a given cell is not blank, the processor leaves that cell alone. On the other hand, if the cell is blank, the processor examines the adjacent cells according to the above algorithm and takes the appropriate action. Note that any changes in a cell at processor time N affect $\Lambda(N + 1)$. This means that changes cannot feed backwards in processor time.
4. Having completed its examination of each cell in $\Lambda(N)$, the processor increases the value of N by one and then goes back to step 3.

7.5.4 Commentary

Figure 7.3 is a portion of $\Lambda(\infty)$, the long-term result of evolving from a specific initial state given by cells with zeros in the figure. It shows a number of significant features.

1. Integers in cells running from left to right at the top of the diagram represent lab time in the array rest frame. Integers in cells running from bottom to top on the left of the diagram represent spatial position in the array rest frame.
2. There is a set of empty cells towards the upper left-hand side of the figure. These cells represent spacetime locations in the laboratory before any state was prepared by the processor.
3. Cells with a display of zero ('0') are those cells which had initial data set up by the processor at processor time $T = 0$. In order to simulate initial data set up on a frame $\mathcal{F}$ moving relative to the rest frame of the array $\mathcal{F}_0$, the zeros run in a zig-zag way. This zig-zag is an approximation to the straight line in $\mathcal{F}_0$, which would represent a special-relativistic spacelike hyperplane

of simultaneity in the $\mathcal{F}$ frame, in the continuum limit. The set of zeros in Figure 7.3 approximates a hyperplane in a frame $\mathcal{F}$ moving with speed 3/5 relative to $\mathcal{F}_0$.

4. To the right of the line of zeros, integers greater than zero are displayed. Each such integer indicates the processor time at which the calculation was completed for that cell.
5. There is a zone to the right of the initial data, known as the causal boundary layer, where the simulation has not settled into its long-term pattern. Causal boundary layers should be avoided when searching for significant patterns in a simulation.
6. Beyond the causal boundary layer, to the right in the diagram, the completion times form a pattern that can be analyzed to reveal time dilation.

The pattern of implication settles down after several clock cycles and assumes a regularity in which a relativistic pattern may be discerned. In Figure 7.3 we show lines that connect cells carrying dates suggestive of times and position coordinates in a frame $\mathcal{F}_{\frac{3}{5}}$ moving with velocity $v = 3/5$ relative to the rest frame $\mathcal{F}_0$ of the original lattice.

7.5.5 Analysis

We analyze the data from four events P, Q, R and S in Figure 7.3. Coordinates measured in terms of lab time t and array position x are given by (t, x), whereas values of processor time T and $\mathcal{F}$ spatial coordinate X are denoted by $[T, X]$. The data read off are given in the following table:

event	(t,x)	$[T,X]$
P	$(0,0)$	$[10,0]$
Q	$(6,10)$	$[10,X_{\mathrm{Q}}]$
R	$(4,-4)$	$[26,X_{\mathrm{R}}]$
S	$(16,16)$	$[26,X_{\mathrm{S}}]$

(7.7)

The spatial coordinate value for X_{p} is set to zero for convenience and does not affect the analysis.

The next step in the analysis is to subtract the initial time $T_{\mathrm{P}} = 10$ from all times T. This does not affect any subsequent analysis, giving the table

event	(t,x)	$[T,X]$
P	$(0,0)$	$[0,0]$
Q	$(6,10)$	$[0,X_{\mathrm{Q}}]$
R	$(4,-4)$	$[16,16]$
S	$(16,16)$	$[16,-16]$

. (7.8)

At this step we have set $X_{\mathrm{R}} = 16$ and $X_{\mathrm{S}} = -16$, because R and S are events reached by 'light' propagating from source P. The speed of light c is set to unity in this simulation.

The next step is to relate the coordinates $[T, X]$ to (n, i) by the equivalent of a Lorentz matrix, viz.,

$$\begin{bmatrix} T \\ X \end{bmatrix} = \begin{bmatrix} a & b \\ c & d \end{bmatrix} \begin{bmatrix} n \\ i \end{bmatrix}. \tag{7.9}$$

Taking the data for events R and S from (7.8) gives $a = d = 5/2$ and $b = c = -3/2$.

The final step is to input the principle of special relativity, which asserts that the laws of physics, excluding gravitation, are the same in all standard inertial frames. Given standard inertial coordinates $\{x^\mu\}$ and $\{X^\mu\}$ for two inertial frames and assuming a Lorentz transformation $X^\mu = \Lambda^\mu{}_\nu x^\nu$, we have the condition

$$\Lambda^\mu{}_\alpha \eta_{\mu\nu} \Lambda^\nu{}_\beta = \eta_{\alpha\beta}, \tag{7.10}$$

where $\eta_{\mu\nu}$ are the components of the metric tensor relative to standard coordinates, given by

$$[\eta_{\mu\nu}] = \begin{bmatrix} 1 & 0 \\ 0 & -1 \end{bmatrix}. \tag{7.11}$$

As it stands, we cannot satisfy condition (7.10) with the values of a, b, c and d as given. What is needed is an additional scaling factor to standardize frames $\mathcal{F}$ and $\mathcal{F}_0$. Such a standardization is needed because what happens at the processor level is specific to the computer and cannot be influenced, whereas the usage of the standard frames protocol leading to (7.10) is a *choice* made by the observer. We could, for example, have decided to use imperial units for frame $\mathcal{F}$ and metric units for frame $\mathcal{F}_0$, and these would not affect the overall conclusions except by introducing an inessential scaling factor. For instance, the speed of light is numerically different in different systems of units: it is only the same in standardized frames.[3]

We find the scaling factor to be $1/2$ for this simulation. When this factor is incorporated into the transformation, we find

$$[\Lambda^\mu{}_\nu] = \begin{bmatrix} 5/4 & -3/4 \\ -3/4 & 5/4 \end{bmatrix}. \tag{7.12}$$

This matrix satisfies condition (7.10). If we identify

$$\Lambda^0{}_0 \equiv \frac{1}{\sqrt{1-v^2}}, \quad \Lambda^0{}_0 \equiv -\frac{1}{\sqrt{1-v^2}} v, \tag{7.13}$$

according to standard Lorentz-transformation matrices, then we find, as expected, $v = 3/5$, which is the speed of frame $\mathcal{F}$ relative to $\mathcal{F}_0$ as encoded in the initial data for this particular simulation.

[3] Actually, our calculation shows that the speed of light would not change if we scaled space and time coordinates by the same amount.

If we denote the scaled coordinates in the processor frame $\mathcal{F}_0$ by $\{t', x'\}$ then we arrive at the final table

$$\begin{array}{cccc} \textbf{event} & (t,x) & [T,X] & \{t',x'\} \\ \mathrm{P} & (0,0) & [0,0] & \{0,0\} \\ \mathrm{Q} & (6,10) & [0,X_{\mathrm{Q}}] & \{0,8\} \\ \mathrm{R} & (4,-4) & [16,-16] & \{8,-8\} \\ \mathrm{S} & (16,16) & [16,-16] & \{8,8\} \end{array} \quad (7.14)$$

7.5.6 Conclusions

The conclusions of this analysis are as follows.

1. Time dilation and Fitzgerald length contraction can be simulated in a CA, provided that care is taken with protocols and the principles of physics.
2. A cellular array should not be interpreted literally in terms of its spatial structure. Cells are merely depositories of information, and the context of those cells must be involved in the interpretation of the physics. The same remark applies to the way humans perceive the Universe: optical data will be registered and processed in the brain, but not in the form of an exact three-dimensional reconstruction within the brain itself. Nevertheless, the best interpretation of such data is in terms of a three-dimensional model.

7.6 Classical register mechanics

In this section we discuss an approach to DT mechanics combining elements of CA theory and classical computation. The strategy is to focus on the observer and the questions that they can ask of a classical SUO.

7.6.1 Binary questions and bits

The most basic form of question is a binary question, one with two possible, mutually exclusive, answers, *yes* or else *no*. We shall assume that such a question is being asked of an SUO that can exist in only one of two answer states, corresponding to these two possible answers. Further, we assume that, when a question is asked of such an SUO[4] as to which state it is in, there is no error or deception involved in the answer, i.e., observations are classically truthful. One of the differences between CM and QM is that in QM we cannot assume that an SUO is always in a definite answer state if we have not yet asked a question and obtained an answer.

We shall represent the two possible answer states by the two-dimensional real column matrices $|0)$ and $|1)$ defined by

[4] This anthropomorphism should not be taken literally.

$$|0) \equiv \begin{bmatrix}1\\0\end{bmatrix}, \quad |1) \equiv \begin{bmatrix}0\\1\end{bmatrix}. \tag{7.15}$$

It is conventional to regard these states as the two alternative states of a single object known as a classical bit, a familiar concept in the theory of classical computation.

Questions have to be posed in such a way as to have only *yes* or *no* answers. With classical bits, two obvious and most often used questions are $\mathsf{Q}_0 \equiv$ 'is the SUO in state $|0)$?' and $\mathsf{Q}_1 \equiv$ 'is the SUO in state $|1)$?'. Other possible questions could be asked, but are beyond the scope of this chapter (Jaroszkiewicz, 2010) and are discussed in Chapter 29.

With possible quantization in mind, we choose to represent Q_0 and Q_1 by linear mappings of the column arrays (7.15) into the real numbers. Linear maps of vectors can be represented by the action of dual vectors, which in this case are represented by the row matrices

$$\mathsf{Q}_0 \mapsto (0| \equiv \begin{bmatrix}1 & 0\end{bmatrix}, \quad \mathsf{Q}_1 \mapsto (1| \equiv \begin{bmatrix}0 & 1\end{bmatrix}. \tag{7.16}$$

An arbitrary question Q_i asked of arbitrary state $|j)$ is then given by the action of the row matrix $(i|$ on the column matrix $|j)$, the result being denoted

$$(i|j) = \delta_{ij}, \tag{7.17}$$

where δ_{ij} is the Kronecker delta.[5] A value $\delta_{ij} = 0$ represents the answer *no* (no signal found), whilst a value $\delta_{ij} = 1$ represents the answer *yes* (signal found).

7.6.2 Bit operators

There are four useful operators we will use. The projection operators (or filters) P and $\bar{P}$ either pass a state or else block it. A passed state is an eigenstate of the projection operator concerned with eigenvalue $+1$, whilst a blocked state is an eigenstate with eigenvalue 0. We have the following definitions of the two filters:

$$\begin{aligned} P|0) &= |0), & \bar{P}|0) &= 0, \\ P|1) &= 0, & \bar{P}|1) &= |1). \end{aligned} \tag{7.18}$$

A representation of the filters is $P = |0)(0|$ and $\bar{P} \equiv |1)(1|$, and a resolution of the identity operator I is $P + \bar{P} = I$.

Two additional useful operators are the transition operators A and $A^\dagger$, defined by

[5] Technically, the component of the 1×1 matrix resulting from the product of a 1×2 matrix and a 2×1 matrix.

$$A \equiv |0)(1|, \quad A^\dagger \equiv |1)(0|. \tag{7.19}$$

These satisfy the algebra

$$AA = A^\dagger A^\dagger = 0, \quad AA^\dagger + A^\dagger A = I. \tag{7.20}$$

7.6.3 The void state

A classical bit has two possible answer states, $|0)$ and $|1)$. However, there is a third kind of state implied by the signal algebra (7.20), viz., the void state $|\emptyset)$, represented by

$$|\emptyset) \mapsto \begin{bmatrix} 0 \\ 0 \end{bmatrix}. \tag{7.21}$$

This is neither a *yes* $\equiv |1)$ state nor a *no* $\equiv |0)$ state. It is in a sense a state of non-existence and requires interpretation. It is one of the bit power-set elements discussed above and represents a very physical concept: the *absence* of a detector. When a bit is in the void state, there is no possibility of extracting signal information from it by the observer because the void state is neither in the ground state $|0)$ nor in the signal state $|1)$. The void state is encountered, for instance, when an attempt is made to excite a bit more than once, i.e., $A^\dagger A^\dagger |0) = |\emptyset)$, which has the interpretation that there is no state that carries two signal excitations.

The void state, as we call it, is just as meaningful as any other state in that it can be observed: an observer can walk into an empty laboratory, see that there is no detection equipment there and assign the state $|\emptyset)$ to that physical state of the laboratory. Admittedly there is nothing practical that can be done with the void state, but nevertheless it should occur in any theory aiming to describe time-dependent apparatus. It is important not to confuse the void state with the ground state. An account is given in Jaroszkiewicz (2010) and in Chapter 29.

7.6.4 Bit registers

Cellular automata generally contain vast numbers of cells. Likewise, we will extend the discussion from a single-bit SUO to one described by a classical register $\mathcal{R}^r$ consisting of a large number r of bits. By large we do not mean a million. If we wanted to simulate the physical space of the visible Universe by assigning a bit to each Planck-scale cube of space, heuristic estimates give a number of the order of 10^{185} bits.

The number r of bits in $\mathcal{R}^r$ will be called the *rank* of that register. The individual bits of such a register are labelled by an integer i running from 1 to r. How this label is assigned to each bit is quite arbitrary, but it is wise to use context. For example, if we have a line of bits then it is sensible to assign an index value $i + 1$ to the bit on one side of the line relative to bit i and index value $i - 1$ to the bit on the other side.

A general state in the register will be denoted in the occupation representation (O-rep) by $|\varepsilon_1\varepsilon_2\ldots\varepsilon_r)$, where ε_i is the occupancy of the *ith* bit, which takes the value 0 or 1. For example, a rank-2 register has four possible states, given in the O-rep by $|00)$, $|01)$, $|10)$ and $|11)$.

A variant of the O-rep is the product representation, in which the individual bits in a state are written out in the form

$$|\varepsilon_1\varepsilon_2\ldots\varepsilon_r) \mapsto |\varepsilon_1)_1 \times |\varepsilon_2)_2 \times \cdots \times |\varepsilon_r)_r. \tag{7.22}$$

In the quantized version of this theory, such a product would be replaced by a tensor product and the classical register becomes a Hilbert space $\mathcal{H}^r$ known as a quantum register. In such a scenario, the classical bit states (7.22) become the 2^r elements of a preferred basis for $\mathcal{H}^r$.

In general, a rank-r classical bit register has 2^r distinct normal states. For large rank, the O-rep becomes inefficient. For instance, suppose we assign one bit for every pixel of a computer screen, assuming a monochrome (black and white) image. A modern flat screen would typically have a resolution of $1920 \times 1060 = 2\,035\,200$ pixels. Then there would be a total of $2^{2\,035\,200}$ possible distinct images, each represented in the O-rep by a string with over two million characters. Clearly, this is very inefficient notation.

A more efficient representation is the computation representation (C-rep), where we make the identification

$$|\varepsilon_1\varepsilon_2\ldots\varepsilon_r) \mapsto \left|\sum_{i=0}^{r-1} \varepsilon_i 2^i\right\rangle. \tag{7.23}$$

We shall use round brackets for O-rep states and angular brackets for C-rep states. For example, for a rank-4 state, we have

$$|1011) \mapsto |1 \times 2^0 + 0 \times 2^1 + 1 \times 2^2 + 1 \times 2^3\rangle = |13\rangle. \tag{7.24}$$

The C-rep states in a rank-r register are labelled by an integer i running from 0 to $2^r - 1$.

There are advantages and disadvantages to each representation, so the best policy is to use whichever is most suitable for a given purpose. For example, the resolution of the identity operator $\mathbb{I}_r$ in the C-rep is more economical than that in the O-rep: for a rank-20 register we have the compact C-rep

$$\mathbb{I}_{20} = \sum_{i=0}^{2^{20}-1} |i\rangle\langle i|, \tag{7.25}$$

the corresponding expansion in the O-rep being too cumbersome to write out here.

7.6.5 The register ground state

Given a classical register $\mathcal{R}^r$, we define the register ground state to be the state $|0\rangle$ in the C-rep, equivalent to $|000\ldots0)$ in the O-rep, a state where each bit in the register is in its *no* state. The ground state is the CA analogue of the vacuum, the state in relativistic quantum field theory representing empty space. The ground state is not the same as the register void state, which represents a completely empty laboratory.

7.6.6 Arbitrary classical register operators

To construct any rank-r register operator, we first label the projection operators P_i, $\bar{P}_i$, the transition operators A_i, $A_i^\dagger$ and the bit identity I_i for the ith bit by the same index i. A general operator in the register $\mathcal{R}^r$ is then a product of r bit operators, each acting on its own bit. For example, for a rank-4 register, we might have

$$A_1^\dagger P_2 A_3 I_4 |12\rangle = A_1^\dagger P_2 A_3 I_4 |0011) \tag{7.26}$$

$$= A_1^\dagger |0)_1 \times P_2 |0)_2 \times A_3 |1)_3 \times I_4 |1)_4$$

$$= |1)_1 \times |0)_2 \times |0)_2 \times |1)_4 = |1001) = |9\rangle. \tag{7.27}$$

In general, we represent classical register operators by open-face symbols, such as $\mathbb{A}_i^+$, $\mathbb{I}^r$, etc. An arbitrary register operator $\mathbb{O}$ may be readily represented in the C-rep in the form

$$\mathbb{O} = \sum_{i=1}^{2^r-1} \sum_{j=1}^{2^r-1} |i\rangle O_{ij} \langle j|, \tag{7.28}$$

where the O_{ij} are either zeros or ones, with certain conditions imposed. These conditions will depend on the modelling. For example, we may impose the condition that, for any j, the sum $\sum_{i=1}^{2^r-1} O_{ij}$ equals one. This means that every state in the register is mapped into precisely one state in the register. This would not be the case in the quantized version of this theory. The difference is that a classical register is not a vector space, whereas a quantum register is a vector space and the vector sum of register states is a linear superposition, which is meaningful in the QM context.

7.6.7 Signal operators, signality and signality classes

Suppose we wish to excite the ith bit in the ground state so that that bit is in state $|1)_i$, with all the other bits still in the *no* state. This can be arranged by applying the signal operator

$$\mathbb{A}_i^\dagger \equiv I_1 \times I_2 \times \cdots \times I_{i-1} \times A_i^\dagger \times I_{i+1} \times \cdots \times I_r \tag{7.29}$$

to the ground state. This leads us to the algebra of the classical signal operators $\mathbb{A}_i$ and $\mathbb{A}_j^\dagger$ (Jaroszkiewicz, 2008). By inspection, we find that the signal operators behave like fermionic operators insofar as a single bit is concerned, i.e.,

$$\mathbb{A}_i\mathbb{A}_i = \mathbb{A}_i^\dagger\mathbb{A}_i^\dagger = 0_r, \quad \mathbb{A}_i\mathbb{A}_i^\dagger + \mathbb{A}_i^\dagger\mathbb{A}_i = \mathbb{I}_r \quad (\text{no sum over } i), \tag{7.30}$$

whilst satisfying bosonic operator commutation relations when different bits are involved. Specifically, commutators involving signal operators from different bits all vanish.

The signal operators can be used to construct signality classes. These are subsets of a classical register that all have the same signality, that is, the same number of *yes* components in their O-rep. For a rank-r classical register, there are $r+1$ signality classes, the kth signality class having C_k^r elements, calculated as follows.

1. Signality class 0: there is only one element, the ground state $|0\rangle$, so $C_0^r = 1$.
2. Signality class 1: there are r elements, each of the form $\mathbb{A}_i^\dagger|0\rangle$, where $i = 1, 2, \ldots, r$, so $C_1^r = r$.
3. Signality class 2: there are

$$C_2^r = \frac{r!}{2!(r-2)!} \equiv \binom{r}{2}$$

   different ways to generate states of the form $\mathbb{A}_i^\dagger\mathbb{A}_j^\dagger|0\rangle$, for $1 \leqslant i \neq j \leqslant r$. This calculation takes into account that different signal operators commute.
4. And so on until the rth signality class, consisting of one element, the fully saturated state $|2^r - 1\rangle$, so $C_r^r = 1$.

We conclude that the kth signality class subset has

$$C_k^r = \frac{r!}{k!(r-k)!} \equiv \binom{r}{k}$$

members. Since each state in the register belongs to one and only one signality class, the total number of elements C^r in the register is

$$C^r = \sum_{k=0}^{r} \binom{r}{k} = (1+1)^r = 2^r$$

as expected.

The signality of a register state is the signality class it belongs to and is useful because it tells us how many particles or excitations there are in that classical register state.

7.6.8 Evolution

Given a rank-r register $\mathcal{R}^r$ that is fixed in time and the computational representation for the states, consider the temporal evolution of an SUO state $|k, n\rangle$ at

DT n to state $|U_n k, n+1\rangle$ after one elementary time-step to time $n+1$, where k and $U_n k$ are integers in the interval $[0, 2^r - 1]$. Denoting this transition as the action of some temporal evolution operator $\mathbb{U}_n$ acting on the initial state $|k, n\rangle$, we write $\mathbb{U}_n|k, n\rangle \equiv |U_n k, n+1\rangle$, $0 \leqslant k, U_n k < 2^r$.

For a given k, there are in principle 2^r possible states $|U_n k, n+1\rangle$ into which it could be mapped, and, because there are 2^r values of k, we conclude that, for a rank-r classical register, there are $(2^r)^{2^r}$ distinct possible evolution operators in this form of mechanics.

Most of the possible evolution operators over a classical register will not be useful. Many of them will correspond to irreversible and/or unphysical dynamical evolution and only a small subset will be of interest. We need to find some principles to guide us in our choice of evolution operator.

Recall that, in CT CM, Hamilton's equations of motion lead to Liouville's theorem. This tells us that, as we track a small volume element in phase space along a classical trajectory, this volume remains constant in magnitude though not necessarily constant in shape or orientation. This leads to the idea that a system of many non-interacting particles moving along classical trajectories in phase space behaves like an incompressible fluid, such a phenomenon being referred to as a Hamiltonian flow.

An important characteristic of Hamiltonian flows is that flow lines never cross. We shall encode this idea into our approach to signal mechanics. There are two versions of this mechanics, one of which does not necessarily conserve signality whilst the other does. We consider the first one now.

Permutation flows

The physical register $\mathcal{R}^r$ contains 2^r lab states denoted by $|k\rangle$, $k = 0, 1, 2, \ldots$, $2^r - 1$. Consider a permutation P of the integers k, such that, under P, $k \rightarrow Pk \in [0, 2^r - 1]$. Define the evolution of the lab state $|k\rangle$ over one time-step by $|k\rangle \rightarrow \mathbb{U}|k\rangle = |Pk\rangle$. Such a process is reversible and will be referred to as a permutation flow.

Example 7.2 Consider $r = 3$. Then there are $2^3 = 8$ possible register states, denoted in the C-rep by $\{|i\rangle : i = 0, 1, 2, \ldots, 7\}$. Consider the permutation $0 \rightarrow 4, 1 \rightarrow 1, 2 \rightarrow 3, 3 \rightarrow 2, 4 \rightarrow 5, 5 \rightarrow 7, 6 \rightarrow 6, 7 \rightarrow 0$, conveniently denoted by $(0457)(23)$. Consider the state $|0\rangle$ in the C-rep. This permutation flow gives $|0\rangle \rightarrow |4\rangle$, which in the O-rep corresponds to $|000) \rightarrow |001)$, which clearly does not conserve signality.

There is a total of $n!$ distinct permutations of n objects, so there are $(2^r)!$ possible distinct permutation flow processes. For large r, the number of permutation flows is a rapidly decreasing fraction of the number $(2^r)^{2^r}$ of all possible forms of register processes.

Signal-conserving flows

Most permutation flows will not conserve signality, as the above example demonstrated. However, we can readily construct the subset of the permutation flows which do conserve signality by using the O-rep. Consider a physical register state $|\Psi_n)$ at time t_n given by $|\Psi_n) = |i_1 i_2 \dots i_r)$ in the O-rep, where $i_j = 0$ or else 1 for $1 \leqslant j \leqslant r$.

Now let P^* be some permutation of the numbers $1, 2, \dots, r$ and write $P^* j$ to represent the number that j changes to under this permutation. Now suppose that $|\Psi_n)$ evolves into the lab state $|\Psi_{n+1})$ at time t_{n+1} given by the rule

$$|\Psi_n) \to |\Psi_{n+1}) \equiv \mathbb{U}|\Psi_n) = |i_{P^*1} i_{P^*2} \dots i_{P^*r}). \tag{7.31}$$

To determine the new occupancy of the jth bit, we just look at the occupancy of the (P^*j)th bit. This may be summarized as the dynamical rule $i_j \to i'_j \equiv i_{P^*j}$. We shall call this form of signal mechanics *signal permutation dynamics*.

In this form of dynamics, signality is automatically conserved.

Example 7.3 Consider $r = 4$ and the permutation $1 \to 3, 3 \to 4, 4 \to 1, 2 \to 2$, conventionally written (134). Then the signality-2 state $|1010)$ is transformed into the signality state $|0011)$. Under the same dynamics, the signality-3 state $|0111)$ is transformed to the signality-3 state $|1101)$.

Note that $|1010) \to |0011)$ is equivalent to $|5\rangle \to |12\rangle$ in the C-rep and $|0111) \to |1101)$ is equivalent to $|14\rangle \to |11\rangle$.

Another way of seeing that signality is conserved is to use the signal creation operators and note that if $|\Psi_n)$ has signality d, then we can write $|\Psi_n) = \mathbb{A}^\dagger_{j_1}\mathbb{A}^\dagger_{j_2}\dots\mathbb{A}^\dagger_{j_d}|0)$, where $1 \leqslant j_1 < j_2 < \dots < j_d \leqslant r$. Then under the above permutation P^* of the integers $1, 2, \dots, r$ the new state at time t_{n+1} takes the form

$$|\Psi_{n+1}) \equiv \mathbb{U}|\Psi_n) = \mathbb{A}^\dagger_{P^*j_1}\mathbb{A}^\dagger_{P^*j_2}\dots\mathbb{A}^\dagger_{P^*j_d}|0). \tag{7.32}$$

Then clearly signality is conserved.

Example 7.4 Returning to Example (7.3), we can write the O-rep state $|1010)$ in the form $|1010) \equiv \mathbb{A}^\dagger_1\mathbb{A}^\dagger_3|0)$. Under the given signal permutation dynamics, we have $1 \to 3, 3 \to 4$, so we deduce $|1010) \equiv \mathbb{A}^\dagger_1\mathbb{A}^\dagger_3|0) \to \mathbb{A}^\dagger_3\mathbb{A}^\dagger_4|0) \equiv |0011)$, as stated.

Likewise, $|0111) \equiv \mathbb{A}^\dagger_2\mathbb{A}^\dagger_3\mathbb{A}^\dagger_4|0) \to \mathbb{A}^\dagger_2\mathbb{A}^\dagger_4\mathbb{A}^\dagger_1|0) \equiv |1101)$, as stated.

There is a certain efficiency in using the signal operators in this way to describe signality-conserving permutations. The efficiency arises because the permutation of signal operators affects only cells that change, and this depends on the signality class.

The total number of distinct permutations of r objects is $r!$, so there are that many distinct forms of signal permutation dynamics for a rank-r classical register. Since there are $(2^r)!$ distinct forms of permutation dynamics, the set of signal permutation dynamics forms a rapidly decreasing fraction of the set of all possible permutation dynamics.

Signal permutation dynamics is the closest we get in this approach to conventional Newtonian particle mechanics, in that the conservation of signality is akin to conservation of particles. Alternatively, we could think of conservation of signality as analogous to conservation of total electric charge. A one-electron system would be modelled in our register approach by a signality-1 state evolving only to other signality-1 states. Similarly, a two-electron state would be modelled by a signality-2 state evolving only to other signality-2 states. Situations where particle number is not conserved, as in the case with photons, correspond to changes in signality, and this can be accommodated in register mechanics with no difficulty whatsoever.

Permutation flows have a number of features that have analogues in standard classical mechanics. First, permutation flows are reversible. Given a permutation P, its inverse P^{-1} always exists, because permutations form a group.

Another feature of permutation dynamics is the existence of orbits or cycles. A permutation of 2^r objects will in general contain cycles, which are subsets of the objects such that only elements within a given cycle replace each other under the permutation. This is the origin of the bracket notation used in the above examples. For example, the permutation $1 \to 3, 2 \to 2, 3 \to 4, 4 \to 1$ is denoted (134), the $2 \to 2$ cycle being left out by implication.

Cycles are relevant here because they will surface in the dynamics of time-independent autonomous systems for which the evolution is given by repeated application of the same permutation. The structure of the cycles does not change in such cases, so each cycle consisting of p elements has a dynamical period p. For example, the identity permutation gives a trivial form of mechanics where nothing changes. It has 2^r cycles each of period 1. At the other end of the spectrum, the permutation denoted by $(0 \to 1 \to 2 \to \ldots \to 2^2 - 1 \to 0)$ has no cycles except itself and has period 2^r. Therefore, any physical register evolving under time-independent, autonomous permutation mechanics must return to its initial lab state no later than after 2^r time steps. This is the DT analogue of the Poincaré recurrence theorem (Poincaré, 1890).

7.6.9 Evolution and measurement

Context plays a vital role in observation. When, for instance, an observer reports that a particle has been observed at position $x = 1.5$, what they mean is that positive signals have been detected in some apparatus for which the contextual interpretation is that a signal in it corresponds to a position at $x = 1.5$. This assignment is based on the context of the experiment: the observer will know on the basis of prior theoretical knowledge what their signals mean, and it is that which gives a 'value' to any measurable quantity.

A good analogy here is with vision. Light stimulates cells on the retina, which, being two-dimensional, can in no way give a faithful three-dimensional representation of any object being perceived per se. Nevertheless, the brain has contextual information, usually in the form of visual signals previously seen, interpreted and stored in short-term memory. This contextual information is used to relate to signals coming from the retina, which are then interpreted in terms of spatial models. It should not come as a surprise, given the complexity of the process, that there can be occasional failures of this process, which are known as optical illusions.

We shall restrict the discussion here to a single run of a classical experiment, taking place over a total of N discrete time steps. We start off in a given initial register state $|\Psi_0\rangle$, which is one of the 2^r possibilities. After N steps we end up with a state denoted $|\Psi_N\rangle$.

We model the measurement process in terms of *weighted relevant questions*, which means the following. Given a rank-r register, there are 2^r possible final states. Now, in principle, the observer should ask the following question of each potential final state $|i\rangle$: 'is $|\Psi_n\rangle$ the state $|i\rangle$?'.

This question is answered by calculating the inner product $\langle i|\Psi_N\rangle$. A result 0 means that the answer to this question is 'no' and a result 1 means that the answer is 'yes'.

We note that the numbers *zero* and *one* are the only solutions to the equation $x^2 = x$, and these are the only possible values of $\langle i|\Psi_N\rangle$. Hence we can make the formal replacement

$$\langle i|\Psi_N\rangle \mapsto \langle \Psi_N|i\rangle\langle i|\Psi_N\rangle, \tag{7.33}$$

which holds for all register states $|i\rangle$ and all possible final states $|\Psi_N\rangle$, since these are also register states.[6]

We note that, from the resolution of the identity

$$\sum_{i=0}^{2^r-1} |i\rangle\langle i| = \mathbb{I}_r, \tag{7.34}$$

summing (7.33) over i gives

$$\begin{aligned}\sum_{i=0}^{2^r-1} \langle \Psi_N|i\rangle\langle i|\Psi_N\rangle &= \langle \Psi_N| \left\{ \sum_{i=0}^{2^r-1} |i\rangle\langle i| \right\} |\Psi_N\rangle \\ &= \langle \Psi_N|\mathbb{I}_r|\Psi_N\rangle = \langle \Psi_N|\Psi_N\rangle = 1. \end{aligned} \tag{7.35}$$

Now suppose that each potential final state $|i\rangle$ is contextually associated with some known 'value' X_i of some observable, where we assume for simplicity that

[6] This is a classical theory, so we do not allow superpositions of register states to represent physical states.

these values are real. Then the actual value X observed at time N would be given by

$$\begin{aligned} X &= X\langle\Psi_N|\Psi_N\rangle && \text{from (7.35)} \\ &= X\langle\Psi_N|\mathbb{I}_r|\Psi_N\rangle && \text{register identity operator} \\ &= X\langle\Psi_N|\left\{\sum_{i=0}^{2^r-1}|i\rangle\langle i|\right\}|\Psi_N\rangle && \text{resolution of identity} \\ &= \langle\Psi_N|\left\{\sum_{i=0}^{2^r-1}|i\rangle X_i\langle i|\right\}|\Psi_N\rangle && \text{by inspection,} \end{aligned} \tag{7.36}$$

which can be written

$$X = \langle\Psi_N|\mathbb{X}|\Psi_N\rangle, \tag{7.37}$$

where $\mathbb{X}$ is the register operator

$$\mathbb{X} \equiv \sum_{i=0}^{2^r-1}|i\rangle X_i\langle i|. \tag{7.38}$$

We see here the emergence of a formalism with the same structure as that used in QM, where analogous expressions arise for expectation values of observables.

In the classical case, we note that all classical observables are strictly diagonal with respect to the register state basis. The signal operators $\mathbb{A}_i^\dagger$ are not diagonal and therefore do not represent classical observables.

The equivalent of n-point functions in quantum field theory can also be discussed, representing questions about correlations between different detectors. For this we use the register projection operators

$$\mathbb{P}_i \equiv I_1 I_2 \dots I_{i-1} P_i I_{i+1} \dots I_r, \quad \bar{\mathbb{P}}_i \equiv I_1 I_2 \dots I_{i-1} \bar{P}_i I_{i+1} \dots I_r, \tag{7.39}$$

where I_j is the identity operator for the jth bit and P_i, $\bar{P}_i$ are the bit projection operators with properties given by (7.18). With the register projection operators we can ask many questions. For example, the question $Q_i(\Psi) \equiv$ 'would the state $|\Psi\rangle$ give a signal in the ith detector if we looked?' has the answer

$$A_i(\Psi) = \langle\Psi|\mathbb{P}_i|\Psi\rangle, \tag{7.40}$$

where, as before, a value zero means 'no' whilst a value one means 'yes'. A more complicated question such as $Q_{ij\bar{k}}(\Psi) \equiv$ 'would the state $|\Psi\rangle$ give a signal in detectors i and j and no signal in detector k?' would have the answer

$$A_{ij\bar{k}}(\Psi) = \langle\Psi|\mathbb{P}_i\mathbb{P}_j\bar{\mathbb{P}}_k|\Psi\rangle. \tag{7.41}$$

It is possible to ask time-dependent questions. Consider a state evolving from discrete time n_0 to some time $n_1 \geqslant n_0$. If the state at time n_0 is $|\Psi, n_0\rangle$, we denote what it has evolved to at time n_1 by

$$|\Psi, n_1\rangle \equiv \mathbb{U}_{n_1 n_0}|\Psi, n_0\rangle, \tag{7.42}$$

where $\mathbb{U}_{mn}$ is the evolution operator from time m to time n.

Now suppose we ask the question $Q_i(\Psi, n_1) \equiv$ 'is there a signal in detector i at time n_1?'. Applying the operator $\bar{\mathbb{P}}_i$ to $|\Psi, n_1\rangle$ either destroys the ith bit state by projecting it into the void state $|\emptyset)_i$ if there is no signal at that time in the ith detector, or else the state is left alone. Now suppose the SUO is left to evolve from time n_1 to time $n_2 \geqslant n_1$ and another question $Q_j(\Psi, n_2)$ is asked. This is done by applying $\bar{\mathbb{P}}_j$ to $|\Psi, n_2\rangle \equiv \mathbb{U}_{n_2 n_1}\bar{\mathbb{P}}_i|\Psi, n_1\rangle$.

Next, suppose the system is left to evolve further until final time $n_3 \geqslant n_2$, at which point we collect all answers. The answer *yes* $\equiv 1$ or *no* $\equiv 0$ to the question 'Given an evolving initial state $|\Psi, n_0\rangle$, is it true that, by time $n_3 \geqslant n_2$, a signal would have been detected at detector i at time $n_1 \geqslant n_0$ and a signal would have been detected at j at time $t_2 \geqslant t_1$?' is given by the value

$$\langle\Psi, n_0|\mathbb{U}^{+}_{n_3 n_0}\mathbb{U}_{n_3 n_2}\bar{\mathbb{P}}_j\mathbb{U}_{n_2 n_1}\bar{\mathbb{P}}_j\mathbb{U}_{n_1 n_0}|\Psi, n_0\rangle. \tag{7.43}$$

It should be clear that the equivalent of the Heisenberg picture can be constructed, in which the prepared state is frozen in discrete time whilst the projection operators are assigned time evolution. Care should be taken here because the evolution operator $\mathbb{U}_{nm}$ is not necessarily unitary. However, if permutation dynamics is involved, then the evolution operator always has an inverse (since permutations form a group), so the equivalent of an inverse operator $\mathbb{U}^{-1}_{nm}$ can be constructed.

It is possible to develop this formalism further to take into account random mixtures of initial states, in which case a classical analogue of the QM density matrix formalism can be set up in a straightforward way (Jaroszkiewicz, 2010).

Part II

Classical discrete time mechanics

8
The action sum

8.1 Configuration-space manifolds

In advanced classical mechanics (CM), a system under observation (SUO) is modelled by a single point, the *system point*, moving in some real r-dimensional space known as *configuration space*. Typically, configuration spaces are real differentiable manifolds chosen to model the mechanical degrees of freedom believed to represent the SUO.

Whilst it is always more elegant to discuss problems in CM over manifolds in a coordinate-free way (Abraham and Marsden, 2008), it is generally more convenient to use coordinates and we shall do so here. Therefore, having chosen a configuration-space manifold $\mathcal{C}$ relevant to the problem at hand, the next step is to set up a system of coordinate frames or patches to cover the region of interest in the manifold.

In CM, the region of interest will contain the trajectory of the system point over the time interval of interest. Over a finite interval of time, a system point will not move in general over the whole of its configuration space. In practice, therefore, the equations of motion can be discussed relative to a single coordinate patch, which need not be global. This is fortunate for CM, because for many manifolds, such as spheres, a single well-behaved coordinate patch covering the whole manifold cannot be constructed.

Another factor helping here is that equations of motion in continuous time (CT) CM are invariably differential-in-time equations,[1] which means that, for any given time, they can always be formulated in the neighbourhood of any given point in configuration space relative to some chosen coordinate patch.

In quantum mechanics (QM) the situation is quite different. The architecture of a typical quantum experiment requires the preparation of a state of an SUO

[1] Except possibly where impulses occur. At such points along a trajectory, position is continuous in time but velocity is not.

at some initial time t_i and then the testing for possible outcomes at a later time $t_\mathrm{f} > t_\mathrm{i}$, with no interference from the observer with the state in between those two times. Now, according to Heisenberg's uncertainty principle (Heisenberg, 1927), a system point cannot be pinned down precisely whilst it is not being observed between t_i and t_f. This is equivalent to the principle in QM that, in the absence of any information to the contrary, the system point cannot be assumed to be well defined during that time interval and could in principle be found at any given point in configuration space (with some probability), **if** the observer broke laboratory protocol and decided to look earlier than time t_f. QM therefore requires a global, i.e., non-local, discussion over the entire configuration-space manifold for any time in the open interval $(t_\mathrm{f}, t_\mathrm{i})$. This is exploited in the Feynman path-integral approach to QM (Feynman and Hibbs, 1965), which will be discussed in Chapter 18.

The global nature of quantum state temporal evolution makes QM inherently much more difficult to discuss than CM. The good news is that the discretization of time does not make this problem any worse. Indeed, because the CT limit is not required in discrete time (DT) QM, the DT approach to path integrals is less problematic than conventional Feynman path integrals, which may be technically ill-defined as continuum limits.

Each coordinate patch consists of some open subset U of $\mathcal{C}$ with a mapping F into some open subset $F(U)$ in $\mathbb{R}^r$. If the system point P is in some open subset U_P of $\mathcal{C}$ then the mapping F gives the coordinates of P relative to that coordinate patch. These coordinates are the real-number elements of the r-tuple $\mathbf{q}_\mathrm{P} \equiv \{q^1_\mathrm{P}, q^2_\mathrm{P}, \ldots, q^r_\mathrm{P}\}$ representing the image under F of P in $F(U_\mathrm{P})$.

The value of this approach in CT CM is that system point coordinates are usually differentiable functions of time subject to differential equations obtained from Newton's second law of motion, or from Euler–Lagrange equations of motion in a more advanced setting. It is possible to encounter SUOs in CT CM where the system point coordinates are continuous but non-differentiable functions of time. Such a situation occurs, for example, in the case of Brownian motion (Brown, 1828; Einstein, 1905b; Smoluchowski, 1906), the study of particles suspended in a liquid. DT mechanics drops even the requirement of temporal continuity, taking us further away from conventional mechanics and requiring us to discuss system point coordinates that cannot be continuous functions of time. Before we can understand how to deal with such situations, however, we need to review the action principle in CT CM.

8.2 Continuous time action principles

In this section we discuss the derivation of the CT equations of motion for a classical system point moving along some trajectory Γ in r-dimensional configuration space. We use an action principle and the calculus of variations, restricting the motion to a single coordinate patch with coordinates $\{q^i\}$.

At time t, the instantaneous position of the system point will be denoted by the r-tuple $\mathbf{q} \equiv (q^1, q^2, \dots, q^r)$, where each coordinate $q^i(t)$ is a differentiable function of time t. The instantaneous velocity relative to these coordinates will be denoted by $\dot{\mathbf{q}} \equiv (\dot{q}^1, \dot{q}^2, \dots, \dot{q}^r)$, where the dots denote differentiation with respect to time.

The time parameter is assumed to run from some initial time t_i to some final time t_f, these times being determined by the requirements of and the constraints imposed on the observer who is monitoring the system. It is implicit that, during the time interval $[t_\mathrm{i}, t_\mathrm{f}]$, the system point trajectory is contained within the given coordinate patch.

The time parameter is regarded here not as an attribute of the system under observation but as a parameter associated with the observer of that system. In relativity, the role of the time parameter raises questions to do with simultaneity, proper time, the principle of equivalence in general relativity, the role of local inertial frames of reference and the interpretation of relativistic QM.

The calculus of variations requires us to consider differentiable system point trajectories. This means that, at any point on such a trajectory, both $\mathbf{q}$ and $\dot{\mathbf{q}}$ exist.

In analytic CM, isolated points along a system point trajectory where the coordinates are not differentiable, such as in the case of *impulses*, which represent impacts, discontinuities in velocity and acceleration can be handled on an *ad hoc* basis. Such cases do not generally represent particular difficulties, but for SUOs where trajectories are nowhere differentiable, such as in models of Brownian motion, special mathematical technologies are needed in order to discuss the mechanics. Such stochastic processes, which we shall not discuss in this book, have even been quantized (Rabei *et al.*, 2006).

8.2.1 Newtonian versus variational paradigms

A fundamental assumption in CM is that a system point moves from its initial configuration $\mathbf{q}_\mathrm{i} \equiv \mathbf{q}(t_\mathrm{i})$ to its final configuration $\mathbf{q}_\mathrm{f} \equiv \mathbf{q}(t_\mathrm{f})$ along a well-defined, unique trajectory, Γ_c, with the subscript 'c' signifying *classical*. We shall call this the classical trajectory and note that it is unique except in special cases such as a particle in a box.

For a particle in a box, specification of the initial event $(\mathbf{q}_\mathrm{i}, t_\mathrm{i})$ and final event $(\mathbf{q}_\mathrm{f}, t_\mathrm{f})$ does not impose any condition on how many bounces against the walls of the box the particle can make in going from initial to final events. This highlights a fundamental difference between the architecture of the Newtonian approach to mechanics and that of the calculus-of-variations approach. In the former, the equations of motion are solved given initial conditions $\mathbf{q}(t_\mathrm{i})$ and $\dot{\mathbf{q}}(t_\mathrm{i})$. In this approach, the classical trajectory is unique. In the latter, the equations of motion are the same, but now they are fitted to end-point configurations. This does not rule out multiple classical solutions, which the particle-in-a-box SUO demonstrates.

There is a subtle relationship between QM and these two paradigms. A Schrödinger wavefunction evolves deterministically from its initial configuration, so evolution is Newtonian in that sense (albeit of first order in time). However, in order to use that evolved solution in a prediction, the observer has to take its inner product with a specified final outcome state, and this has the flavour of the variational paradigm.

8.2.2 Finding the classical trajectory

Our objective now will be to establish a procedure for finding the (assumed) true or classical trajectory Γ_c, given the initial system point $\mathbf{q}_i$ and the final system point $\mathbf{q}_f$.

In analytic or Lagrangian mechanics, classical trajectories are usually derived via an action principle that is based on the calculus of variations. In this approach, the first step is to write down a suitable Lagrangian, $L = L(\mathbf{q}, \dot{\mathbf{q}})$, a carefully constructed function of the instantaneous coordinates and instantaneous coordinate velocities along a given differentiable trajectory Γ in configuration space.

An important feature of Lagrangian-based mechanics is that it seems sufficient to restrict Lagrangians to functions of position and velocity. Such Lagrangians will be referred to as *canonical Lagrangians*. All fundamental quantum field theories of current interest in physics are canonical. Non-canonical Lagrangians, involving higher derivatives of the coordinates such as accelerations, have indeed been considered in CM (Tapia, 1988). However, quantization becomes much more problematical for such theories.

One reason for the ubiquity of canonical Lagrangians may be to do with the structure of manifolds. Every point in a manifold has an associated tangent space and its dual space, known as cotangent space. These vector spaces provide local vector spaces for velocity and momentum vectors, respectively. These are quantities that can be measured in principle even in QM, albeit with some difficulty on account of the non-locality of the measuring processes concerned. Accelerations and higher derivatives are inherently more difficult to measure in the laboratory, and the mathematical spaces associated with them require correspondingly more complicated mathematical structures (Marsden and West, 2001).

Having chosen a canonical Lagrangian, the next step is to write down the *action integral* $A_{fi}[\Gamma]$, which is given by

$$A_{fi}[\Gamma] \equiv \int_{t_i}^{t_f} L(\mathbf{q}, \dot{\mathbf{q}}) dt. \tag{8.1}$$

This depends on the specific trajectory Γ in configuration space being discussed. An action integral is a real-valued functional, a mapping into the real numbers from the space of all suitable differentiable trajectories in configuration space running from the initial position $\mathbf{q}_i$ to the final position $\mathbf{q}_f$.

Given an action integral $A_{\text{fi}}[\Gamma]$, the next step is to use it to determine the true or classical trajectory Γ_{c} from the set of all possible paths. The traditional method is to invoke Hamilton's principle, which asserts that Γ_{c} is a stationary point of the action integral. However, it is more useful to use the Weiss action principle (Sudarshan and Mukunda, 1983; Weiss, 1936), an enhanced version of Hamilton's principle that allows us to consider variations of the trajectory where the end points are not fixed. The Weiss action principle asserts that, for arbitrary infinitesimal variations away from the true path, the change in the action integral depends only on end-point contributions.

A brief sketch of the method is as follows. Starting with a given trajectory Γ running from $(\mathbf{q}_{\text{i}}, t_{\text{i}})$ to $(\mathbf{q}_{\text{f}}, t_{\text{f}})$, we construct a new trajectory Γ' such that, at time t in the interval $[t_{\text{i}}, t_{\text{f}}]$, the coordinates change according to the rule

$$q^i(t) \to q^{i\prime}(t) \equiv q^i(t) + \epsilon u^i(t), \tag{8.2}$$

where $\mathbf{u} \equiv (u^1, u^2, \ldots, u^r)$ are smooth variations of the trajectory and ϵ is an infinitesimal parameter that will be taken to zero at the end of the calculation, Figure 8.1.

Under such a change, the action integral along the new path is given to first order in ϵ by

$$A_{\text{fi}}[\Gamma'] = A_{\text{fi}}[\Gamma] + \epsilon \left[\mathbf{u} \cdot \frac{\partial L}{\partial \dot{\mathbf{q}}}\right]_{t_{\text{i}}}^{t_{\text{f}}} + \epsilon \int_{t_{\text{i}}}^{t_{\text{f}}} \mathbf{u} \cdot \left\{\frac{\partial L}{\partial \mathbf{q}} - \frac{d}{dt}\left(\frac{\partial L}{\partial \dot{\mathbf{q}}}\right)\right\} dt + O(\epsilon^2), \tag{8.3}$$

after integration by parts. Here we use the notion

$$\mathbf{u} \cdot \frac{\partial L}{\partial \mathbf{q}} \equiv \sum_{i=1}^{r} u^i \frac{\partial L}{\partial q^i}. \tag{8.4}$$

This leads us to define the functional derivative of the action integral with respect to $\mathbf{u}$:

$$\begin{aligned} \mathcal{D}_{\mathbf{u}} A_{\text{fi}}[\Gamma] &\equiv \lim_{\epsilon \to 0} \frac{A_{\text{fi}}[\Gamma'] - A_{\text{fi}}[\Gamma]}{\epsilon} \\ &= \left[\mathbf{u} \cdot \frac{\partial L}{\partial \dot{\mathbf{q}}}\right]_{t_{\text{i}}}^{t_{\text{f}}} + \int_{t_{\text{i}}}^{t_{\text{f}}} \mathbf{u} \cdot \left\{\frac{\partial L}{\partial \mathbf{q}} - \frac{d}{dt}\left(\frac{\partial L}{\partial \dot{\mathbf{q}}}\right)\right\} dt. \end{aligned} \tag{8.5}$$

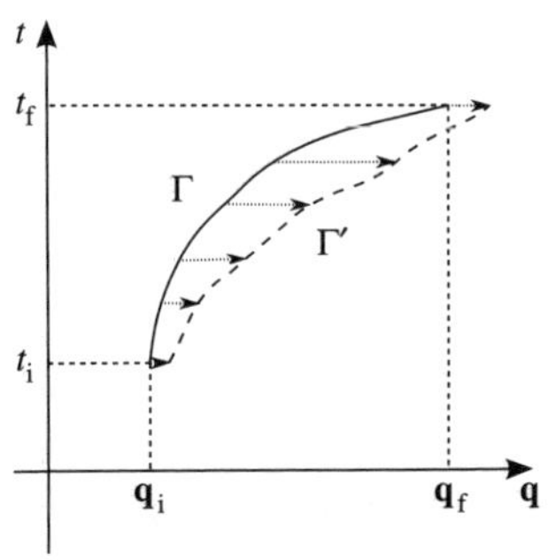

Figure 8.1. A variation $\Gamma \to \Gamma'$ of path in configuration space.

The Weiss action principle asserts that the functional derivative (8.5), evaluated on the true trajectory $\Gamma_{\rm c}$, depends only on the end-point contributions, for arbitrary path variation $\mathbf{u}$, i.e.,

$$\mathcal{D}_{\mathbf{u}} A_{\rm fi}[\Gamma_{\rm c}] \underset{\rm c}{=} \left[\mathbf{u} \cdot \frac{\partial L}{\partial \dot{\mathbf{q}}} \right]_{t_{\rm i}}^{t_{\rm f}}, \tag{8.6}$$

where we use the notation $\underset{\rm c}{=}$ to denote an equality that holds only on the true or classical trajectory. Hence we deduce

$$\int_{t_{\rm i}}^{t_{\rm f}} \mathbf{u} \cdot \left\{ \frac{\partial L}{\partial \mathbf{q}} - \frac{d}{dt}\left(\frac{\partial L}{\partial \dot{\mathbf{q}}} \right) \right\} dt \underset{\rm c}{=} 0 \tag{8.7}$$

for any arbitrary variation $\mathbf{u}$. Since this has to hold for any variation, we deduce that the integrand has to vanish pointwise along the trajectory $\Gamma_{\mathbf{c}}$, i.e., we arrive at the Euler–Lagrange equations of motion

$$\boxed{\frac{d}{dt}\left(\frac{\partial L}{\partial \dot{q}^i} \right) \underset{\rm c}{=} \frac{\partial L}{\partial q^i}, \qquad i = 1, 2, \ldots, r.} \tag{8.8}$$

We chose to use the Weiss action principle because it leads us directly to Noether's theorem, a theorem of great importance in mechanics and field theory. Suppose we have chosen a variation of path $\mathbf{v}$ that is a *near symmetry* of the Lagrangian. What we mean by this is that under the point transformation

$$q^i(t) \to q^{i\prime}(t) = q^i(t) + \epsilon v^i(t) \tag{8.9}$$

the Lagrangian changes according to the rule

$$L(\mathbf{q}, \dot{\mathbf{q}}) \to L(\mathbf{q}', \dot{\mathbf{q}}') = L(\mathbf{q}, \dot{\mathbf{q}}) + O(\epsilon^2), \tag{8.10}$$

i.e., such that there is no term that is of first order in ϵ. Then clearly

$$\mathcal{D}_{\mathbf{v}} A_{\rm fi}[\Gamma] = 0. \tag{8.11}$$

Hence, from the Weiss action principle (8.6), we deduce

$$\mathbf{v} \cdot \left. \frac{\partial L}{\partial \dot{\mathbf{q}}} \right|_{t_{\rm f}} \underset{\rm c}{=} \mathbf{v} \cdot \left. \frac{\partial L}{\partial \dot{\mathbf{q}}} \right|_{t_{\rm i}}. \tag{8.12}$$

The conclusion is that the quantity $\mathbf{v} \cdot \partial L / \partial \dot{\mathbf{q}}$ is conserved along a classical trajectory.

There are two special cases to consider.

1. The point transformation may be an *exact symmetry* of the Lagrangian, i.e.,

$$L(\mathbf{q}, \dot{\mathbf{q}}) \to L(\mathbf{q}', \dot{\mathbf{q}}') = L(\mathbf{q}, \dot{\mathbf{q}}), \tag{8.13}$$

2. The point transformation may be a *dynamical symmetry* of the Lagrangian, i.e., is a near or exact symmetry of the Lagrangian only on the true classical trajectory.

8.3 The discrete time action principle

We are now in a position to adapt the results of the previous section to DT mechanics. We do not need to change the configuration-space manifold, the only change of significance in this respect being that we replace continuous paths in configuration space by discrete sets of points.

In the DT scenario, we imagine that the system point is observed at a finite number of points in configuration space. These points will be labelled by an integer n, which runs from some initial value M to some final value N, with $M < N$. We identify $\mathbf{q}_M$ with $\mathbf{q}_\mathrm{i}$ and $\mathbf{q}_N$ with $\mathbf{q}_\mathrm{f}$, the initial and final points discussed in the previous section.

In classical DT mechanics, the $N+1-M$ points on a DT trajectory correspond to those places where the system point is known to occur. If t_n is the observer's time for which the system point is at $\mathbf{q}_n$ then, if the system point were actually moving along a continuous trajectory, we would make the identification $\mathbf{q}_n \equiv \mathbf{q}(t_n)$.

An important condition imposed on the labelling of such discrete trajectories is that $t_n < t_{n+1}$, i.e., that increasing the integer n corresponds to an increase in the observer's laboratory time. Alternatively, we could take $t_n > t_{n+1}$ for all n. What we will exclude in this chapter is the possibility that the temporal ordering changes along the trajectory. Such a possibility arises in CT mechanics. Specifically, the Feynman–Stueckelberg interpretation of anti-particles (Stueckelberg, 1941; Bjorken and Drell, 1964) takes negative-energy solutions of particle wave equations as positive-energy particles moving backwards in time. We shall encounter in Chapters 10 and 14 approaches to classical DT mechanics where the possibility that $t_{n-1} > t_n < t_{n+1}$ occurs.

A fundamental point has to be addressed now, concerning the values of the time differences $\mathbb{D}_n t_n \equiv t_{n+1} - t_n$. There are three distinct possibilities, leading to three different types of DT mechanics.

1. **Type 1: regular fixed time intervals**

   In such a DT mechanics, we have the condition

$$\mathbb{D}_n t_n = T, \tag{8.14}$$

   where T has the same fixed value for $M \leqslant n < N$ given by

$$T = \frac{t_N - t_M}{N - M}. \tag{8.15}$$

2. **Type 2: irregular fixed time intervals**

   In such a DT mechanics, the temporal intervals $\mathbb{D}_n t_n \equiv t_{n+1} - t_n$ are fixed but not necessarily equal, and subject to the conditions

$$\begin{aligned} &\text{(i)} \quad \mathbb{D}_n t_n > 0, \quad n = M, M+1, \ldots, N-1, \\ &\text{(ii)} \quad \sum_{n=M}^{N-1} \mathbb{D}_n t_n = t_N - t_M. \end{aligned} \tag{8.16}$$

3. **Type 3: dynamically determined time intervals**
   In this form of DT mechanics, the temporal intervals $\mathbb{D}_n t_n$ become dynamical variables and are determined by the dynamics. Whether they are subject to condition (i) for Type-2 DT mechanics depends on the modelling.

We are now in a position to develop a variational approach to DT mechanics for Type-1 and Type-2 discretizations. Given an initial system point $\mathbf{q}_M$ and a final system point $\mathbf{q}_N$, consider a chain Γ of system point nodes $\{\mathbf{q}_M, \mathbf{q}_{M+1}, \ldots, \mathbf{q}_N\}$. For each n such that $M \leqslant n < N$, define the *system function*

$$F^n \equiv F(\mathbf{q}_n, \mathbf{q}_{n+1}, n), \tag{8.17}$$

where F^n is some chosen function modelling the system under observation. The system function plays the role of the Lagrangian in CT dynamics. Knowing F^n, we can work out many important details about the dynamics.

Given Γ and the system function, we now calculate the *action sum*, defined by

$$A_{NM}[\Gamma] \equiv \sum_{n=M}^{N-1} F^n. \tag{8.18}$$

This is the DT analogue of the action integral in CT mechanics.

The system function is in fact more closely related to *Hamilton's principal function*, an integral over time of a CT Lagrangian, than to the Lagrangian itself, i.e.,

$$F^n \sim \int_{t_n}^{t_{n+1}} L\, dt, \tag{8.19}$$

where we integrate over a classical trajectory, i.e., a solution of the CT equations of motion. This is discussed in more detail in Chapter 12.

Now consider a variation $\{\mathbf{u}_n\}$ of path such that

$$\mathbf{q}_n \longrightarrow \mathbf{q}'_n \equiv \mathbf{q}_n + \varepsilon \mathbf{u}_n, \qquad M \leqslant n \leqslant N. \tag{8.20}$$

where ε is a real parameter and the

$$\mathbf{u}_n \equiv \{u_n^1, u_n^2, \ldots, u_n^r\} \tag{8.21}$$

are a set of real quantities assumed independent of ε. Define the *functional derivative*

$$\mathcal{D}_{\mathbf{u}} A_{MN}[\Gamma] \equiv \lim_{\varepsilon \longrightarrow 0} \left\{ \frac{A_{MN}[\Gamma'] - A_{MN}[\Gamma]}{\varepsilon} \right\}, \tag{8.22}$$

where Γ' is the new path. Now, for all systems of interest, the system function will be a real analytic function of the coordinates, so we find

$$\mathcal{D}_{\mathbf{u}} A_{MN}[\Gamma] = \sum_{n=M}^{N-1} \left\{ \mathbf{u}_n \cdot \frac{\partial}{\partial \mathbf{q}_n} + \mathbf{u}_{n+1} \cdot \frac{\partial}{\partial \mathbf{q}_{n+1}} \right\} F^n$$

$$= \mathbf{u}_M \cdot \frac{\partial F^M}{\partial \mathbf{q}_M} + \mathbf{u}_N \cdot \frac{\partial F^{N-1}}{\partial \mathbf{q}_N} + \sum_{n=M+1}^{N-1} \mathbf{u}_n \cdot \frac{\partial}{\partial \mathbf{q}_n} \{F^n + F^{n-1}\}. \tag{8.23}$$

8.4 The discrete time equations of motion

We will now apply the DT analogue of the Weiss action principle (Sudarshan and Mukunda, 1983) for CT. We suppose that the true or classical trajectory Γ_c from given $\mathbf{q}_M$ to $\mathbf{q}_N$ is such that

$$\mathcal{D}_{\mathbf{u}} A_{MN}[\Gamma_c] \underset{c}{=} \mathbf{u}_M \cdot \frac{\partial F^M}{\partial \mathbf{q}_M} + \mathbf{u}_N \cdot \frac{\partial F^{N-1}}{\partial \mathbf{q}_N}, \tag{8.24}$$

for all variations $\mathbf{u}_n$, $M \leqslant n \leqslant N$. Equivalently, we assert that

$$\sum_{n=M+1}^{N-1} \mathbf{u}_n \cdot \frac{\partial}{\partial \mathbf{q}_n} \{F^n + F^{n-1}\} \underset{c}{=} 0 \quad \forall \mathbf{u}_n. \tag{8.25}$$

The conclusion is that Γ_c must be such that

$$\boxed{\frac{\partial}{\partial \mathbf{q}_n} \{F^n + F^{n-1}\} \underset{c}{=} \mathbf{0}}, \quad M < n < N, \tag{8.26}$$

which we shall refer to as Cadzow's equations of motion. These are the DT equations of motion for node variables (Cadzow, 1970; Marsden and West, 2001). They can be understood as analogues of conservation-of-momentum equations in CT mechanics if we make the following definitions of node momenta:

$$\mathbf{p}_n^{(-)} \equiv -\frac{\partial}{\partial \mathbf{q}_n} F^n, \quad \mathbf{p}_{n-1}^{(+)} \equiv \frac{\partial}{\partial \mathbf{q}_n} F^{n-1}. \tag{8.27}$$

Then (8.26) is equivalent to

$$\mathbf{p}_n^{(-)} \underset{c}{=} \mathbf{p}_{n-1}^{(+)}. \tag{8.28}$$

Although the momenta (8.27) are link variables, the equations (8.28) mean that it is meaningful to assign a unique node momentum $\mathbf{p}_n$ at the nth temporal node, such that $\mathbf{p}_n \underset{c}{=} \mathbf{p}_n^{(-)}$ (Marsden and West, 2001).

8.5 The discrete time Noether theorem

We now discuss an important theorem, which, in the form given by Noether, has wide applicability in CT mechanics.

Suppose that we are given a DT trajectory $\Gamma \equiv \{\mathbf{q}_n\}$ and deform it by a variation $\Gamma \to \Gamma_{\mathbf{u}} \equiv \{\mathbf{q}_n + \varepsilon \mathbf{u}_n\}$ such that the variation $\{\mathbf{u}_n\}$ is an exact symmetry of the system function, i.e.,

$$F^n(\mathbf{q}_n + \varepsilon \mathbf{u}_n, \mathbf{q}_{n+1} + \varepsilon \mathbf{u}_{n+1}, n) = F^n(\mathbf{q}_n, \mathbf{q}_{n+1}, n) \tag{8.29}$$

for any real ε and for $M \leqslant n < N$. Alternatively, we could perform a more restricted variation, where the equality holds only on a classical trajectory, i.e.,

$$F^n(\mathbf{q}_n + \varepsilon \mathbf{u}_n, \mathbf{q}_{n+1} + \varepsilon \mathbf{u}_{n+1}, n) \underset{c}{=} F^n(\mathbf{q}_n, \mathbf{q}_{n+1}, n) \tag{8.30}$$

for any $\varepsilon \in \mathbb{R}$ and for $M \leqslant n < N$. In this case, we refer to the symmetry as a *dynamical symmetry* of the system function. Then, by performing a Taylor expansion in the arbitrary parameter ε and comparing coefficients of ε, we deduce that, on the true trajectory Γ_c,

$$\left\{ \mathbf{u}_n \cdot \frac{\partial}{\partial \mathbf{q}_n} + \mathbf{u}_{n+1} \cdot \frac{\partial}{\partial \mathbf{q}_{n+1}} \right\} F^n \underset{c}{=} 0, \tag{8.31}$$

for $M \leqslant n < N$. Now, applying the equation of motion (8.26), we deduce

$$\mathbf{u}_n \cdot \frac{\partial}{\partial \mathbf{q}_n} F^n \underset{c}{=} \mathbf{u}_{n+1} \cdot \frac{\partial}{\partial \mathbf{q}_{n+1}} F^{n+1}, \tag{8.32}$$

for $M \leqslant n < N - 1$. By inspection, we see that (8.32) implies that the quantity

$$Q^n_{\mathbf{u}} \equiv \mathbf{u}_n \cdot \frac{\partial}{\partial \mathbf{q}_n} F^n \tag{8.33}$$

is conserved, i.e.,

$$Q^{n+1}_{\mathbf{u}} \underset{c}{=} Q^n_{\mathbf{u}}. \tag{8.34}$$

Such invariants of the motion were discussed by Maeda (1981).

More generally, suppose that $\{\mathbf{u}_n\}$ is a *near symmetry*, i.e., for

$$\mathbf{q}_n \longrightarrow \mathbf{q}'_n \underset{c}{=} \mathbf{q}_n + \varepsilon \mathbf{u}_n \tag{8.35}$$

we find

$$F^n(\mathbf{q}_n + \varepsilon \mathbf{u}_n, \mathbf{q}_{n+1} + \varepsilon \mathbf{u}_{n+1}, n) \underset{c}{=} F^n(\mathbf{q}_n, \mathbf{q}_{n+1}, n) + \varepsilon G_{n+1} - \varepsilon G_n + O(\varepsilon^2),$$

where $G_n \equiv G(\mathbf{q}_n, \mathbf{q}_{n+1})$ is some function of the indicated variables. Then we deduce that the quantity

$$Q^n_{\mathbf{u}} \equiv \mathbf{u}_n \cdot \frac{\partial F^n}{\partial \mathbf{q}_n} + G_n \tag{8.36}$$

is a constant of the motion. Such an invariant will be referred to as a Logan invariant (Logan, 1973).

8.6 Conserved quantities via the discrete time Weiss action principle

In the previous section we derived conserved quantities from various types of symmetry directly, using the DT equations of motion. A more elegant approach, related to the DT Schwinger action principle which we shall use in the quantization of DT field theories, is to use the DT Weiss action principle. This is the DT analogue of the CT Weiss action principle discussed in Section 8.2.

The DT Weiss action principle is given by (8.24). Now suppose that the variation $\{\mathbf{u}_n : M \leqslant n \leqslant N\}$ is at worst a near symmetry of the action sum. Then the functional derivative (8.22) is zero, so we deduce that the right-hand side of (8.24) is zero, i.e.,

$$\mathbf{u}_M \cdot \frac{\partial F^M}{\partial \mathbf{q}_M} \underset{c}{=} -\mathbf{u}_N \cdot \frac{\partial F^{N-1}}{\partial \mathbf{q}_N}. \tag{8.37}$$

If now the DT equations of motion hold for $n = M$ and $n = N$, then we deduce that the left-hand side of (8.37) is conserved, as was obtained directly in the previous section.

9
Worked examples

In this chapter we apply the principles of discrete time (DT) mechanics discussed in the previous chapter to some important examples of Type-1 DT mechanical systems.

9.1 The complex harmonic oscillator

Consider a system under observation (SUO) described by complex coordinates z_n, z_n^* and system function

$$F^n(z_n, z_{n+1}) \equiv \alpha\big(z_n^* z_n + z_{n+1}^* z_{n+1}\big) - \beta\big(z_n^* z_{n+1} + z_n z_{n+1}^*\big), \tag{9.1}$$

where α and β are real with $\beta \neq 0$. From (8.26) the equations of motion are

$$\frac{\partial}{\partial z_n}\{F^n + F^{n-1}\} \underset{c}{=} 0, \quad \frac{\partial}{\partial z_n^*}\{F^n + F^{n-1}\} \underset{c}{=} 0, \quad M < n < N, \tag{9.2}$$

which gives

$$z_{n+1} \underset{c}{=} 2\eta z_n - z_{n-1}, \quad M < n < N, \tag{9.3}$$

and its complex conjugate, where $\eta \equiv \alpha/\beta$.

The system function (9.1) is invariant with respect to the global phase transformation

$$\begin{aligned} z_n &\longrightarrow z_n' \equiv \exp(i\varepsilon\theta)z_n, & z_{n+1} &\longrightarrow z_{n+1}' \equiv \exp(i\varepsilon\theta)z_{n+1}, \\ z_n^* &\longrightarrow z_n^{*\prime} \equiv \exp(-i\varepsilon\theta)z_n^*, & z_{n+1}^* &\longrightarrow z_{n+1}^{*\prime} \equiv \exp(-i\varepsilon\theta)z_{n+1}^*, \end{aligned} \tag{9.4}$$

where ε and θ are real. For infinitesimal ε we have

$$z_n \longrightarrow z_n' = z_n + i\varepsilon\theta z_n + O(\varepsilon^2), \tag{9.5}$$

etc., from which we can take $u_n = i\theta z_n$. Hence the conserved charge is

$$Q_u^n = \left\{u_n \frac{\partial}{\partial z_n} + u_n^* \frac{\partial}{\partial z_n^*}\right\} F^n, \tag{9.6}$$

which leads to the prediction that the quantity

$$Q^n \equiv i\left(z_n^* z_{n+1} - z_n z_{n+1}^*\right) \tag{9.7}$$

is conserved under the equations of motion (9.3), i.e.

$$Q^{n+1} \underset{c}{=} Q^n, \tag{9.8}$$

which is readily verified.

There is a *Logan invariant* analogous to the conserved Hamiltonian in the corresponding continuous time (CT) theory. At this point the above notation becomes somewhat cumbersome and a better notation is to work with the column variable $\mathbf{Z}_n$ defined by

$$\mathbf{Z}_n \equiv \begin{bmatrix} z_n \\ z_{n+1} \end{bmatrix}. \tag{9.9}$$

Then the system function (9.1) may be written in the form

$$F^n = \mathbf{Z}_n^+ \mathbf{F} \mathbf{Z}_n, \tag{9.10}$$

where $\mathbf{F}$ is the matrix

$$\mathbf{F} \equiv \begin{bmatrix} \alpha & -\beta \\ -\beta & \alpha \end{bmatrix}, \tag{9.11}$$

and the equation of motion (9.3) then takes the form

$$\mathbf{Z}_{n+1} \underset{c}{=} \mathbf{E}\mathbf{Z}_n, \quad \mathbf{E} \equiv \begin{bmatrix} 0 & 1 \\ -1 & 2\eta \end{bmatrix}. \tag{9.12}$$

The charge (9.7) may be written in the form

$$Q_n \equiv \mathbf{Z}_n^+ \mathbf{Q} \mathbf{Z}_n, \quad \mathbf{Q} \equiv \begin{bmatrix} 0 & i \\ -i & 0 \end{bmatrix}, \tag{9.13}$$

and then conservation follows immediately from the result $\mathbf{E}^+\mathbf{Q}\mathbf{E} = \mathbf{Q}$.

A Logan invariant is constructed by considering the change to order ε of the system function under the transformation $\mathbf{Z}_n \to \mathbf{Z}_n + \delta\mathbf{Z}_n$, where

$$\delta\mathbf{Z}_n \underset{c}{=} \varepsilon \mathbf{L} \mathbf{Z}_n, \quad \mathbf{L} \equiv \begin{bmatrix} \alpha & -\beta \\ \beta & -\alpha \end{bmatrix}, \tag{9.14}$$

which gives

$$\delta F^n \underset{c}{=} 2\varepsilon \mathbf{Z}_n^+ \begin{bmatrix} \alpha^2 - \beta^2 & 0 \\ 0 & \beta^2 - \alpha^2 \end{bmatrix} \mathbf{Z}_n. \tag{9.15}$$

By inspection, we deduce that transformation (9.14) is a near symmetry of the system function, which is actually a difference, i.e.,

$$\delta F^n = \mathbb{D}_n v_n, \quad v_n \equiv \varepsilon\{\beta^2 - \alpha^2\}|z_n|^2. \tag{9.16}$$

This means that we can construct a Logan invariant, denoted C^n, using the rule

$$C^n \propto \frac{\partial F^n}{\partial z_n}\delta z_n + \frac{\partial F^n}{\partial z_n^*}\delta z_n^* + v_n. \tag{9.17}$$

Taking out unnecessary constant factors, we take as our Logan invariant

$$C^n \equiv \mathbf{Z}_n^+\mathbf{C}\mathbf{Z}_n, \quad \mathbf{C} \equiv \begin{bmatrix} \beta & -\alpha \\ -\alpha & \beta \end{bmatrix}. \tag{9.18}$$

This is conserved because

$$C^{n+1} = \mathbf{Z}_{n+1}^+\mathbf{C}\mathbf{Z}_{n+1} \underset{c}{=} (\mathbf{Z}_n^+\mathbf{E}^+)\mathbf{C}(\mathbf{E}\mathbf{Z}_n) = \mathbf{Z}_n^+(\mathbf{E}^+\mathbf{C}\mathbf{E})\mathbf{Z}_n = \mathbf{Z}_n^+\mathbf{C}\mathbf{Z}_n = C^n, \tag{9.19}$$

using the result $\mathbf{E}^+\mathbf{C}\mathbf{E} = \mathbf{C}$.

9.2 The anharmonic oscillator

In CT mechanics, the anharmonic oscillator is a single point-particle system of mass m moving in three-dimensional space under the action of a modified harmonic oscillator potential. Configuration space is three-dimensional Euclidean space $\mathbb{E}^3$, which permits us to use a single global coordinate patch $\{x^1, x^2, x^3\}$, where superscripts represent distinct coordinates, not exponents. The Lagrangian is given by (Jaroszkiewicz, 1995a)

$$L_{\mathrm{AO}}(\mathbf{r}, \dot{\mathbf{r}}) \equiv \frac{1}{2}m\dot{\mathbf{r}}\cdot\dot{\mathbf{r}} - \frac{1}{2}m\omega^2\mathbf{r}\cdot\mathbf{r} - \frac{\lambda}{4}(\mathbf{r}\cdot\mathbf{r})^2, \tag{9.20}$$

where $\mathbf{r} \equiv (x^1, x^2, x^3)$.

The DT anharmonic oscillator in three spatial dimensions (Jaroszkiewicz, 1994a) has the system function

$$F^n \equiv -\beta\mathbf{r}_n\cdot\mathbf{r}_{n+1} + \frac{\alpha}{2\gamma}\ln\{1+\gamma\mathbf{r}_n\cdot\mathbf{r}_n\} + \frac{\alpha}{2\gamma}\ln\{1+\gamma\mathbf{r}_{n+1}\cdot\mathbf{r}_{n+1}\} \tag{9.21}$$

in Cartesian coordinates, where α, β and γ are constants with $\gamma \neq 0$. To see in what sense this system function represents the anharmonic oscillator (9.20), we need to take limits carefully. If T is the chronon, then making the replacements

$$\beta \to \frac{m}{T} + \frac{m\omega^2}{2}T, \quad \alpha \to \frac{m}{T}, \quad \gamma \to \frac{\lambda T^2}{2m} \tag{9.22}$$

and assuming a trajectory of the form

$$\mathbf{r}_n \equiv \mathbf{r}, \quad \mathbf{r}_{n+1} = \mathbf{r} + \dot{\mathbf{r}}T + \frac{1}{2}\ddot{\mathbf{r}}T^2 + O(T^3) \tag{9.23}$$

allows us to find

$$\lim_{T\to 0}\frac{F^n}{T} = L_{\mathrm{AO}}(\mathbf{r}, \dot{\mathbf{r}}). \tag{9.24}$$

The system function (9.21) gives the DT equation of motion

$$\mathbf{r}_{n+1} \underset{c}{=} \frac{2\eta}{1+\gamma \mathbf{r}_n \cdot \mathbf{r}_n}\mathbf{r}_n - \mathbf{r}_{n-1}, \quad \eta \equiv \alpha/\beta. \tag{9.25}$$

The formal correspondence of (9.25) with the equation of motion (6.30) motivates the terminology *anharmonic oscillator* for the SUO discussed in Section 6.2.1.

When the anharmonic parameter γ in (9.25) tends to zero, we recover the DT harmonic oscillator equation

$$\mathbf{r}_{n+1} \underset{c}{=} 2\eta \mathbf{r}_n - \mathbf{r}_{n-1}, \quad \eta \equiv \alpha/\beta. \tag{9.26}$$

If we make the replacements (9.22), assume a trajectory of the form (9.23) and take the limit $T \to 0$, we recover the expected Newtonian equation for the anharmonic oscillator, viz.,

$$m\ddot{\mathbf{r}} \underset{c}{=} -m\omega \mathbf{r} - \lambda(\mathbf{r}\cdot\mathbf{r})\mathbf{r}. \tag{9.27}$$

If now we make the infinitesimal transformations

$$\mathbf{r}_n \longrightarrow \mathbf{r}'_n \equiv \mathbf{r}_n + \varepsilon \mathbf{u}_n, \tag{9.28}$$

where ε is infinitesimal and

$$\mathbf{u}_n \equiv \eta \mathbf{r}_n - (1+\gamma \mathbf{r}_n \cdot \mathbf{r}_n)\mathbf{r}_{n+1}, \tag{9.29}$$

then we find that the transformation is a near symmetry of the system function, with

$$v_n = \beta \mathbf{r}_n \cdot \mathbf{r}_n - \frac{\eta\alpha \mathbf{r}_n \cdot \mathbf{r}_n}{1+\gamma \mathbf{r}_n \cdot \mathbf{r}_n}. \tag{9.30}$$

From the DT Weiss action principle (Sudarshan and Mukunda, 1983; Weiss, 1936) we obtain the conserved quantity

$$C^n \equiv \frac{1}{2}\beta(\mathbf{r}_n \cdot \mathbf{r}_n + \mathbf{r}_{n+1}\cdot \mathbf{r}_{n+1}) - \alpha \mathbf{r}_n \cdot \mathbf{r}_{n+1} + \frac{1}{2}\beta\gamma(\mathbf{r}_n \cdot \mathbf{r}_n)(\mathbf{r}_{n+1}\cdot \mathbf{r}_{n+1}). \tag{9.31}$$

This is a three-dimensional version of the invariant discussed in Section 6.2. If again we make the replacements (9.22), assume a trajectory of the form (9.23) and take the limit $T \to 0$, we find

$$\lim_{T\to 0} \frac{C^n}{T} = \frac{1}{2}m\dot{\mathbf{r}}\cdot\dot{\mathbf{r}} + \frac{1}{2}m\omega^2 \mathbf{r}\cdot\mathbf{r} + \frac{\lambda}{4}(\mathbf{r}\cdot\mathbf{r})^2, \tag{9.32}$$

which corresponds to conserved energy in the Newtonian CT theory.

The system function (9.21) is invariant with respect to spatial rotations,

$$\begin{aligned} \mathbf{r}_n &\longrightarrow \mathbf{r}'_n \equiv R(\varepsilon\boldsymbol{\omega})\mathbf{r}_n, \\ \mathbf{r}_{n+1} &\longrightarrow \mathbf{r}'_{n+1} \equiv R(\varepsilon\boldsymbol{\omega})\mathbf{r}_{n+1}, \end{aligned} \tag{9.33}$$

where $R(\varepsilon\boldsymbol{\omega})$ is an orthogonal matrix and $\boldsymbol{\omega}$ is a fixed three-vector. If we write

$$\mathbf{r}'_n \equiv \mathbf{r}_n + \varepsilon\boldsymbol{\omega} \times \mathbf{r}_n + O(\varepsilon^2) \tag{9.34}$$

then we find a conserved 'angular-momentum' vector

$$\mathbf{L}_n \equiv \mathbf{r}_n \times \mathbf{r}_{n+1} \underset{c}{=} \mathbf{L}_{n+1}. \tag{9.35}$$

9.3 Relativistic-particle models

Next, we discuss two models of a special relativistic particle of rest mass m 'skipping' over continuous four-dimensional Minkowski spacetime. We shall work in a single inertial frame $\mathcal{F}$ with standard Cartesian coordinates $\{x^\mu\}$, taking our metric tensor η to have diagonal components $(+1, -1, -1, -1)$ and zero elsewhere.

The architecture of these models is analogous to that of a stone skipping over the surface of a pond, the places where the stone touches the pond constituting the discrete points being discussed.

In Model 1, successive appearances are separated by equal intervals of the particle's 'proper time', whilst in Model 2 successive proper time intervals themselves are treated as dynamical variables subject to equations of motion. The latter model is related to Lee's approach to DT mechanics, which will be discussed in the next chapter.

9.3.1 Relativistic-particle Model 1

We start by recalling that, for CT relativistic-particle mechanics, an appropriate Lagrangian for a free relativistic particle of rest mass m is given by

$$L(x^\mu, \dot{x}^\mu) = -mc^2\sqrt{\dot{x}^2}, \tag{9.36}$$

where c is the speed of light and we use the metric tensor components $\eta_{\mu\nu}$ to lower indices, i.e., we define $\dot{x}_\mu = \eta_{\mu\nu}\dot{x}^\nu$ and $\dot{x}^2 \equiv \dot{x}^\mu \dot{x}_\mu$. Here the dots denote differentiation with respect to the path parameter, which is chosen to be proper time, and we use the summation convention.

In some respects, the DT approach to such a model system is conceptually more satisfactory than in CT mechanics, because in the DT approach we do not need to specify what the continuous path parameter means. In our first model, we discretize the right-hand side of (9.36) in the obvious way:

$$-mc^2\sqrt{\dot{x}^2} \to -mc^2\frac{\sqrt{(x_{n+1}-x_n)^2}}{T}. \tag{9.37}$$

The appropriate action sum (Jaroszkiewicz, 1994a) contains Lagrange multipliers λ_n that enforce the condition that the proper time T between successive appearances is constant along the trajectory,

$$A_{MN}[\Gamma] \equiv -\sum_{n=M}^{N-1}\left\{mc\sqrt{(x_{n+1}-x_n)^2} + \lambda_n[(x_{n+1}-x_n)^2 - c^2T^2]\right\}. \tag{9.38}$$

The procedure is to treat the coordinates x_n^μ as node variables and the Lagrange multipliers as link variables. From (9.38) we read off the system function

$$F^n(x_n, x_{n+1}, \lambda_n) = -mc\sqrt{(x_{n+1} - x_n)^2} + \lambda_n[(x_{n+1} - x_n)^2 - c^2T^2]. \quad (9.39)$$

Taking the link-variable equations of motion enforces the conditions

$$(x_{n+1} - x_n)^2 \underset{c}{=} c^2T^2, \quad M \leqslant n < N. \quad (9.40)$$

For the node variables we find

$$\frac{\partial}{\partial x_n^\mu} F^n = mc\frac{(x_{n+1} - x_n)_\mu}{\sqrt{(x_{n+1} - x_n)^2}} - 2\lambda_n(x_{n+1} - x_n)_\mu, \quad M < n < N, \quad (9.41)$$

$$\frac{\partial}{\partial x_n^\mu} F^{n-1} = mc\frac{(x_{n-1} - x_n)_\mu}{\sqrt{(x_{n-1} - x_n)^2}} + 2\lambda_n(x_n - x_{n-1})_\mu, \quad M < n < N, \quad (9.42)$$

giving the node equations of motion

$$\{m - 2\lambda_n T\}(x_{n+1} - x_n)^\mu \underset{c}{=} \{m - 2\lambda_{n-1}T\}(x_n - x_{n-1})^\mu, \quad (9.43)$$

taking the positive square root in (9.40). The equations reduce to

$$(x_n - x_{n+1})^\mu \underset{c}{=} \mp(x_n - x_{n-1})^\mu, \quad M < n < N, \quad (9.44)$$
$$m/T + 2\lambda_n \underset{c}{=} \pm(m/T + 2\lambda_{n-1}),$$

with the $\pm$ indicating that there are two possible branches to the solutions.

The first sign gives the expected free-particle (linear) equation of motion

$$x_{n+1}^\mu \underset{c}{=} 2x_n^\mu - x_{n-1}^\mu, \quad \lambda_n \underset{c}{=} \lambda_{n-1}, \quad (9.45)$$

whilst the second sign gives the peculiar and physically unacceptable solution

$$x_{n+1}^\mu \underset{c}{=} x_{n-1}^\mu, \quad \lambda_n \underset{c}{=} -\lambda_{n-1} - \frac{m}{T}. \quad (9.46)$$

9.3.2 Relativistic-particle Model 2

An alternative approach to relativistic-particle mechanics is to introduce an auxiliary dynamical degree of freedom e, such that its equation of motion enforces the mass-shell constraint. Consider the Lagrangian

$$L = \frac{\dot{x}^\mu \dot{x}_\mu}{2e} + \frac{em^2c^2}{2}. \quad (9.47)$$

Then the conjugate momentum is

$$p_\mu = \frac{\dot{x}_\mu}{e}, \quad (9.48)$$

with equation of motion

$$\frac{d}{d\tau}p_\mu \underset{c}{=} 0. \tag{9.49}$$

The equation of motion for the auxiliary variable e gives

$$\dot{x}^\mu \dot{x}_\mu \underset{c}{=} e^2 m^2 c^2. \tag{9.50}$$

The choice $e = m^{-1}$ is a consistent one, leading both to the kinematically correct one

$$\dot{x}^\mu \dot{x}_\mu = c^2 \tag{9.51}$$

and to the mass-shell constraint

$$p^\mu p_\mu = mc^2. \tag{9.52}$$

From the given Lagrangian (9.47) we construct the system function

$$F_n = \frac{(x_{n+1} - x_n)^2}{2e_n T} + \frac{m^2 c^2}{2} e_n T, \tag{9.53}$$

where e_n is a link variable.

Now the DT equations of motion become

$$-\frac{x^\mu_{n+1} - x^\mu_n}{e_n} + \frac{x^\mu_n - x^\mu_{n-1}}{e_{n-1}} \underset{c}{=} 0 \tag{9.54}$$

and

$$(x_{n+1} - x_n)^2 \underset{c}{=} e_n^2 m^2 c^2 T^2. \tag{9.55}$$

The dynamical choice $e_n = m^{-1}$ then reproduces the results of Model 1 in the previous section.

After we have discussed Lee's approach to DT CM, which considers Type-3 chronons (which are dynamically variable), we will appreciate that the combination $e_n T$ in the system function (9.53) is equivalent to having a dynamical chronon. For this model, however, there is insufficient information given to solve completely for the e_n, requiring us to add in by hand extra information, such as $e_n = m^{-1}$. This corresponds to what happens in the CT version of Model 2, given by the Lagrangian (9.47). There we encounter a primary constraint for the momentum conjugate to e (Dirac, 1964). The stability analysis of this constraint then leads to a secondary first-class constraint, namely the mass-shell constraint (9.52). There is therefore a gauge symmetry in the model. In the DT theory, the choice $e_n = m^{-1}$ is analogous to a choice of gauge in the CT theory.

10

Lee's approach to discrete time mechanics

10.1 Lee's discretization

Time is a continuous parameter in standard Schrödinger wave mechanics, and quantum wavefunctions are differentiable functions of that parameter. We shall refer to this as *temporal differentiability*. Temporal differentiability is assumed in Dirac's more abstract formulation of quantum mechanics, where state vectors in some abstract Hilbert space and Hermitian operators over that space depend on continuous time (CT) (Dirac, 1958). The same is assumed in relativistic quantum field theories, where quantum field operators in the Heisenberg picture are differentiable over time and space.

Temporal differentiability is a necessary condition for the existence of the quantum wave equations and operator field equations found in such theories. Unfortunately, these differentiable equations are usually impossible to solve exactly, so various techniques such as perturbation theory and computer simulation are employed to make suitable approximations.

Numerical techniques on their own, however, are insufficient to answer questions of principle, such as the meaning of quantized spacetime. A more principled approach to the solution of quantum differential equations, namely the path integral (PI), was developed, exploited and popularized principally by Feynman (Feynman and Hibbs, 1965). In some situations, such as quantum gravity, the PI was found to be virtually the only technology available to discuss the quantum physics. We discuss this approach in greater detail in Chapter 18.

The PI approach uses the fact that solving a differential equation is formally equivalent to integrating it. Feynman showed how to replace the problem of solving the Schrödinger differential equation by the integration of the so-called *Feynman kernel*, denoted by $F(\mathbf{x}, t_f; \mathbf{y}, t_i)$. For a single-particle system under observation (SUO), the Feynman kernel represents the relative quantum amplitude for the particle to be found at position $\mathbf{x}$ at a final time t_f, given that it was at position $\mathbf{y}$ at initial time t_i. The Feynman kernel is the focus of our discussion in Chapter 18.

Two points are relevant to us in our exploration of discrete time (DT). First, a Feynman kernel is a function of two times: the initial time and the final time, and such a function can just as easily be a function in DT mechanics as in CT mechanics. Second, Feynman used a process of temporal discretization in his approach to the calculation of the kernel (Feynman, 1948; Feynman and Hibbs, 1965). He divided the time interval $[t_\mathrm{i}, t_\mathrm{f}]$ into a number N of equal steps of duration $T \equiv (t_\mathrm{f} - t_\mathrm{i})/N$ and considered the quantum amplitude for the system to propagate step by step from its initial to its final configuration. The CT limit $N \to \infty$ is taken at the end of the calculation. Feynman's method is a development of Dirac's use of Lagrangians in a discussion of the relationship of quantum amplitudes to CT trajectories in classical mechanics (CM) (Dirac, 1933).

Feynman's original approach to path integrals can be classified as Type-1 temporal discretization, i.e., fixed intervals of equal length. In 1983, T. D. Lee considered a variant of Feynman's approach that can be classified as Type-3 discretization(Lee, 1983). Recall from our definition in Section 8.3 that, in Type-3 discretization, intermediate node times t_n become dynamical variables on a par with the original configuration-space variables.

On comparing Lee's mechanics and the analogous CT equations, we are reminded of Dirac's treatment of the action-integral time parameter in his theory of constraint mechanics (Dirac, 1964). In particular, both of these theories are reminiscent of the generally covariant way in which spacetime and matter are coupled in the Einstein–Hilbert action approach to general relativity (Hilbert, 1915). We discuss this further in Section 10.3.

In this chapter we restrict our attention to classical mechanics. Following Lee's approach, we first make a Type-3 partition $\{t_M, t_{M+1}, \ldots, t_N\}$ of the interval $[t_\mathrm{i}, t_\mathrm{f}]$, with $t_M \equiv t_\mathrm{i}$, $t_N \equiv t_\mathrm{f}$ and $\mathbf{q}_n \equiv \mathbf{q}(t_n)$. Then we define the extended coordinates $\mathbf{Q}_n \equiv \{\mathbf{q}_n, t_n\}$ and write the system function in the form $F^n \equiv F(\mathbf{Q}_n, \mathbf{Q}_{n+1})$ with action sum

$$A_{MN}[\mathcal{Q}] \equiv \sum_{n=M}^{N-1} F^n, \tag{10.1}$$

where $\mathcal{Q} \equiv \{\mathbf{Q}_M, \mathbf{Q}_{M+1}, \ldots, \mathbf{Q}_N\}$ denotes a specific DT path in extended configuration space. Now treat the partition time values t_n, $M \leqslant n \leqslant N$, as dynamical variables as well as the original $\mathbf{q}_n$. Under an infinitesimal variation $\mathcal{Q} \to \mathcal{Q}'$ of path such that $\mathbf{Q}_n \to \mathbf{Q}'_n \equiv \mathbf{Q}_n + \delta\mathbf{Q}_n$, we have the first-order variation

$$\begin{aligned}
\delta A_{MN}[\mathcal{Q}] &= \sum_{n=M}^{N-1} \left\{ \delta\mathbf{Q}_n \cdot \frac{\partial}{\partial \mathbf{Q}_n} + \delta\mathbf{Q}_{n+1} \cdot \frac{\partial}{\partial \mathbf{Q}_{n+1}} \right\} F^n \\
&= \delta\mathbf{Q}_M \cdot \frac{\partial}{\partial \mathbf{Q}_M} F^M + \delta\mathbf{Q}_N \cdot \frac{\partial}{\partial \mathbf{Q}_N} F^{N-1}
\end{aligned}$$

$$+\sum_{n=M+1}^{N-1} \delta\mathbf{Q}_n \cdot \frac{\partial}{\partial \mathbf{Q}_n}\{F^n + F^{n-1}\}. \tag{10.2}$$

Applying the DT Weiss action principle gives the equations of motion

$$\frac{\partial}{\partial \mathbf{Q}_n}\{F^n + F^{n-1}\} \underset{c}{=} \mathbf{0}, \quad M < n < N, \tag{10.3}$$

which reduce to the two sets of equations

$$\frac{\partial}{\partial t_n}\{F^n + F^{n-1}\} \underset{c}{=} 0, \quad \frac{\partial}{\partial \mathbf{q}_n}\{F^n + F^{n-1}\} \underset{c}{=} \mathbf{0}, \quad M < n < N. \tag{10.4}$$

10.2 The standard particle system

The standard particle system in CT mechanics consists of r equal-mass point particles moving in three-dimensional Euclidean physical space under the action of some time-dependent interparticle potential. Taking standard Cartesian configuration-space coordinates $\mathbf{q} \equiv (q^1, q^2, \ldots, q^r)$ with t denoting Newton's absolute time parameter, we may write a Lagrangian L for the system in the form

$$L = L(\mathbf{q}, \mathbf{v}, t), \tag{10.5}$$

where

$$\mathbf{v} \equiv \dot{\mathbf{q}} = \left(\frac{d}{dt}q^1, \frac{d}{dt}q^2, \ldots, \frac{d}{dt}q^r\right). \tag{10.6}$$

Now consider a Type-3 temporal discretization $\{t_\mathrm{i} \equiv t_M, t_{M+1}, \ldots, t_N \equiv t_\mathrm{f}\}$ of the evolution interval $[t_\mathrm{i}, t_\mathrm{f}]$ and make the discretization replacements

$$\mathbf{q} \to \mathbb{A}_n \mathbf{q}_n, \quad \mathbf{v} \to \mathbf{v}_n \equiv \frac{\mathbb{D}_n \mathbf{q}_n}{\mathbb{D}_n t_n}, \quad t \to \mathbb{A}_n t_n, \qquad M \leqslant n < N, \tag{10.7}$$

where $\mathbb{D}_n$ and $\mathbb{A}_n$ are the difference and average operators, respectively, discussed in Chapter 4, and take as system function

$$F^n \equiv L^n \, \mathbb{D}_n t_n, \tag{10.8}$$

where

$$L^n \equiv L(\mathbb{A}_n \mathbf{q}_n, \mathbf{v}_n, \mathbb{A}_n t_n). \tag{10.9}$$

It is worth noting that, starting from a single-particle Lagrangian of the form

$$L = \frac{1}{2}m\dot{x}^2 - V(x), \tag{10.10}$$

Lee assumed a temporal discretization of the form

$$L \to L^n \equiv \frac{1}{2}m\left(\frac{\mathbb{D}_n x_n}{\mathbb{D}_n t_n}\right)^2 - \mathbb{A}_n V(x_n), \qquad M \leqslant n < N \tag{10.11}$$

(Lee, 1983), where $\mathbb{A}_n$ is the forwards averaging operator (4.16), rather than the discretization

$$L \to L^n \equiv \frac{1}{2}m\left(\frac{\mathbb{D}_n x_n}{\mathbb{D}_n t_n}\right)^2 - V(\mathbb{A}_n x_n) \tag{10.12}$$

discussed by us here, but this does not alter the overall thrust of the argument.

Applying Lee's equations (10.4) to the system function (10.8) gives two sets of equations as follows.

1. **For the** t_n**,**

$$\mathbb{D}_n E^n \underset{c}{=} -\mathbb{A}_n \frac{\partial F^n}{\partial \mathbb{A}_n t_n}, \qquad M \leqslant n < N-1, \tag{10.13}$$

where

$$E^n \equiv \frac{\partial L^n}{\partial \mathbf{v}_n} \cdot \mathbf{v}_n - L^n \tag{10.14}$$

is a DT analogue of the energy.

Proof Using $F^n \equiv L(\mathbb{A}_n \mathbf{q}_n, \mathbf{v}_n, \mathbb{A}_n t_n)\mathbb{D}_n t_n$ gives

$$\frac{\partial F^n}{\partial t_n} = -L^n + \frac{\partial L^n}{\partial \mathbf{v}_n} \cdot \frac{\mathbb{D}_n \mathbf{q}_n}{\mathbb{D}_n t_n} + \frac{1}{2}\,\mathbb{D}_n t_n \frac{\partial L^n}{\partial \mathbb{A}_n t_n}, \tag{10.15}$$

whilst $F^{n-1} = L(\mathbb{A}_n \mathbf{q}_{n-1}, \mathbf{v}_{n-1}, \mathbb{A}_n t_{n-1})\mathbb{D}_n t_{n-1}$ gives

$$\frac{\partial F^{n-1}}{\partial t_n} = L^{n-1} - \frac{\partial L^{n-1}}{\partial \mathbf{v}_{n-1}} \cdot \frac{\mathbb{D}_n \mathbf{q}_{n-1}}{\mathbb{D}_n t_{n-1}} + \frac{1}{2}\,\mathbb{D}_n t_{n-1} \frac{\partial L^{n-1}}{\partial \mathbb{A}_n t_{n-1}}. \tag{10.16}$$

Hence

$$\begin{aligned} \frac{\partial F^n}{\partial t_n} + \frac{\partial F^{n-1}}{\partial t_n} &= -L^n + \frac{\partial L^n}{\partial \mathbf{v}_n} \cdot \mathbf{v}_n + \frac{1}{2}\,\mathbb{D}_n t_n \frac{\partial L^n}{\partial \mathbb{A}_n t_n} \\ &\quad + L^{n-1} - \frac{\partial L^{n-1}}{\partial \mathbf{v}_{n-1}} \cdot \mathbf{v}_{n-1} + \frac{1}{2}\,\mathbb{D}_n t_{n-1} \frac{\partial L^{n-1}}{\partial \mathbb{A}_n t_{n-1}} \\ &= -\mathbb{D}_n L^{n-1} + \mathbb{D}_n \left\{ \frac{\partial L^{n-1}}{\partial \mathbf{v}_{n-1}} \cdot \mathbf{v}_{n-1} \right\} \\ &\quad + \mathbb{A}_n \left\{ \mathbb{D}_n t_{n-1} \frac{\partial L^{n-1}}{\partial \mathbb{A}_n t_{n-1}} \right\}, \end{aligned} \tag{10.17}$$

which leads immediately to (10.13).

The significance of this result is that, if there is no explicit time dependence in the original Lagrangian (10.5), then the 'discrete energy' E^n is conserved modulo Lee's DT equations.

2. **For the** $\mathbf{q}_n$**,** the equations of motion give

$$\mathbb{A}_n \left\{ \frac{\partial F^n}{\partial \mathbb{A}_n \mathbf{q}_n} \right\} \underset{c}{=} \mathbb{D}_n \frac{\partial L^n}{\partial \mathbf{v}_n}, \qquad M \leqslant n < N-1. \tag{10.18}$$

Proof Using $F^n = \mathbb{D}_n t_n \, L(\mathbb{A}_n \mathbf{q}_n, \mathbf{v}_n, \mathbb{A}_n t_n)$ gives

$$\frac{\partial F^n}{\partial \mathbf{q}_n} = \frac{1}{2} \, \mathbb{D}_n t_n \, \frac{\partial L^n}{\partial \mathbb{A}_n \mathbf{q}_n} - \frac{\partial L^n}{\partial \mathbf{v}_n}, \tag{10.19}$$

whilst $F^{n-1} = \mathbb{D}_n t_{n-1} \, L(\mathbb{A}_n \mathbf{q}_{n-1}, \mathbf{v}_{n-1}, \mathbb{A}_n t_{n-1})$ gives

$$\frac{\partial F^{n-1}}{\partial \mathbf{q}_n} = \frac{1}{2} \, \mathbb{D}_n t_{n-1} \, \frac{\partial L^{n-1}}{\partial \mathbb{A}_n \mathbf{q}_{n-1}} + \frac{\partial L^{n-1}}{\partial \mathbf{v}_{n-1}}. \tag{10.20}$$

Hence

$$\begin{aligned} \frac{\partial F^n}{\partial \mathbf{q}_n} + \frac{\partial F^{n-1}}{\partial \mathbf{q}_n} &= \frac{1}{2} \, \mathbb{D}_n t_n \, \frac{\partial L^n}{\partial \mathbb{A}_n \mathbf{q}_n} - \frac{\partial L^n}{\partial \mathbf{v}_n} \\ &\quad + \frac{1}{2} \, \mathbb{D}_n t_{n-1} \, \frac{\partial L^{n-1}}{\partial \mathbb{A}_n \mathbf{q}_{n-1}} + \frac{\partial L^{n-1}}{\partial \mathbf{v}_{n-1}} \\ &= \mathbb{A}_n \left\{ \mathbb{D}_n t_{n-1} \, \frac{\partial L^{n-1}}{\partial \mathbb{A}_n \mathbf{q}_{n-1}} \right\} - \mathbb{D}_n \left\{ \frac{\partial L^{n-1}}{\partial \mathbf{v}_{n-1}} \right\}, \end{aligned} \tag{10.21}$$

giving the stated result (10.18).

10.3 Discussion

Lee's DT mechanics bears a remarkable structural similarity to the discussion given by Dirac on the arbitrary reparametrization of time in CT CM (Dirac, 1964). We review Dirac's treatment briefly now.

Consider the canonical action integral

$$A_{\text{fi}}[\mathbf{q}] \equiv \int_{t_{\text{i}}}^{t_{\text{f}}} L(\mathbf{q}, \mathbf{v}, t) dt, \tag{10.22}$$

where $\mathbf{q} \equiv \mathbf{q}(t)$ and $\mathbf{v} \equiv (d/dt)\mathbf{q}$. Now reparametrize the parameter t by making it a monotonically increasing, differentiable function of a new parameter λ, which runs from $\lambda = 0$ to $\lambda = 1$. We define

$$t \equiv t(\lambda), \quad \dot{t} \equiv \frac{d}{d\lambda} t(\lambda) > 0, \quad \mathbf{Q}(\lambda) \equiv \mathbf{q}(t(\lambda)), \quad \dot{\mathbf{Q}}(\lambda) = \frac{d}{d\lambda}\mathbf{Q}(\lambda) = \dot{t}\mathbf{v}. \tag{10.23}$$

Then

$$\tilde{A}_{\text{fi}}[\mathbf{Q}] \equiv A_{\text{fi}}[\mathbf{q}] = \int_0^1 \tilde{L}(\mathbf{Q}, t, \dot{\mathbf{Q}}, \dot{t}) d\lambda, \tag{10.24}$$

where

$$\tilde{L}(\mathbf{Q}, t, \dot{\mathbf{Q}}, \dot{t}) \equiv \dot{t} L \left(\mathbf{Q}, \frac{\dot{\mathbf{Q}}}{\dot{t}}, t \right). \tag{10.25}$$

Now we consider arbitrary variations in $\mathbf{Q}$ and t separately. For $\mathbf{Q}$ we have the equations of motion

$$\frac{d}{d\lambda} \left\{ \frac{\partial}{\partial \dot{\mathbf{Q}}} \tilde{L} \right\} \mathop{=}_{c} \frac{\partial}{\partial \mathbf{Q}} \tilde{L}, \tag{10.26}$$

which gives the standard Euler–Lagrange equation of motion

$$\frac{d}{dt}\left\{\frac{\partial}{\partial \mathbf{v}}L\right\} \underset{c}{=} \frac{\partial}{\partial \mathbf{q}}L \tag{10.27}$$

as expected. This is the CT analogue of (10.18).

On the other hand, the equation of motion for t gives

$$\frac{d}{d\lambda}\left\{\frac{\partial}{\partial t}\tilde{L}\right\} \underset{c}{=} \frac{\partial}{\partial t}\tilde{L}, \tag{10.28}$$

which gives

$$\frac{d}{dt}\left\{L - \frac{\partial L}{\partial \mathbf{v}} \cdot \mathbf{v}\right\} \underset{c}{=} \frac{\partial L}{\partial t}, \tag{10.29}$$

which is equivalent to the standard equation

$$\frac{d}{dt}H(\mathbf{p}, \mathbf{q}, t) \underset{c}{=} -\frac{\partial L}{\partial t} \tag{10.30}$$

and is the CT analogue of (10.13).

Dirac developed constraint mechanics, the basis for his discussion on the reparametrization of time, with general relativity (GR) in mind (Dirac, 1964). In GR, there is a powerful principle known as general covariance. This principle states that the laws of physics including gravitation should be form invariant with respect to arbitrary reparametrizations of coordinate patch. Dirac's reparametrization of time can be viewed as a simplified discussion of generalized covariance.

10.4 Return to the relativistic point particle

With Lee's DT mechanics and constraint mechanics in mind, we return to a discussion of Model 1, the first relativistic point particle model discussed in Chapter 9. In that model, we introduced Lagrange multipliers to enforce the condition that the particle appeared at regular intervals of proper time.

Suppose, however, we had decided not to use Lagrange multipliers as in Model 1 or the technique of Model 2, which introduces an auxiliary variable e to encode constraints in an otherwise unconstrained Lagrangian. Instead, suppose we decided to try a variable proper time interval approach in the manner of Lee, writing

$$F^n \equiv -mc\, \mathbb{D}_n\tau_n \sqrt{\frac{(\mathbb{D}_n x_n)^\mu}{\mathbb{D}_n\tau_n}\frac{(\mathbb{D}_n x_n)_\mu}{\mathbb{D}_n\tau_n}}, \tag{10.31}$$

where now the node variable τ_n is the proper time recorded by the particle at each node and is regarded as a proper dynamical variable. A problem arises, because, by inspection, we see that $\mathbb{D}_n\tau_n$ cancels out above and below in (10.31).

Hence proper time effectively disappears from the system function (10.31), giving the system function $F^n \equiv -mc\sqrt{(\mathbb{D}_n x_n)^2}$. Therefore, we can expect problems to do with the physical interpretation.

To see this, we find that the momenta $p^{(-)}_{n\mu}$ and $p^{(+)}_{n-1\mu}$ defined in Section 8.4 satisfy the equivalent of a primary constraint (Dirac, 1964). These momenta are given by

$$p^{(-)}_{n\mu} \equiv -\frac{\partial F^n}{\partial x^\mu_n} = mc\frac{\mathbb{D}_n x_{n\mu}}{\sqrt{(\mathbb{D}_n x_n)^2}}, \quad p^{(+)}_{n-1\mu} \equiv \frac{\partial F^{n-1}}{\partial x^\mu_n} = mc\frac{\mathbb{D}_n x_{n-1\mu}}{\sqrt{(\mathbb{D}_n x_{n-1})^2}}, \tag{10.32}$$

from which we see $p^{(-)2}_n = p^{(+)2}_{n-1} = m^2c^2$, which is the mass-shell constraint.

The equations of motion are $p^{(-)}_{n\mu} \underset{c}{=} p^{(+)}_{n-1\mu}$, which means

$$\frac{\mathbb{D}_n x_{n\mu}}{\sqrt{(\mathbb{D}_n x_n)^2}} \underset{c}{=} \frac{\mathbb{D}_n x_{n-1\mu}}{\sqrt{(\mathbb{D}_n x_{n-1})^2}}. \tag{10.33}$$

Unfortunately, solutions to these equations are not unique. For example, introduce a timelike unit vector u^μ such that $u^\mu u_\mu = 1$ and define

$$X^\mu_{n+1} = X^\mu_n + \alpha u^\mu, \quad X^\mu_n = X^\mu_{n-1} + \beta u^\mu, \tag{10.34}$$

where α and β are arbitrary positive constants. Then the $\{X^\mu_n\}$ satisfy equations (10.33) with no constraints on α or β.

This DT analysis of the relativistic point particle suggests the following interpretation of the constraint issues arising in relativistic particle mechanics. Proper time is an internal attribute of a particle SUO moving in relativistic spacetime. That is why different particles can evolve with different proper time rates and this is the basis of the twin 'paradox'.

Coordinate time, $x^0 \equiv t$, however, is more like an attribute of the observer's detectors, situated at rest in some inertial frame of reference and external to the particle. If there is no coupling between the observer's frame (which represents the laboratory and all the detectors in it) and the internal mechanics of the particle (as represented by its proper time), then the observer cannot expect to extract information about that internal mechanics, hence the non-uniqueness of the constants α and β.

On the other hand, the system function (9.39) of Model 1 explicitly couples in the proper time intervals, so that the laboratory can indeed register those events in spacetime where the internal clock of the particle has clicked the passage of a chronon of proper time. These events are analogous to the splashes that a skipping stone makes on the surface of a pond.

11

Elliptic billiards

11.1 The general scenario

Elliptic billiards is the name given to a particular class of discrete time (DT) mechanical system in which a particle moves in continuous time (CT) and space but is observed only at a discrete set of times whenever it bounces off a fixed surface. It is therefore a form of *stroboscopic mechanics* but, unlike conventional stroboscopic mechanics where the time intervals between successive observations are determined by the observer and are usually of equal duration, the time between successive observations in elliptic billiards is variable, being determined by the dynamical behaviour of the system under observation. We shall first discuss this form of mechanics using a purely geometrical approach (Moser and Veselov, 1991) and then we shall show that the same system can be analysed in terms of Lee mechanics, which was studied in the previous chapter.

We will discuss the situation when the particle under observation is confined to the interior of some container such as an ellipsoid (hence the title of this chapter), being observed only when it bounces elastically off the surface of that container, Figure 11.1. The container can in principle be any closed shape, i.e., not necessarily a box with rectangular sides. In the scenario we are interested in in this chapter, the particle is assumed to move inertially, i.e., it moves uniformly between successive impacts on the surface of the container.

Both the geometrical approach and the dynamical approach show that the particle bounces off the surface in the equivalent of an elastic collision. This will be discussed in detail.

In the first instance we follow Moser and Veselov's analysis of elliptic billiards (Moser and Veselov, 1991). Consider a closed surface in $\mathbb{R}^d$ defined via a positive symmetric $d \times d$ matrix M, such that the coordinates $\mathbf{x}^{\mathrm{T}} = [x^1, x^2, \ldots, x^d]$ of points on the elliptic surface satisfy the condition

$$\mathbf{x}^{\mathrm{T}}\mathsf{M}\mathbf{x} = 1. \tag{11.1}$$

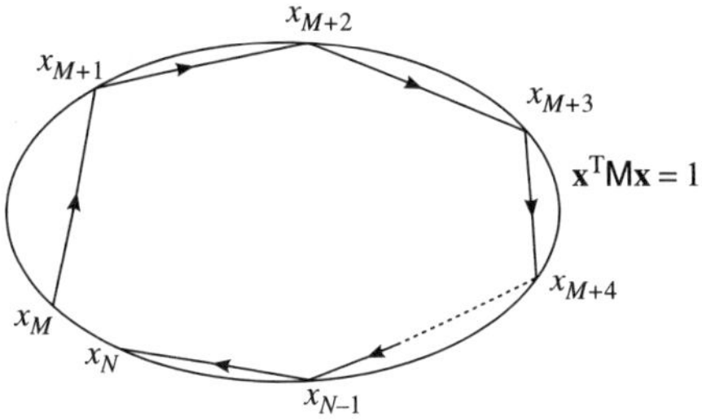

Figure 11.1. A particle trajectory inside a closed surface such as an ellipsoid defined by $\mathbf{x}^{\mathrm{T}}\mathsf{M}\mathbf{x} = 1$.

Example 11.1 The surface of an ellipsoid in $\mathbb{R}^3$ with its principal axes aligned along the standard Cartesian coordinate axes satisfies the equation

$$\frac{x^2}{a^2} + \frac{y^2}{b^2} + \frac{z^2}{c^2} = 1, \tag{11.2}$$

so we recover (11.1) if we define

$$x^1 \equiv x, \quad x^2 \equiv y, \quad x^3 \equiv z \tag{11.3}$$

and

$$\mathsf{M} \equiv \begin{bmatrix} a^{-2} & 0 & 0 \\ 0 & b^{-2} & 0 \\ 0 & 0 & c^{-2} \end{bmatrix}. \tag{11.4}$$

The particle is assumed to move uniformly inside the surface between successive impacts on that surface. The observer monitors only those impacts, which therefore occur at a discrete number of places $\{\mathbf{x}_M, \mathbf{x}_{M+1}, \ldots, \mathbf{x}_N\}$ and at discrete times $\{t_M, t_{M+1}, \ldots, t_N\}$, for integers $M < N$.

11.2 Elliptic billiards via the geometrical approach

The geometrical approach determines the points $\{\mathbf{x}_n : M < n < N\}$ of impact between given initial and final points $\mathbf{x}_M$ and $\mathbf{x}_N$ by minimizing the Euclidean length of the total path between those points, i.e., we minimize the path functional

$$S_{NM} \equiv \sum_{n=M}^{N-1} |\mathbb{D}_n \mathbf{x}_n| + \frac{1}{2} \sum_{n=M}^{N} \nu_n (\mathbf{x}_n^{\mathrm{T}} \mathsf{M} \mathbf{x}_n - 1). \tag{11.5}$$

Here the ν_n are Lagrange multipliers, which enforce the condition that the points of impact lie on the given surface.

Taking infinitesimal variations in the path functional with respect to the impact positions and the Lagrange multipliers gives

$$\delta S_{NM} = \sum_{n=M}^{N-1} \frac{(\mathbb{D}_n \delta \mathbf{x}_n)^{\mathrm{T}}(\mathbb{D}_n \mathbf{x}_n)}{|\mathbb{D}_n \mathbf{x}_n|} + \frac{1}{2}\sum_{n=M}^{N} \delta\nu_n(\mathbf{x}_n^{\mathrm{T}}\mathsf{M}\mathbf{x}_n - 1) + \sum_{n=M}^{N} \nu_n\, \delta\mathbf{x}_n^{\mathrm{T}}\, \mathsf{M}\mathbf{x}_n$$
$$= \delta\mathbf{x}_N^{\mathrm{T}}\left\{\frac{\mathbb{D}_N \mathbf{x}_{N-1}}{|\mathbb{D}_N \mathbf{x}_{N-1}|} + \nu_N \mathsf{M}\mathbf{x}_N\right\} - \delta\mathbf{x}_M^{\mathrm{T}}\left\{\frac{\mathbb{D}_M \mathbf{x}_M}{|\mathbb{D}_M \mathbf{x}_M|} - \nu_M \mathsf{M}\mathbf{x}_M\right\}$$
$$+ \sum_{n=M+1}^{N-1} \delta\mathbf{x}_n^{\mathrm{T}}\left\{\frac{\mathbb{D}_n \mathbf{x}_{n-1}}{|\mathbb{D}_n \mathbf{x}_{n-1}|} - \frac{\mathbb{D}_n \mathbf{x}_n}{|\mathbb{D}_n \mathbf{x}_n|} + \nu_n \mathsf{M}\mathbf{x}_n\right\}$$
$$+ \frac{1}{2}\sum_{n=M}^{N} \delta\nu_n(\mathbf{x}_n^{\mathrm{T}}\mathsf{M}\mathbf{x}_n - 1). \tag{11.6}$$

Hence, assuming that the true trajectory lies on an extremum of the path functional, we deduce from the independence of the $\delta\nu_n$ variations

$$\mathbf{x}_n^{\mathrm{T}}\mathsf{M}\mathbf{x}_n \underset{c}{=} 1, \quad M \le n \le N, \tag{11.7}$$

and from the independence of the $\delta\mathbf{x}_n$ variations

$$\frac{\mathbb{D}_n \mathbf{x}_n}{|\mathbb{D}_n \mathbf{x}_n|} - \frac{\mathbb{D}_n \mathbf{x}_{n-1}}{|\mathbb{D}_n \mathbf{x}_{n-1}|} \underset{c}{=} \nu_n \mathsf{M}\mathbf{x}_n, \qquad M < n < N. \tag{11.8}$$

The ν_n are readily determined by using the surface constraints (11.7):

$$\mathbf{x}_n^{\mathrm{T}}\left\{\frac{\mathbb{D}_n \mathbf{x}_n}{|\mathbb{D}_n \mathbf{x}_n|} - \frac{\mathbb{D}_n \mathbf{x}_{n-1}}{|\mathbb{D}_n \mathbf{x}_{n-1}|}\right\} \underset{c}{=} \nu_n \mathbf{x}_n^{\mathrm{T}}\mathsf{M}\mathbf{x}_n = \nu_n, \quad M < n < N. \tag{11.9}$$

In (11.9), the left-hand side involves positions at discrete times $n-1, n$ and $n+1$, so there arises the question of whether these equations can indeed be solved in terms of the end-point positions alone, i.e., on $\mathbf{x}_M$ and $\mathbf{x}_N$. In fact, the equations can be rearranged as follows: introduce new quantities $\mathbf{w}_n$ and $\mu_n > 0$ given by

$$\mathbb{D}_n \mathbf{x}_n \equiv \mu_n \mathbf{w}_n, \quad M \le n < N, \tag{11.10}$$

with the condition that $\mathbf{w}_n^{\mathrm{T}}\mathbf{w}_n = 1,\ M \le n < N$. Then the equations of motion (11.8) are equivalent to

$$\mathbb{D}_n \mathbf{w}_{n-1} = \nu_n \mathsf{M}\mathbf{x}_n, \quad M < n < N. \tag{11.11}$$

The μ_n are determined as follows. The symmetry of M and the constraint conditions $\mathbf{x}_n^{\mathrm{T}}\mathsf{M}\mathbf{x}_n = 1,\ M \le n \le N$, lead to the identity

$$[\mathbb{A}_n \mathbf{x}_n]^{\mathrm{T}}\mathsf{M}\mathbb{D}_n \mathbf{x}_n = 0, \quad M \le n < N, \tag{11.12}$$

which, with the above definitions, gives

$$[\mathbb{A}_n \mathbf{x}_n]^{\mathrm{T}}\mathsf{M}\mathbf{w}_n = 0, \quad M \le n < N. \tag{11.13}$$

Rearranging this in the form

$$(\mathbb{D}_n \mathbf{x}_n + 2\mathbf{x}_n)^{\mathrm{T}}\, \mathsf{M}\mathbf{w}_n = 0, \quad M \le n < N, \tag{11.14}$$

leads to the result (Moser and Veselov, 1991)

$$\mu_n = -\frac{2\mathbf{x}_n^{\mathrm{T}}\mathsf{M}\mathbf{w}_n}{\mathbf{w}_n^{\mathrm{T}}\mathsf{M}\mathbf{w}_n}, \quad M \leq n < N, \tag{11.15}$$

where the quantity we denote by $\mathbf{w}_n$ is denoted by Moser and Veselov by $\mathbf{y}_{n+1}$.

To obtain an expression for the ν_n, first we rewrite the equation of motion (11.11) in the form

$$\mathbf{w}_n \underset{c}{=} \mathbf{w}_{n-1} + \nu_n \mathsf{M}\mathbf{x}_n, \tag{11.16}$$

and 'square' both sides, giving

$$1 = 1 + 2\nu_n \mathbf{w}_{n-1}^{\mathrm{T}}\mathsf{M}\mathbf{x}_n + \nu_n^2(\mathsf{M}\mathbf{x}_n)^{\mathrm{T}}\mathsf{M}\mathbf{x}_n. \tag{11.17}$$

Now by inspection of (11.16), it is readily seen that a value $\nu_n = 0$ would mean that there was no deflection as the particle hit the surface of the container, so we reject this possibility. Hence taking $\nu_n \neq 0$ in (11.17) gives (Moser and Veselov, 1991)

$$\nu_n = -\frac{2\mathbf{w}_{n-1}^{\mathrm{T}}\mathsf{M}\mathbf{x}_n}{(\mathsf{M}\mathbf{x}_n)^{\mathrm{T}}\mathsf{M}\mathbf{x}_n}, \quad M < n < N. \tag{11.18}$$

11.2.1 Elastic bounces: the geometrical approach

Some of the fundamental properties of classical geometry, whilst very useful, can cause significant issues to arise when it comes to physics. For example, the geometrical distance between two points in Euclidean space does not depend on the ordering in which we take those points: the distance from A to B is the same as the distance from B to A.

However, in the real world, this symmetry need not hold. For instance, we may be travelling along a road that has a long diversion, due to roadworks, affecting only one side of the road, whilst not affecting travellers going in the opposite direction. In such a case the effective distance between two places on that road may be very different depending on the direction in which we are going.

This issue became acute early on in the twentieth century, once it had been realized by Minkowski (1908) that special relativity could be discussed geometrically. The subsequent development of general relativity by Einstein (1915) compounded the problem, because it then seemed to many physicists that the whole of physics could be described in terms of geometry. Unfortunately, the temporal symmetry of such a description runs counter to the general observation that the Universe runs along irreversible lines, as evinced by the observed Hubble expansion of distant galaxies, the second law of thermodynamics, and the fact that probabilities in quantum mechanics involve a fundamental irreversibility associated with the acquisition of information and memory.

This issue arises in the case under discussion here, because the geometrical approach to elliptic billiards does not involve any explicit notion of time. By symmetry, we would expect that angles of incidence should equal angles of reflection, precisely because the geometrical approach does not tell us which is which.

Nevertheless, there seems something unsatisfactory about using a geometrical argument to derive what is essentially a dynamical result. Therefore, the concept of an elastic bounce needs to be discussed more carefully.

We define an *elastic geometrical bounce* to be a deflection of a line incident at a point P on a surface, satisfying the following conditions:

(i) the normal $\hat{\mathbf{n}}_{\mathrm{P}}$ to the surface at a point P of impact lies in the plane defined by the incident unit vector $\hat{\mathbf{u}}$ and the outgoing unit vector $\hat{\mathbf{v}}$; and
(ii) the magnitude of the angle θ_{i} between $\hat{\mathbf{u}}$ and $\hat{\mathbf{n}}$ equals the magnitude of the angle θ_{f} between $\hat{\mathbf{v}}$ and $\hat{\mathbf{n}}$.

We define the angle θ_{ab} between unit vectors $\mathbf{a}$ and $\mathbf{b}$ at a point P in a Euclidean signature space by

$$\cos(\theta_{ab}) \equiv \mathbf{g}_{\mathrm{P}}(\mathbf{a}, \mathbf{b}),$$

where $\mathbf{g}_{\mathrm{P}}$ is the metric tensor, with evaluation at P. What is missing in this description, of course, is any reference to the *speed* of the particle.

We prove that elliptic billiards consists of elastic geometrical bounces as follows.

From basic vector calculus, the normal to the surface at any point of impact $\mathbf{x}_n$, $M < n < N$, is proportional to $\mathsf{M}\mathbf{x}_n$. From definition (11.10), the incident unit vector is $\mathbf{w}_{n-1}$ and the outgoing unit vector is $\mathbf{w}_n$. Then, from the equation of motion (11.11), we see that condition (i) is satisfied. The cosine of the angle of incidence (in three dimensions) between $\mathbf{w}_{n-1}$ and the normal is proportional to $\mathbf{w}_{n-1}^{\mathrm{T}}\mathsf{M}\mathbf{x}_n$, and similarly for the angle of reflection. Now taking the 'inner product' of both sides of the equation of motion with $\mathsf{M}\mathbf{x}_n$ gives

$$\mathbf{w}_n^{\mathrm{T}}\mathsf{M}\mathbf{x}_n = \mathbf{w}_{n-1}^{\mathrm{T}}\mathsf{M}\mathbf{x}_n + \nu_n(\mathsf{M}\mathbf{x}_n)^{\mathrm{T}}\mathsf{M}\mathbf{x}_n, \quad M < n < N. \tag{11.19}$$

Squaring both sides then gives

$$\left(\mathbf{w}_n^{\mathrm{T}}\mathsf{M}\mathbf{x}_n\right)^2 = \left(\mathbf{w}_{n-1}^{\mathrm{T}}\mathsf{M}\mathbf{x}_n\right)^2, \quad M < n < N, \tag{11.20}$$

using expressions (11.18). This and the fact that the normal to the surface is in the plane of the incident and reflected vectors proves that each bounce is indeed elastic.

11.3 Elliptic billiards via Lee mechanics

We may derive equivalent equations for elliptic billiards using Lee's DT mechanics formalism, which was discussed in the previous chapter, as follows. First, we note that, according to CT mechanics, a particle moving inside a container and subject to no external forces will behave as a free particle between successive bounces off the confining surface. Therefore, during that free part of the motion, the CT Lagrangian can be taken to be the free-particle Lagrangian

$$L(\mathbf{x}, \dot{\mathbf{x}}, t) = \frac{1}{2} m \dot{\mathbf{x}} \cdot \dot{\mathbf{x}}. \tag{11.21}$$

Knowing the classical trajectory between events $(\mathbf{x}_n, t_n)$ and $(\mathbf{x}_{n+1}, t_{n+1})$ immediately allows us to determine Hamilton's principal function S^n, which is defined by

$$S^n \equiv \int_{t_n}^{t_{n+1}} dt\ L. \tag{11.22}$$

Here we integrate over the free-particle trajectory from $\mathbf{x}_n$ to $\mathbf{x}_{n+1}$. We readily find

$$S^n = \frac{m[\mathbb{D}_n \mathbf{x}_n]^{\mathrm{T}} \mathbb{D}_n \mathbf{x}_n}{2\, \mathbb{D}_n t_n}, \tag{11.23}$$

where $\mathbb{D}_n t_n \equiv t_{n+1} - t_n > 0$.

The system function F^n is now taken to be given by S^n, so that the action sum to be considered is

$$A_{MN} = \sum_{n=M}^{N-1} S^n + \sum_{n=M}^{N} \frac{1}{2} \lambda_n \left(\mathbf{x}_n^{\mathrm{T}} \mathsf{M} \mathbf{x}_n - 1 \right), \tag{11.24}$$

where the λ_n are Lagrange multipliers and M is a real symmetric matrix.

We now determine the equations of motion on the basis that the intermediate times t_n, $M < n < N$, and the Lagrange multipliers λ_n, $M \leq n \leq N$, as well as the $\mathbf{x}_n, M \leq n \leq N$, are to be considered as dynamical variables. Then the equations of motion are found to be

$$m \left(\frac{\mathbb{D}_n \mathbf{x}_n}{\mathbb{D}_n t_n} \right) - m \left(\frac{\mathbb{D}_n \mathbf{x}_{n-1}}{\mathbb{D}_n t_{n-1}} \right) \underset{c}{=} \lambda_n \mathsf{M} \mathbf{x}_n, \quad M < n < N, \tag{11.25}$$

$$\frac{m}{2} \left(\frac{\mathbb{D}_n \mathbf{x}_n}{\mathbb{D}_n t_n} \right)^{\mathrm{T}} \left(\frac{\mathbb{D}_n \mathbf{x}_n}{\mathbb{D}_n t_n} \right) \underset{c}{=} E, \quad M \leq n < N, \tag{11.26}$$

$$\mathbf{x}_n^{\mathrm{T}} \mathsf{M} \mathbf{x}_n \underset{c}{=} 1, \quad M \leq n \leq N. \tag{11.27}$$

We may relate these equations to the equations derived from the geometrical approach, as follows. First, despite taking the intermediate times t_n, $M < n < N$, as variable, physics requires solutions to the Lee equations of motion to be causal relative to the clocks in the observer's laboratory, i.e., we require $t_n < t_{n+1}$. Hence, from (11.26), we have

$$\mathbb{D}_n t_n = \sqrt{\frac{m}{2E}} |\mathbb{D}_n \mathbf{x}_n|, \tag{11.28}$$

where the 'energy' E is given from the initial conditions by

$$E = \frac{1}{2m} \left(\frac{\mathbf{x}_1 - \mathbf{x}_0}{t_1 - t_0} \right)^{\mathrm{T}} \left(\frac{\mathbf{x}_1 - \mathbf{x}_0}{t_1 - t_0} \right). \tag{11.29}$$

From this we see why the two approaches are equivalent: the spatial interval between successive impacts on the container surface is directly proportional to the time elapsed between those impacts. Hence we find

$$\lambda_n = \sqrt{2mE} \nu_n, \tag{11.30}$$

which demonstrates the formal equivalence of the two approaches.

11.4 Complex-plane billiards

To illustrate the formalism, we consider elliptic billiards on a plane, Figure 11.2, with the confining perimeter given by the unit circle

$$x^2 + y^2 = 1. \tag{11.31}$$

It is efficient to encode the two degrees of freedom x and y into the complex plane by defining the complex variable $z \equiv x + iy$. Then the confining perimeter is given by $\bar{z}z = 1$.

For convenience, we take the mass of the particle to be unity. Then the DT Lee mechanics equations of motion are

$$\begin{aligned} \frac{\mathbb{D}_n z_n}{\mathbb{D}_n t_n} - \frac{\mathbb{D}_n z_{n-1}}{\mathbb{D}_n t_{n-1}} &\underset{c}{=} \lambda_n z_n, && n > 0, \\ \bar{z}_n z_n &\underset{c}{=} 1, && n \geqslant 0, \\ \frac{|\mathbb{D}_n z_n|^2}{|\mathbb{D}_n t_n|^2} &\underset{c}{=} \frac{|\mathbb{D}_n z_{n+1}|^2}{|\mathbb{D}_n t_{n+1}|^2}, && n > 0, \end{aligned} \tag{11.32}$$

with initial events (z_0, t_0) and (z_1, t_1) given. Then

$$\mathbb{D}_n t_{n=0} = t_1 - t_0, \tag{11.33}$$

$$\frac{1}{2}\frac{|\mathbb{D}_n z_{n=0}|^2}{|\mathbb{D}_n t_{n=0}|^2} = \frac{1}{2}\frac{|z_1 - z_0|^2}{|t_1 - t_0|^2} \equiv E. \tag{11.34}$$

Now define.

$$p_n \equiv \frac{\mathbb{D}_n z_n}{\mathbb{D}_n t_n}, \quad n \geqslant 0.$$

Then

$$\begin{aligned} p_n - p_{n-1} &\underset{c}{=} \lambda_n z_n, && n > 0, \\ \bar{z}_n z_n &\underset{c}{=} 1, && n \geqslant 0, \\ |p_n|^2 &\underset{c}{=} 2E, && n \geqslant 0, \end{aligned} \tag{11.35}$$

with p_0 determined from the initial conditions.

Taking the first set of equations, we write $p_n \underset{c}{=} p_{n-1} + \lambda_n z_n$, so

$$|p_n|^2 \underset{c}{=} |p_{n-1}|^2 + \lambda_n(\bar{p}_{n-1} z_n + p_{n-1}\bar{z}_n) + \lambda_n^2 |z_n|^2 \tag{11.36}$$

or

$$\lambda_n = -(\bar{p}_{n-1} z_n + p_{n-1}\bar{z}_n), \quad n > 0. \tag{11.37}$$

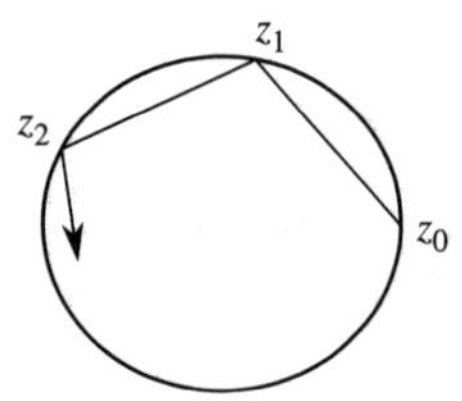

Figure 11.2. Circular billiards in the complex plane.

Clearly, λ_1 can be determined from the initial conditions. We find

$$\begin{aligned} p_n &\underset{\mathrm{c}}{=} p_{n-1} - (\bar{p}_{n-1} z_n + p_{n-1} \bar{z}_n) z_n \\ &= p_{n-1} - \bar{p}_{n-1} z_n^2 - p_{n-1} |z_n|^2, \end{aligned} \tag{11.38}$$

i.e., $p_n = -\bar{p}_{n-1} z_n^2$, $n > 0$. This obviously satisfies the equations of motion

$$|p_n|^2 \underset{\mathrm{c}}{=} |p_{n-1}|^2. \tag{11.39}$$

Now $\mathbb{D}_n z_n = p_n \, \mathbb{D} t_n$ gives $z_{n+1} = z_n + p_n \, \mathbb{D} t_n$, which implies

$$|z_{n+1}|^2 = |z_n|^2 + (\bar{p}_n z_n + p_n \bar{z}_n) \mathbb{D} t_n + |p_n|^2 (\mathbb{D} t_n)^2, \tag{11.40}$$

or

$$\begin{aligned} \mathbb{D} t_n &= -\frac{\bar{p}_n z_n + p_n \bar{z}_n}{2E} \\ &= -\frac{(-p_{n-1} \bar{z}_n^2) z_n + (-\bar{p}_{n-1} z_n^2) \bar{z}_n}{2E} \\ &= \frac{p_{n-1} \bar{z}_n + \bar{p}_{n-1} z_n}{2E}, \qquad n > 0. \end{aligned} \tag{11.41}$$

Hence

$$\begin{aligned} z_{n+1} &= z_n + p_n \frac{p_{n-1} \bar{z}_n + \bar{p}_{n-1} z_n}{2E} \\ &= z_n - \bar{p}_{n-1} z_n^2 \frac{p_{n-1} \bar{z}_n + \bar{p}_{n-1} z_n}{2E} \\ &= z_n - \frac{|p_{n-1}|^2 z_n + \bar{p}_{n-1}^2 z_n^3}{2E} z_{n+1} = -\frac{\bar{p}_{n-1}^2 z_n^3}{2E}, \qquad n > 0. \end{aligned} \tag{11.42}$$

We note that, for this system,

$$\begin{aligned} \mathbb{D} t_{n+1} &= \frac{p_n \bar{z}_{n+1} + \bar{p}_n z_{n+1}}{2E} \\ &= \frac{(-\bar{p}_{n-1} z_n^2)\big(-p_{n-1}^2 \bar{z}^3/(2E)\big) + (-p_{n-1} \bar{z}_n^2)(-\bar{p}_{n-1}^2 z_n^3/(2E))}{} \\ &= \frac{p_{n-1} \bar{z}_n + \bar{p}_{n-1} z_n}{2E} = \mathbb{D} t_n, \end{aligned} \tag{11.43}$$

as expected.

Worked example

As an application of the above, consider the initial conditions:

$$\begin{aligned} z_0 &= 1, & t_0 &= 0, \\ z_1 &= e \equiv e^{i\theta}, & t_1 &= 1, \end{aligned} \tag{11.44}$$

where θ is a constant. Then

$$p_0 = \frac{z_1 - z_0}{t_1 - t_0} = e - 1, \quad E = \frac{1}{2} |p_0|^2 = 1 - \cos\theta. \tag{11.45}$$

From the above theory we find the solution

$$p_n = e^n (e - 1), \quad z_n = e^n, \quad \mathbb{D} t_n = 1, \qquad n \geqslant 0, \tag{11.46}$$

which can be readily proved by induction.

12

The construction of system functions

12.1 Phase space

In Chapter 8 we discussed the principles involved in the construction of canonical Lagrangians, action integrals and the derivation of the Euler–Lagrange equations of motion in continuous time (CT) classical mechanics (CM). Then we looked at the analogous situation in discrete time (DT) CM, where the notions of system function and action sum were introduced and used to work out DT equations of motion.

Given the central role of system functions in our approach, the natural question is the following: by what principles do we construct system functions?

The problem we face is that DT CM cannot be logically derived from CT CM, no more than quantum mechanics (QM) can be derived from CM. We have to take a leap and jump into the unknown. Depending on our choice of discretization, we may end up with a system function that does what we expect or one that leads to novel solutions to its equations of motion. This is not necessarily a bad thing.

How we view CT will influence how we discretize time. In Chapter 15, we shall discuss the geometrical approach of Marsden and others to DT CM. Generally, their methodology is aimed at finding variational integrators representing good approximations to CT, because that is their agenda. Our aim is fundamentally different: we want to explore the notion that DT may be a better model of time than CT, rather than an approximation to CT. In this respect we are of the same mindset as Caldirola and the Italian School, as discussed in Section 1.4.2.

To understand the system function and later the quantization of DT systems, it will be useful to review phase space and its relation to the Lagrangian. We shall discuss the standard formulation in terms of coordinates.

Given a canonical Lagrangian of the form $L(\mathbf{q}, \dot{\mathbf{q}}, t)$, the action integral $A_{\mathrm{fi}}[\Gamma]$ is the integral of the Lagrangian evaluated over a given canonical trajectory Γ of the system point in configuration space $\mathcal{C}$, over a given run-time interval $[t_{\mathrm{i}}, t_{\mathrm{f}}]$:

$$A_{\mathrm{fi}}[\Gamma] \equiv \int_{t_{\mathrm{i}}}^{t_{\mathrm{f}}} L(\mathbf{q}, \dot{\mathbf{q}}, t)\Big|_{\Gamma} dt. \tag{12.1}$$

Here $\mathbf{q} \equiv \{q^1, q^2, \ldots, q^r\}$ are suitable coordinates covering some region of interest in configuration space $\mathcal{C}$, which is assumed to be an r-dimensional manifold. Phase space $\mathcal{P}$ is a $2r$-dimensional manifold, identified with the cotangent bundle $T^*\mathcal{C}$ (Abraham and Marsden, 2008). The conventional approach is to use a coordinate patch for $\mathcal{P}$ with half of the $2r$ coordinates given by $\mathbf{q}$ and the other half given by 'canonical momentum' coordinates $\mathbf{p} \equiv \{p_1, p_2, \ldots, p_r\}$.

The standard route to phase space is to define momentum coordinates by a Legendre transformation from configuration-velocity space via the rule

$$\mathbf{p} \equiv \frac{\partial}{\partial \dot{\mathbf{q}}} L, \tag{12.2}$$

a relationship that requires detailed analysis in the case where primary and secondary constraints occur (Dirac, 1964).

The Hamiltonian $H(\mathbf{p}, \mathbf{q}, t)$ is a function over phase space obtained by standard rules in the case where there are no constraints (Goldstein *et al.*, 2002) and by closing the algebra of constraints as discussed by Dirac (1964) in the case of primary constraints. The Hamiltonian is used to derive Hamilton's equations of motion

$$\dot{\mathbf{p}} \underset{c}{=} -\frac{\partial}{\partial \mathbf{q}} H, \quad \dot{\mathbf{q}} \underset{c}{=} \frac{\partial H}{\partial \mathbf{p}}. \tag{12.3}$$

12.2 Hamilton's principal function

Now the principal objective of CT CM is to determine the classical or true trajectory Γ_c. Assuming we could solve the equations of motion (8.8) and thereby find Γ_c, Hamilton's principal function $S_c(\mathbf{q}_f, t_f; \mathbf{q}_i, t_i)$ is just the value of the action integral evaluated over that trajectory, i.e.,

$$S_c(\mathbf{q}_f, t_f; \mathbf{q}_i, t_i) \equiv A_{\mathrm{fi}}[\Gamma_c] = \int_{t_{\mathrm{i}}}^{t_{\mathrm{f}}} L(\mathbf{q}, \dot{\mathbf{q}}, t)\Big|_{\Gamma_c} dt. \tag{12.4}$$

Our notion here shows explicitly that S_c depends on the end-point events $(\mathbf{q}_i, t_i)$ and $(\mathbf{q}_f, t_f)$.

Hamilton's principal function plays a fundamental role in transformation theory and geometrical mechanics (Goldstein, 1964; Leech, 1965). Transformation theory is the study of coordinate transformations that leave the canonical formalism intact whilst geometrical mechanics gives a classical wave equation related to particle trajectories. To gain an understanding of the relationship between our DT equations and CT mechanics, we will now investigate infinitesimal variations in the classical path Γ_c and its effect on S_c.

We take the same approach as in our discussion in Section 10.3. We introduce a new path parameter λ running from 0 to 1, allowing us to reparametrize space and time coordinates along a given trajectory such that

$$t \to t(\lambda), \quad \mathbf{q} \to \mathbf{Q}(\lambda) \equiv \mathbf{q}(t(\lambda)), \tag{12.5}$$

with the end points fixed as follows:

$$t(0) \equiv t_{\mathrm{i}}, \quad t(1) \equiv t_{\mathrm{f}}. \tag{12.6}$$

Assuming that these new coordinates are differentiable functions of λ, Hamilton's principal function takes the form

$$S_{\mathrm{c}}(\mathbf{q}_{\mathrm{f}}, t_{\mathrm{f}}; \mathbf{q}_{\mathrm{i}}, t_{\mathrm{i}}) = \int_{\lambda=0}^{\lambda=1} \dot{t} L\left(\mathbf{Q}, \frac{\dot{\mathbf{Q}}}{\dot{t}}, t\right)\bigg|_{\Gamma_{\mathrm{c}}} d\lambda, \tag{12.7}$$

where the dots indicate derivatives with respect to the path parameter λ. Note that we exclude the possibility $\dot{t} = 0$ anywhere in the interval $0 \leqslant \lambda \leqslant 1$, as this would mean that the new path parameter λ was not a sensible replacement for time t. Generally we shall take $\dot{t} > 0$.

By definition, the integral in (12.7) is taken along the true or classical trajectory Γ_{c}. Now consider an infinitesimal variational departure from this trajectory, so that $\Gamma_{\mathrm{c}} \to \Gamma'_{\mathrm{c}}$, such that

$$\mathbf{Q} \to \mathbf{Q}' \equiv \mathbf{Q} + \delta\mathbf{Q}, \quad t \to t' \equiv t + \delta t. \tag{12.8}$$

Then by integrating by parts and applying the equations of motion, we find to first order in the variation

$$\delta S_{\mathrm{c}} = -\delta t_{\mathrm{f}}\, H_{\mathrm{f}} + \delta t_{\mathrm{i}}\, H_{\mathrm{i}} + \delta\mathbf{q}_{\mathrm{f}} \cdot \mathbf{p}_{\mathrm{f}} - \delta\mathbf{q}_{\mathrm{i}} \cdot \mathbf{p}_{\mathrm{i}}, \tag{12.9}$$

where subscripts denote evaluation at the particular run-time end points concerned.

From this we deduce

$$\begin{aligned} \frac{\partial S_{\mathrm{c}}}{\partial t_{\mathrm{f}}} &= -H_{\mathrm{f}}, & \frac{\partial S_{\mathrm{c}}}{\partial t_{\mathrm{i}}} &= H_{\mathrm{i}}, \\ \frac{\partial S_{\mathrm{c}}}{\partial \mathbf{q}_{\mathrm{f}}} &= \mathbf{p}_{\mathrm{f}}, & \frac{\partial S_{\mathrm{c}}}{\partial \mathbf{q}_{\mathrm{i}}} &= -\mathbf{p}_{\mathrm{i}}. \end{aligned} \tag{12.10}$$

We may use these equations to derive the famous Hamilton–Jacobi equation as follows. First, suppress the initial conditions in the notation, since these are assumed fixed, and write for the final end points

$$\begin{aligned} (\mathbf{p}, \mathbf{q}, t) &\equiv (\mathbf{p}_{\mathrm{f}}, \mathbf{q}_{\mathrm{f}}, t_{\mathrm{f}}), \\ H(\mathbf{p}, \mathbf{q}, t) &\equiv H_{\mathrm{f}}, \\ S(\mathbf{q}, t) &\equiv S_{\mathrm{c}}(\mathbf{q}, t; \mathbf{q}_{\mathrm{i}}, t_{\mathrm{i}}). \end{aligned} \tag{12.11}$$

Then equations (12.10) lead to the Hamilton–Jacobi equation

$$\frac{\partial S}{\partial t} + H(\nabla S, \mathbf{q}, t) = 0. \tag{12.12}$$

This equation looks like a classical wave equation. Its relationship with the Schrödinger equation has been much discussed in the literature (Goldstein *et al.*, 2002; Leech, 1965).

Cadzow's equation of motion (8.26) in DT mechanics can be given an interpretation in terms of Hamilton's principal function. Assume that CT mechanics is valid and consider an interval of time $[t_0, t_N]$ with initial and final configurations $\mathbf{q}_0 \equiv \mathbf{q}(t_0)$ and $\mathbf{q}_N(t_N)$ specified. Next, assume that we could solve the Euler–Lagrange equations of motion, thereby determining the true or classical trajectory Γ_c. Now create a partition of this interval by defining a set of intermediate times such that

$$t_0 < t_1 < \ldots < t_N. \tag{12.13}$$

For each subinterval $[t_n, t_{n+1}]$, construct Hamilton's principal function S^n by application of the rule

$$S^n \equiv S_c(\mathbf{q}_{n+1}, t_{n+1}; \mathbf{q}_n, t_n). \tag{12.14}$$

In other words, we solve the CT equations of motion over each interval $[t_n, t_{n+1}]$ subject to the boundary conditions that $\mathbf{q}_n \equiv \mathbf{q}(t_n)$ and $\mathbf{q}_{n+1} \equiv \mathbf{q}(t_{n+1})$, and hence evaluate the action integral over that interval, thereby giving S^n. From (12.10) we can write down the following expressions for upper and lower momenta for each subinterval:

$$\mathbf{p}_n^{(-)} \equiv -\frac{\partial}{\partial \mathbf{q}_n} S^n, \qquad \mathbf{p}_{n+1}^{(+)} \equiv \frac{\partial}{\partial \mathbf{q}_{n+1}} S^n. \tag{12.15}$$

Now, continuity of momentum across any boundary t_n between successive links of the partition means $\mathbf{p}_n^{(-)} \underset{c}{=} \mathbf{p}_{n-1}^{(+)}$, which, from (12.15), requires the condition

$$\frac{\partial}{\partial \mathbf{q}_n}\{S^n + S^{n-1}\} \underset{c}{=} 0. \tag{12.16}$$

This looks just like Cadzow's equation (8.26). In other words, an interpretation of the DT equation of motion is that it is a statement of the conservation of canonical momentum at temporal nodes.

The fundamental problem is this: we do not know the true classical trajectory, so we cannot actually work out any of the S^n. If we could work out any S^n, then presumably we could work out Hamilton's principal function for any interval, and hence we would have a complete solution to the classical problem.

The strategy in DT CM is to find some reasonable approximation Γ_{VP} to the true path Γ_c, evaluate the action integral $A_{fi}[\Gamma_{VP}]$ function and then take that as a system function. Here the subscript VP stands for 'virtual path', which we explain presently.

There is no unique approximation Γ_{VP} to Γ_{c}: we need to make an informed choice for any virtual path. Whatever we choose, we will *define* our system function F^n to be given by the integral

$$F^n \equiv \int_{t_n}^{t_{n+1}} L\Big|_{\Gamma_{\mathrm{VP}}} dt. \tag{12.17}$$

A dimensional analysis of (12.17) shows that a system function has the physical dimensions of an action, i.e., an angular momentum, rather than the dimensions of energy, which a Lagrangian always has.

A question arises at this point: over which virtual path should the Lagrangian be integrated in (12.17)? We shall discuss this in more detail in the next section. We can be sure of one thing: we cannot use the true or classical CT path because we do not know it. Indeed, if we believe that time really is discrete, then CT paths do not actually exist, which is why we refer to the path Γ_{VP} in (12.17) as virtual.

12.3 Virtual-path construction of system functions

We now focus on our choice of virtual path. We emphasize that this is a choice of discretization scheme, its principal advantages being that (i) it provides an algorithmic approach to temporal discretization; and (ii), in the limit $T \to 0$, the distinction between the Lagrangian and the system function should disappear, in the sense that we expect

$$\lim_{t_{\mathrm{f}} \to t_{\mathrm{i}}} \frac{\int_{t_{\mathrm{i}}}^{t_{\mathrm{f}}} L\Big|_{\Gamma_{\mathrm{VP}}} dt}{t_{\mathrm{f}} - t_{\mathrm{i}}} = L\Big|_{t_{\mathrm{i}}}. \tag{12.18}$$

The virtual-path method is based on the observation that, if time really is discrete on some fundamental level, then dynamical variables can only be associated with either the discrete times themselves, in which case they are referred to as *node variables*, or the links between successive nodes, in which case they are referred to as *link variables*.

For a CT field $\varphi(t, \mathbf{x})$ that is associated with nodes in the discretized theory, the virtual-path discretization algorithm is to make the replacement

$$\varphi(t, \mathbf{x}) \to \varphi_{n\lambda}(\mathbf{x}) \equiv \lambda\varphi_{n+1}(\mathbf{x}) + \bar{\lambda}\varphi_n(\mathbf{x}), \tag{12.19}$$

where the virtual-path parameter λ runs from zero to unity and $\bar{\lambda} \equiv 1 - \lambda$. This particular virtual path is a simple linear interpolation between the end-point values of the dynamical variable. When we come to discuss gauge invariance, this changes radically.

For a field variable $\phi(t, \mathbf{x})$ associated with the link running from t_n to t_{n+1} the corresponding virtual-path rule is

$$\phi(t, \mathbf{x}) \to \phi_n(\mathbf{x}). \tag{12.20}$$

Temporal derivatives of node fields are discretized according to the rule

$$\frac{\partial}{\partial t}\varphi(t,\mathbf{x}) \to \frac{1}{T}\frac{\partial}{\partial \lambda}\varphi_{n\lambda}(\mathbf{x}) = \frac{\mathbb{D}_n\varphi_n(\mathbf{x})}{T}, \qquad (12.21)$$

whilst spatial derivatives are discretized according to the rule

$$\nabla\varphi(t,\mathbf{x}) \to \nabla\varphi_{n\lambda}(\mathbf{x}) = \lambda\ \nabla\varphi_{n+1}(\mathbf{x}) + \bar{\lambda}\ \nabla\varphi_n(\mathbf{x}). \qquad (12.22)$$

We note that temporal derivatives of link variables generally do not occur, a fact connected with the existence of primary constraints involving such variables in CT mechanics (Dirac, 1964).

When all fields in a Lagrange density have been discretized according to these rules, i.e., once we have made the replacement

$$\mathcal{L}(\varphi, \phi, \partial_t\varphi, \nabla\varphi, \nabla\phi) \to \mathcal{L}_{n\lambda} \equiv \mathcal{L}(\varphi_{n\lambda}, \phi_n, T^{-1}\ \mathbb{D}_n\varphi_n, \nabla\varphi_{n\lambda}, \nabla\phi_n), \qquad (12.23)$$

the system function density $\mathcal{F}^n$ is found by integrating $\mathcal{L}_{n\lambda}$ over λ and multiplying by the chronon T, i.e.,

$$\mathcal{F}^n = T\langle\mathcal{L}_{n\lambda}\rangle, \qquad (12.24)$$

where we use angular brackets to denote integration over λ from 0 to 1. Specifically, for any virtual path f_λ we define

$$\langle f_\lambda\rangle \equiv \int_0^1 d\lambda\ f_\lambda. \qquad (12.25)$$

The same algorithm is used for point mechanics as for field mechanics, the only difference being that we do not have spatial derivatives in the former.

Example 12.1 Point-particle mechanics
For a given node degree of freedom q, the virtual path from time n to time $n+1$ gives

$$q_{n\lambda} \equiv \lambda q_{n+1} + \bar{\lambda} q_n, \quad \bar{\lambda} \equiv 1-\lambda, \quad 0 \leqslant \lambda \leqslant 1. \qquad (12.26)$$

Hence, given a canonical Lagrangian $L(\dot{q}, q, t)$, the system function F^n is given by

$$L(\dot{q}, q) \to F^n = T_n\langle L\,(T_n^{-1}\ \mathbb{D}_n q_n, q_{n\lambda}, t_{n\lambda})\rangle, \qquad (12.27)$$

where $t_{n\lambda} \equiv \lambda t_{n+1} + \bar{\lambda} t_n$.

For a point particle in a one-dimensional time-dependent potential, the Lagrangian

$$L = \frac{1}{2}m\dot{x}^2 - V(x,t) \qquad (12.28)$$

goes over to the system function

$$F^n = \frac{m(\mathbb{D}_n x_n)^2}{T_n} - T_n\int_0^1 d\lambda\ V(x_{n\lambda}, t_{n\lambda}). \qquad (12.29)$$

This discretization is based on a particular interpretation of the potential V in (12.28). We shall see in later chapters that, when the potential is of electromagnetic origin, then a different virtual path may be encountered. For instance, an alternative virtual path for the potential would be to make the replacement

$$V(x,t) \to V_{n\lambda} \equiv \lambda V(x_{n+1}, t_{n+1}) + \bar{\lambda} V(x_n, t_n), \tag{12.30}$$

rather than

$$V(x,t) \to V_{n\lambda} \equiv V(x_{n\lambda}, t_{n\lambda}). \tag{12.31}$$

It really is a matter of interpretation of the physical origin of the potential V.

In the long-term development of DT mechanics, we would naturally expect a quantized field-theoretical approach to be the one to use, in which case we would find virtual paths related to (12.30) to be the ones encountered naturally in the theory, rather than the more *ad hoc* (12.31). Essentially, spatial coordinates are not dynamical degrees of freedom in field theory, whereas in point-particle mechanics they are.

In applications of this approach, we frequently encounter integrals over the virtual-path parameter λ of the form $\langle \lambda^p \bar{\lambda}^q \rangle$, where p and q are integers. We use the result

$$\langle \lambda^p \bar{\lambda}^q \rangle \equiv \int_0^1 d\lambda\ \lambda^p (1-\lambda)^q = B(p+1, q+1) = \frac{\Gamma(1+p)\Gamma(1+q)}{\Gamma(2+p+q)}, \tag{12.32}$$

where $B(x,y)$ is the standard beta function (Arfken, 1985). Hence

$$\langle \lambda^p \bar{\lambda}^q \rangle = \frac{p!q!}{(1+p+q)!}, \tag{12.33}$$

which is symmetric in p and q. Useful applications of this result are

$$\langle 1 \rangle = 1, \quad \langle \lambda \rangle = \langle \bar{\lambda} \rangle = \frac{1}{2}, \quad \langle \lambda^2 \rangle = \langle \bar{\lambda}^2 \rangle = \frac{1}{3}, \quad \langle \lambda \bar{\lambda} \rangle = \frac{1}{6}. \tag{12.34}$$

13

The classical discrete time oscillator

13.1 The discrete time oscillator

The harmonic oscillator is perhaps the most important system under observation (SUO) in mathematical physics: its mathematics underpins the physics of quantum optics and relativistic quantum field theory (QFT), from free-particle fields to full-blown string theory. Remarkably, even the Coulombic interaction responsible for the structure of atoms in three dimensions can be modelled in terms of two coupled harmonic oscillators, each moving in a plane (Cornish, 1984). The continuous time (CT) harmonic oscillator has the great merit of being completely solvable from every theoretical direction in classic mechanics (CM) or quantum mechanics (QM).

The DT harmonic oscillator is also completely solvable classically and quantum mechanically. However, the range of dynamical behaviour is greater in DT than in CT, there being found three possible modes of behaviour in the former theory in contrast to the one mode in the latter. None of the three DT modes is strictly periodic except under very special circumstances when parameters and initial conditions take on special values. One of the three DT modes, the *elliptic mode*, is bounded, whilst the other two modes, the *parabolic* and *hyperbolic* modes, are unbounded. In DT QFT, the onset of the hyperbolic regime introduces a natural cutoff for particle momentum, which may be of potential significance as a natural regularization mechanism. This is discussed in later chapters.

In general, trajectories for the DT harmonic oscillator appear chaotic. Despite this we shall still refer to this SUO as an *oscillator* because its description in terms of ladder operators in the quantum case corresponds almost exactly to the quantized CT harmonic oscillator, albeit with some important differences.

The DT oscillator highlights the facts that DT mechanics is not equivalent to CT mechanics, being in many ways a richer theory:

> If we appeal to Cantor's hierarchy of the infinite, then the collection of all possible discontinuous changes is a higher order of infinity than the

number of continuous changes. By building our study of the physical world ultimately upon the assumption of the ubiquity of continuous changes at the most macroscopic level, we may be making not merely a gross simplification but an infinite simplification.

(Barrow, 1992)

The appearance of non-oscillatory solutions in DT mechanics can be attributed to the existence of a unit of time, the chronon, denoted by T. The pure number $\kappa \equiv T\omega$, where ω is the natural frequency parameter of the CT oscillator, sets the scale or barrier which separates the elliptic and hyperbolic modes of behaviour, with the parabolic mode marking the transition between these extremes.

13.2 The Newtonian oscillator

In the next section we will discuss a generic system function for the DT oscillator. In this section we will focus our attention on the origin of this generic form, viz., an SUO consisting of a Newtonian (non-relativistic) point particle of mass m moving in CT in one spatial dimension with coordinate x, under the action of a harmonic restoring force. The CT Lagrangian is given by

$$L(x, \dot{x}) = \frac{m}{2}\dot{x}^2 - \frac{m}{2}\omega^2 x^2, \tag{13.1}$$

where $x \equiv x(t)$ is the instantaneous position of a particle of mass m, $\dot{x}$ is the instantaneous rate of change of x and ω is a constant related to the frequency of oscillation of the system. We shall take $\omega > 0$.

Every possible solution to the Euler–Lagrange equation of motion

$$\ddot{x} \underset{c}{=} -\omega^2 x \tag{13.2}$$

is a trigonometric function of the form $x(t) = A\cos(\omega t) + B\sin(\omega t)$, where A and B are arbitrary. There are no divergent oscillator trajectories in CT.

The arbitrary constants A and B represent the two pieces of input data required in the general solution to (13.2). These data can be written into the general solution in various ways. In particular, we may express them in terms of boundary conditions in the form of the initial position $x_0 \equiv x(t_0)$ at time t_0 and the final position $x_1 \equiv x(t_1)$ at time $t_1 > t_0$. We find

$$x(t) = \frac{\sin(\omega t_1 - \omega t)x_0 + \sin(\omega t - \omega t_0)x_1}{\sin(\omega t_1 - \omega t_0)}, \tag{13.3}$$

provided that $\omega(t_1 - t_0)$ is not a multiple of π. The latter condition is required in order to ensure that the denominator is not zero in the representation (13.3).

Hamilton's principal function S_c is the action integral evaluated over Γ_c, the true or classical trajectory, from the initial time to the final time (Goldstein *et al.*, 2002). Inserting the classical solution (13.3) into the Lagrangian (13.1) and then evaluating the action integral from t_0 to $t_1 \equiv t_0 + T$ gives

$$S_c(x_1, t_1; x_0, t_0) = \frac{m\omega}{2\sin(\omega T)}\{(x_1^2 + x_0^2)\cos(\omega T) - 2x_1 x_0\}, \quad \sin(\omega T) \neq 0. \tag{13.4}$$

A significant feature of (13.4) is that it appears to diverge as ωT approaches any integer multiple of π. Since the motion is bounded and there is no singularity in the potential, the inescapable conclusion must be that there is a corresponding zero in the numerator in (13.4) at precisely the same values, so that in the limit $\omega T \to n\pi$, for n any integer (including zero), $S_c(x_1, t_0 + T; x_0, t_0)$ is well defined. At such limit points of time, we deduce from the numerator of (13.4) that $x_1 = (-1)^n x_0$, which means that there is strict periodicity in the system with a period that is independent of the starting position. We refer to this phenomenon as harmonic recurrence.

13.3 Temporal discretization of the Newtonian oscillator

We now discuss the temporal discretization of the SUO with the Lagrangian (13.1) via the virtual-path approach discussed in Chapter 12. We will work with a Type-1 temporal lattice, i.e., fixed t_n with $t_n \equiv nT$ and n an integer. The dynamical degrees of freedom are node variables denoted by $x_n \equiv x(t_n)$. The virtual path from t_n to t_{n+1} is defined by

$$x_{n\lambda} \equiv \lambda x_{n+1} + \bar{\lambda} x_n, \qquad \bar{\lambda} \equiv 1 - \lambda. \tag{13.5}$$

Then the system function F^n is given by $F^n \equiv T\langle L_{n\lambda}\rangle$, where

$$L_{n\lambda} = \frac{m}{2}\left(T^{-1}\frac{\partial}{\partial\lambda}x_{n\lambda}\right)^2 - \frac{m\omega^2}{2}x_{n\lambda}^2. \tag{13.6}$$

This gives

$$F^n = \frac{m(x_{n+1} - x_n)^2}{2T} - \frac{Tm\omega^2}{6}(x_{n+1}^2 + x_{n+1}x_n + x_n^2). \tag{13.7}$$

which is most conveniently rearranged into the form

$$F^n = \frac{m(3 - \omega^2 T^2)}{6T}(x_n^2 + x_{n+1}^2) - \frac{m(6 + \omega^2 T^2)}{6T}x_n x_{n+1}. \tag{13.8}$$

The DT equation of motion (8.26) then gives

$$\frac{x_{n+1} - 2x_n + x_{n-1}}{T^2} \underset{c}{=} -\omega^2\frac{x_{n+1} + 4x_n + x_{n-1}}{6}, \tag{13.9}$$

which can be rearranged in the explicit form

$$x_{n+1} \underset{c}{=} 2\frac{6 - 2\kappa^2}{6 + \kappa^2}x_n - x_{n-1}, \tag{13.10}$$

where $\kappa \equiv \omega T$.

We shall see from our analysis of the generalized oscillator discussed in the next section that bounded behaviour occurs provided $\kappa < 2\sqrt{3}$.

13.4 The generalized oscillator

The DT equation of motion (13.9) is a particular version of a generic equation that we find reappearing in various guises throughout this book, so it is economical to look at the general situation. By inspection of the Newtonian particle system function (13.8) we are led to define the generic oscillator system function

$$F^n = \frac{1}{2}\alpha\left(x_n^2 + x_{n+1}^2\right) - \beta x_n x_{n+1}, \qquad \beta > 0, \tag{13.11}$$

where α and β are real constants, with β taken positive. By reading off these coefficients from (13.8) we make the identifications

$$\alpha \equiv \frac{m(6 - 2\omega^2T^2)}{6T}, \qquad \beta \equiv \frac{m(6 + \omega^2T^2)}{6T} \tag{13.12}$$

when we have a Newtonian harmonic oscillator.

We note that these constants are separately divergent in the CT limit $T \to 0$. This is a throwback from the formal divergence in Hamilton's principal function (13.4) for the CT oscillator as $\omega T \to n\pi$, $n = 0, \pm 1$, from which we deduced the phenomenon of recurrence. In the DT case there is no actual divergence in the model per se, because the CT limit $T \to 0$ is not part of the DT dynamics if we believe time is really discrete. The only reason for taking the CT limit is for purposes of comparison between DT and CT.

The system function (13.11) is invariant with respect to the interchange $x_n \leftrightarrow x_{n+1}$, reflecting the fact that an oscillator is a non-dissipative system and so its equations of motion should be time-reversal-invariant. These equations of motion are found to be

$$x_{n+1} \underset{c}{=} 2\eta x_n - x_{n-1}, \qquad \eta \equiv \frac{\alpha}{\beta}, \tag{13.13}$$

and are indeed time-reversal-invariant. It is the value of the parameter η in (13.13) which decides the mode of the solution, i.e., whether the system evolves in the elliptic, parabolic or hyperbolic mode.

13.5 Solutions

We can solve the equations of motion (13.13) in several ways. The easiest approach, and the one that is useful in the quantization discussed in Chapter 17, is as follows. We take a linear combination of two successive positions, introducing a complex constant μ and defining the auxiliary variable

$$a_n \equiv x_n - \mu x_{n+1}, \tag{13.14}$$

which may be complex, depending on the values of α and β in the system function (13.11). The constant μ is fine tuned so that

$$a_n \underset{c}{=} \mu a_{n-1} \tag{13.15}$$

under the equation of motion (13.13). This implies

$$a_n \underset{c}{=} \mu^n a_0. \tag{13.16}$$

The equation of motion (13.13), definition (13.14) and requirement (13.15) lead to the condition

$$\mu = \eta \pm \sqrt{\eta^2 - 1}. \tag{13.17}$$

We define the two branches: $\mu^{(+)} \equiv \eta + \sqrt{\eta^2 - 1}$ and $\mu^{(-)} \equiv \eta - \sqrt{\eta^2 - 1}$. The constants $\mu^{(+)}$ and $\mu^{(-)}$ are not independent, being related by the constraint $\mu^+\mu^- = 1$. They are real for $|\eta| > 1$ and complex conjugates when $|\eta| < 1$. The auxiliary variable a is now replaced by the pair $a^{(\pm)} \equiv x_n - \mu^{(\pm)} x_{n-1}$.

13.6 The three regimes

The complete solution to the problem is now readily obtained from (13.14) and (13.16). It is given by

$$x_n \underset{c}{=} \frac{(\mu^+)^n - (\mu^-)^n}{\mu^+ - \mu^-} x_1 + \frac{(\mu^-)^{n-1} - (\mu^+)^{n-1}}{\mu^+ - \mu^-} x_0, \tag{13.18}$$

for $\eta^2 \neq 1$. For the case when $\eta^2 < 1$ we write $\eta = \cos\theta$ for some positive parameter θ. We take $\sqrt{1 - \eta^2} = \sin\theta$ and then

$$x_n \underset{c}{=} \frac{\sin(n\theta) x_1 - \sin((n-1)\theta) x_0}{\sin\theta}. \tag{13.19}$$

The limit $\theta \to 0$ corresponds to the free particle case $\eta = 1$ and gives, by l'Hôpital's rule, the free-particle trajectory

$$x_n \underset{c}{=} n(x_1 - x_0) + x_0, \quad n = 0, 1, 2, \ldots. \tag{13.20}$$

For $\eta^2 > 1$ we write $\eta = \cosh\theta$, again assuming that η is positive, and then

$$x_n \underset{c}{=} \frac{\sinh(n\theta) x_1 - \sinh((n-1)\theta) x_0}{\sinh\theta}, \quad \theta > 0. \tag{13.21}$$

A similar result holds for $\eta < -1$.

The crucial result is that bounded behaviour occurs when $\eta^2 < 1$, whereas unbounded behaviour occurs when $\eta^2 > 1$.

The readily solved case when $\eta^2 = 1$ corresponds to the free particle and will be referred to as the *parabolic* case. The system will be said to be *hyperbolic* when $\eta^2 > 1$ and *elliptic* when $\eta^2 < 1$.

For the case $\eta^2 < 1$ it is useful to define $\mu = \eta + i\sqrt{1-\eta^2}$ and

$$a_n \equiv \mu^n[x_{n+1} - \mu x_n], \quad a_n^* \equiv \mu^{-n}[x_{n+1} - \mu^{-1} x_n], \tag{13.22}$$

the advantage being that these are constants of the motion, viz.,

$$a_n \underset{c}{=} a_{n-1}, \quad a_n^* \underset{c}{=} a_{n-1}^*. \tag{13.23}$$

These are useful in the construction of particle states in QFT because they correspond to annihilation and creation operators in the Heisenberg picture.

13.7 The Logan invariant

A Logan invariant for the DT harmonic oscillator can be found using two different approaches.

The first is to exploit the properties of the a_n^+ and a_n^- constructs defined above. From (13.15) and the fact that $\mu^+\mu^- = 1$, the construct

$$C^n \equiv \frac{1}{2} a_n^+ a_n^- \tag{13.24}$$

is an invariant. Specifically,

$$C^n \equiv \frac{1}{2}\beta(x_n^2 + x_{n+1}^2) - \alpha x_n x_{n+1}. \tag{13.25}$$

The other approach is to find a near symmetry of the system function. Consider the transformation of coordinates

$$x_n \to x_n' \equiv x_n + \epsilon u_n, \tag{13.26}$$

where u_n depends on x_n and x_{n+1}, and the real parameter ϵ is considered an infinitesimal. Then the change to first order in ϵ of the system function (13.11) is

$$\delta F^n = \epsilon u_n\{\alpha x_n - \beta x_{n+1}\} + \epsilon u_{n+1}\{\alpha x_{n+1} - \beta x_n\}. \tag{13.27}$$

Now consider the linear form $u_n \equiv A x_n + B x_{n+1}$, where A and B are constants. Then, on the true trajectory,

$$u_{n+1} \underset{c}{=} (A + 2\eta B)x_{n+1} - B x_n. \tag{13.28}$$

Hence

$$\begin{aligned} \delta F^n \underset{c}{=} & \epsilon(\alpha A + \beta B)x_n^2 - 2\epsilon\{\beta A + \alpha B\}x_n x_{n+1} \\ & + \epsilon\{-\beta B + \alpha A + 2\alpha\eta B\}x_{n+1}^2. \end{aligned} \tag{13.29}$$

Now setting $A = -\eta B$ means that the cross terms vanish and we find

$$\delta F^n = \epsilon B\{\alpha\eta - \beta\}\mathbb{D}_n\{x_n^2\}. \tag{13.30}$$

From our discussion of the Logan invariant in Chapter 8 we take $v_n = \epsilon B\{\alpha\eta - \beta\}x_n^2$ in (8.36), leading to the Logan invariant

$$C^n = -\epsilon B\{\beta(x_n^2 + x_{n+1}^2) - 2\alpha x_n x_{n+1}\}, \tag{13.31}$$

consistent with (13.25).

13.8 The oscillator in three dimensions

Taking standard Cartesian coordinates $\mathbf{r} \equiv (x, y, z)$, the Lagrangian for this SUO is

$$L(\mathbf{r}, \dot{\mathbf{r}}) \equiv \frac{1}{2}m\dot{\mathbf{r}} \cdot \dot{\mathbf{r}} - \frac{1}{2}m\omega^2\mathbf{r} \cdot \mathbf{r}. \tag{13.32}$$

We define the virtual paths

$$\mathbf{r}_{n\lambda} \equiv \lambda\mathbf{r}_{n+1} + \bar{\lambda}\mathbf{r}_n. \tag{13.33}$$

The system function F^n is then given by $F^n \equiv T\langle L(\mathbf{r}_{n\lambda}, \mathbb{T}_n\mathbf{r}_n)\rangle$, i.e.,

$$\begin{aligned} F^n &\equiv T\int_0^1 \left\{\frac{1}{2}m\frac{\mathbb{D}_n\mathbf{r}_n}{T} \cdot \frac{\mathbb{D}_n\mathbf{r}_n}{T} - \frac{1}{2}m\omega^2\mathbf{r}_{n\lambda} \cdot \mathbf{r}_{n\lambda}\right\} d\lambda \\ &= \frac{m(\mathbf{r}_{n+1} - \mathbf{r}_n)^2}{T} - \frac{m\omega^2 T}{6}(\mathbf{r}_{n+1}^2 + \mathbf{r}_{n+1} \cdot \mathbf{r}_n + \mathbf{r}_n^2) \\ &= \frac{1}{2}\alpha(\mathbf{r}_n^2 + \mathbf{r}_{n+1}^2) - \beta\mathbf{r}_n \cdot \mathbf{r}_{n+1}, \end{aligned} \tag{13.34}$$

where α and β are given by (13.12). This system function gives the DT equation of motion

$$\mathbf{r}_{n+1} \underset{c}{=} 2\eta\mathbf{r}_n - \mathbf{r}_{n-1}, \quad \eta \equiv \alpha/\beta. \tag{13.35}$$

The analysis follows that of the one-dimensional oscillator discussed above, with elliptic, parabolic and hyperbolic regimes decided by the value of the parameter η as before.

If now we make the infinitesimal transformations $\mathbf{r}_n \longrightarrow \mathbf{r}'_n \equiv \mathbf{r}_n + \varepsilon\mathbf{u}_n$, where ε is infinitesimal and $\mathbf{u}_n \equiv \eta\mathbf{r}_n - \mathbf{r}_{n+1}$, then we find that the transformation is a near symmetry of the system function, with $v_n = \beta\mathbf{r}_n \cdot \mathbf{r}_n - \eta\alpha\mathbf{r}_n \cdot \mathbf{r}_n$. From the DT Weiss action principle (Sudarshan and Mukunda, 1983; Weiss, 1936) we obtain the conserved quantity

$$C^n \equiv \frac{1}{2}\beta(\mathbf{r}_n \cdot \mathbf{r}_n + \mathbf{r}_{n+1} \cdot \mathbf{r}_{n+1}) - \alpha\mathbf{r}_n \cdot \mathbf{r}_{n+1}. \tag{13.36}$$

If again we make the replacements (13.12), assume a trajectory of the form (9.23) and take the limit $T \rightarrow 0$, we find

$$\lim_{T\rightarrow 0}\frac{C^n}{T} = \frac{1}{2}m\dot{\mathbf{r}} \cdot \dot{\mathbf{r}} + \frac{1}{2}m\omega^2\mathbf{r} \cdot \mathbf{r}, \tag{13.37}$$

which corresponds to conserved energy in the Newtonian CT theory, as expected.

The system function (9.21) is invariant with respect to spatial rotations,

$$\begin{aligned} \mathbf{r}_n &\longrightarrow \mathbf{r}'_n \equiv R(\varepsilon\boldsymbol{\omega})\mathbf{r}_n, \\ \mathbf{r}_{n+1} &\longrightarrow \mathbf{r}'_{n+1} \equiv R(\varepsilon\boldsymbol{\omega})\mathbf{r}_{n+1}, \end{aligned} \tag{13.38}$$

where $R(\varepsilon\boldsymbol{\omega})$ is an orthogonal matrix and $\boldsymbol{\omega}$ is a fixed three-vector. If we write

$$\mathbf{r}'_n \equiv \mathbf{r}_n + \varepsilon\boldsymbol{\omega} \times \mathbf{r}_n + O(\varepsilon^2) \tag{13.39}$$

then we find a conserved 'angular-momentum' vector

$$\mathbf{L}_n \equiv \mathbf{r}_n \times \mathbf{r}_{n+1} \underset{c}{=} \mathbf{L}_{n+1}. \tag{13.40}$$

13.9 The anharmonic oscillator

The anharmonic oscillator has been encountered twice already. In Section 6.2 we reverse engineered a particular quartic function for a one-dimensional system and found the DT equation of motion (6.30), which we referred to as 'the' anharmonic oscillator. Then in Section 9.2 we asserted that a system function for the three-dimensional anharmonic oscillator was given by (9.21) with equation of motion (9.25).

If, on the other hand, we were to apply the virtual-path discretization procedure employed up to now, we would get a very different system function. Specifically, on taking the virtual paths (13.33) and applying the rule $F^n \equiv T\langle L(\mathbf{r}_{n\lambda}, \mathbb{T}_n\mathbf{r}_n)\rangle$ to the CT anharmonic oscillator Lagrangian (9.20), we would find a system function very different from (9.21). In particular, the virtual-path system function, which we do not show here, depends only on powers of the coordinates, up to the fourth power, and there is no logarithmic dependence of the kind shown in (9.21) What is going on?

The explanation is as follows. Because DT is not a logical consequence of CT, it is quite possible to encounter very different DT system functions that have the same CT limit. The two very different system functions for the same CT SUO that we have discovered are based on different discretization principles. The virtual-path approach discussed in the previous section is an algorithmic one that always yields a system function, no matter how complex the CT Lagrangian, provided that this Lagrangian is a finite polynomial in the variables. Whilst this is not the case for the inverse-square potential, it is the case for the anharmonic oscillator and for all the field theories we are going to discuss.

On the other hand, the origin of the logarithmic system function (9.21) is based on reverse engineering and goes as follows. Assume a system function $F^n(\mathbf{r}_n, \mathbf{r}_{n+1})$ of the time-reversal-invariant form

$$F^n(\mathbf{r}_n, \mathbf{r}_{n+1}) \equiv -\beta\mathbf{r}_n \cdot \mathbf{r}_{n+1} + \frac{1}{2}W(\mathbf{r}_n) + \frac{1}{2}W(\mathbf{r}_{n+1}), \tag{13.41}$$

where $W(\mathbf{r}_n)$ is to be determined. In Chapter 17 on quantization, we shall refer to such system functions as being in 'normal' form.[1] Then the DT equation of motion based on such a system function is

$$\mathbf{r}_{n+1} \underset{c}{=} \beta^{-1} \nabla_n W(\mathbf{r}_n) - \mathbf{r}_{n-1}. \tag{13.42}$$

Assuming an equation of motion of the form (9.25) then leads to a first-order differential equation for W, which can be integrated to give the system function (9.21).

The advantage of the logarithmic system function is that it comes with a conserved quantity, given by (9.31). Unfortunately, finding an invariant seems impossible for most other SUOs. In that situation, using a virtual-path approach in a Type-3 setting, i.e., with a variable chronon, as in Lee's approach discussed in Chapter 10, leads to a conserved quantity analogous to energy.

[1] Specifically, if a system function looks like (13.41) in some coordinate patch, then those coordinates will be called *normal*.

14

Type-2 temporal discretization

14.1 Introduction

In this chapter we discuss an approach to temporal discretization that gives us a method of generating Type-2 discrete time (DT) mechanics, i.e., non-dynamical (i.e., fixed) irregular intervals between successive temporal nodes (Klimek, 1993). This approach generalizes the formalism of q-calculus discussed in Chapter 3.

We recall that the q-derivative $D_q f(t)$ of a real- or complex-valued function $f(t)$ of a real variable t is defined by

$$D_q f(t) \equiv \frac{f(qt) - f(t)}{qt - t}, \quad qt \neq t. \tag{14.1}$$

Here q is a real parameter controlling the departure from the standard derivative, which is given by the limit $q \to 1$, if the limit exists. We note, however, that definition (14.1) does not require f to be a continuous function of t.

The q-derivative has a dual character. On the one hand, it exists, provided only that f is defined at t and qt, assuming $qt \neq t$. If f is defined over the positive reals for instance, then the q-derivative exists everywhere in $(0, \infty)$, given $q \neq 1$. This means that the q-derivative can be defined over a continuum, given the right conditions. On the other hand, the q-derivative involves a discrete time-step for its definition, so it involves discreteness from that point of view. This dual nature gives q-calculus an interesting flavour. Many interesting theorems and formulae in q-calculus retain this dual nature.

Assuming f is well defined wherever required, consider a given starting time t_0. Then in q-calculus there is an implied associated sequence $\{t_n\}$ of temporal node values starting at t_0 given by $t_n \equiv q^n t_0$, $n = 0, 1, 2, \ldots$, assuming we wish to go forwards in the discrete time label n. Successive links, i.e., intervals of time between successive nodes, are of increasing value if $q > 1$ or decreasing value if $q < 1$, assuming that q is positive. We shall call the infinite sequence $\{t_n : n = 0, 1, 2, \ldots, \infty\}$ a standard temporal sequence and the sequence

$\{t_n : n = M, M+1, \ldots, N\}$ a finite temporal sequence for finite integers M, N such that $M < N$.

Our main interest in this book is to describe the physical Universe in as conventional a fashion as possible, with the sole exception that time is discretized. Therefore, we shall be interested only in dynamics based on conventional notions of time. Specifically, this means that a dynamical state associated with a time t_{n+1} will be regarded as physically later than a state associated with a time t_n if $t_{n+1} > t_n$. Here the concept of 'later' will be defined relative to any observer acquiring information as the Universe expands. We shall call this the 'positive-time' direction. This means, for instance, that, in DT mechanics based on q-calculus, we will take $t_{n+1} = q^{-1}t_n$, since the formalism is generally based on taking $q < 1$. A positive-time sequence will be any sequence $\{t_n\}$ such that $t_{n+1} > t_n$.

14.2 q-Mechanics

In this section we shall review the setting up of a one-dimensional particle model that is based on q-calculus (Klimek, 1993). Consider a real dynamical variable $x(t)$ evolving in CT from $t = a > 0$ to $t = b > a$. We take as 'q-Lagrangian' a function L of $x_t \equiv x(t)$ and of $v_t \equiv D_q x(t)$, and consider a q-action integral $A_q[x]$ defined by

$$A_q[x] \equiv \int_a^b L(x_t, v_t) d_q t. \tag{14.2}$$

Now consider a variation of path given by

$$x_t \to x'_t \equiv x_t + \varepsilon u_t, \tag{14.3}$$

where ε is a real parameter and u_t is an arbitrary real function of t. We shall assume all functions are well defined for positive t. q-Lagrangians will be assumed to be analytic functions of x and v, so an expansion in powers of ε gives

$$A_q[x + \varepsilon u] = A_q[x] + \varepsilon \int_a^b \{u_t \, \partial_x L_t + D_q u_t . p_t\} d_{qt} + O(\varepsilon^2), \tag{14.4}$$

where $\partial_x L_t \equiv (\partial/\partial x) L(x_t, v_t)$ and $p_t \equiv (\partial/\partial v) L(x_t, v_t)$.

At this point, we recall the fundamental difference between the Leibniz rule $(fg)' \equiv f'g + fg'$ in standard calculus and its discrete analogues, two forms of which are given by (3.15). We choose to rewrite the second term on the right-hand side of (14.4) in the form

$$D_q u_t . p_t = D_q[\bar{p}_t u_t] - u_t \, D_q \bar{p}_t, \tag{14.5}$$

where $\bar{p}_t \equiv p_{q^{-1}t}$. Note that

$$D_q \bar{p}_t = D_q \bar{p}(t) \equiv \frac{\bar{p}(qt) - \bar{p}(t)}{qt - 1} = \frac{\bar{p}_{qt} - \bar{p}_t}{qt - t} = \frac{p_t - p_{q^{-1}t}}{qt - t}, \tag{14.6}$$

but

$$D_q p_{q^{-1}t} = D_q p(q^{-1}t) \equiv \frac{p(qq^{-1}t) - p(q^{-1}t)}{qq^{-1}t - q^{-1}t} = \frac{q\{p_t - p_{q^{-1}t}\}}{qt - t} = q\ D_q \bar{p}_t, \quad (14.7)$$

which is why we have introduced the $\bar{p}$ notation.

The functional derivative $\mathcal{D}_u A_q[x]$ is given by

$$\mathcal{D}_u A_q[x] \equiv \lim_{\varepsilon \to 0} \frac{A_q[x + \varepsilon u] - A_q[x]}{\varepsilon} \quad (14.8)$$

$$= \int_a^b \{u_t\ \partial_x L_t + D_q[\bar{p}_t u_t] - u_t\ D_q \bar{p}_t\} d_{qt} \quad (14.9)$$

$$= [\bar{p}_t u_t]_a^b + \int_a^b u_t \{\partial_x L_t - D_q \bar{p}_t\} d_{qt}, \quad (14.10)$$

where we use property (3.27) of q-integration.

If now we apply the q-analogue of the Weiss action principle (Weiss, 1936), we arrive at the q-mechanics equation of motion

$$\partial_x L_t \underset{c}{=} D_q \bar{p}_t. \quad (14.11)$$

This equation of motion is a CT difference equation, but we may relate it to the DT equations of motion (8.26) as follows.

Pick a value of $t > 0$ and define $t_n \equiv t$, and then $t_{n-1} \equiv qt$, $t_{n+1} \equiv q^{-1}t, x_n \equiv x(t)$, $x_{n-1} \equiv x(qt)$, and $x_{n+1} \equiv x(q^{-1}t)$. Next, write

$$F^{n-1}(x_n, x_{n-1}) \equiv (t - qt) L(x_t, D_q x_t) = T_{n-1} L\left(x_n, \frac{x_n - x_{n-1}}{T_{n-1}}\right), \quad (14.12)$$

where $T_{n-1} \equiv t - qt$. Hence

$$\frac{\partial}{\partial x^n} F^{n-1} = T_{n-1}\ \partial_x L_t + \partial_v L_t = T_{n-1}\ \partial_x L_t + p_t = T_{n-1}\ \partial_x L_t + \bar{p}_{qt},$$

$$\frac{\partial}{\partial x^n} F^n = -\partial_v L_{q^{-1}t} = -p_{q^{-1}t} = -\bar{p}_t, \quad (14.13)$$

so we find

$$\frac{\partial}{\partial x_n}\{F^n + F^{n-1}\} = T_{n-1}\ \partial_x L_t + \bar{p}_{qt} - \bar{p}_t \underset{c}{=} 0, \quad (14.14)$$

which is equivalent to (14.11).

Example 14.1 We take $L = (D_q x_t)^2/(2m)$, which is the free-particle q-Lagrangian. Then

$$\partial_x L_t = 0, \quad p_t \equiv \partial_v L_t = D_q x_t / m. \quad (14.15)$$

Hence

$$\bar{p}_t \equiv p_{q^{-1}t} = \frac{x_{q^{-1}t} - x_t}{m(1-q)t}. \quad (14.16)$$

Then the equation of motion (14.11) gives $D_q \bar{p}_t \underset{c}{=} 0$, which reduces to

$$q x_t \underset{c}{=} (q+1) x_{qt} - x_{q^2 t}. \tag{14.17}$$

On the other hand, the corresponding system function is

$$F^n \equiv \frac{(x_{n+1} - x_n)^2}{2mT_n}, \tag{14.18}$$

from which the equation of motion (8.26) gives

$$q x_{n+1} \underset{c}{=} (q+1) x_n - x_{n-1}, \tag{14.19}$$

which is consistent with (14.17).

The equation of motion (14.19) is readily solved using the conservation of q-momentum for this system. We rearrange (14.19) into the form

$$q x_{n+1} - x_n \underset{c}{=} q x_n - x_{n-1}, \tag{14.20}$$

which means that the quantity $a \equiv q x_n - x_{n-1}$ is an invariant. Hence

$$x_n = \frac{a}{q} + \frac{x_{n-1}}{q}, \tag{14.21}$$

which immediately gives

$$x_n = a\left[\frac{1}{q} + \frac{1}{q^2} + \cdots + \frac{1}{q^n}\right] + \frac{x_0}{q^n}. \tag{14.22}$$

Hence the solution for $n = 1, 2, \ldots$ is

$$x_n = \frac{(1-q^n)}{q^n(1-q)} a + \frac{x_0}{q^n} = \frac{(1-q^{n+1})x_0 - (1-q^n)x_{-1}}{q^n(1-q)}. \tag{14.23}$$

14.2.1 q-Energy

In her analysis of q-mechanics (Klimek, 1993), Klimek noted that the q-analogue of energy was not conserved. This is to be expected in any Type-2 discretization, unless a suitable Logan invariant that corresponds in some way to CT energy can be found. This is the case for the Type-1 oscillator studied in previous chapters in the CT limit, for instance.

The problem can be understood given our analysis of Lee's Type-3 mechanics studied in Chapter 10: the temporal intervals $T_n \equiv t_{n+1} - t_n = qT_{n+1}$ in Klimek's approach are fixed. A possible solution is to allow q itself to be a time-dependent variable, so that the temporal intervals become dynamical variables.

However, this would mean that we were no longer dealing with conventional q-calculus, in which it is assumed that q is fixed. Dynamical chronons were discussed in Chapter 10. In the next section we shall investigate Klimek's generalization of q-calculus, which stops short of taking that radical step.

14.3 Phi-functions

The generalization of q-calculus taken here follows Klimek's methodology, with the introduction of a special function φ of continuous time t called a phi-function (Klimek, 1993). Any phi-function φ will be a real-valued function of t, usually with domain $\mathbb{R}$, such that

$$\varphi(t) > t, \qquad \forall t. \tag{14.24}$$

Phi-functions can be used to generate positive-time sequences as illustrated by Figure 14.1. Specifically, for a given initial time t_0, we define $t_1 \equiv \varphi(t_0)$, $t_2 \equiv \varphi(t_1)$ and so on. Then the basic property (14.24) of a phi-function ensures that we will have a positive-time sequence.

We will assume that all functions are well defined over the real line, so we shall ignore the warning we gave at the end of Section 3.4 about the domain of definition. For q-calculus, it is normally assumed that $0 < q < 1$, although the case $q > 1$ has been touched upon (Oney, 2007). We shall take $\varphi(t) > t$ to make the intuition easier: this choice gives $t_{n+1} \equiv \varphi(t_n) > t_n$. Recall that, in the case of q-mechanics above, $q < 1$ requires us to work in a reverse direction and define $t_{n+1} \equiv q^{-1} t_n$.

Given a phi-function φ, it is convenient to introduce an associated invasive phi-operator $\hat{\varphi}$, such that, for any function f of time t, we have

$$\hat{\varphi} f(t) = f(\varphi(t)). \tag{14.25}$$

From this definition we note that $\hat{\varphi}\{f(t)g(t)\} = f(\varphi(t))g(\varphi(t))$.

Successive applications of a phi-operator are indicated by powers. For example,

$$\hat{\varphi}^2 f(t) = \hat{\varphi}(\hat{\varphi} f(t)) = f(\varphi(\varphi(t))), \tag{14.26}$$

and so on. We shall denote multiple applications of a phi-operator on the function $g(t) = t$ by the notation $\varphi^{[n]}(t) \equiv \hat{\varphi}^n t$, and then we have, for an arbitrary function f of t, the rule

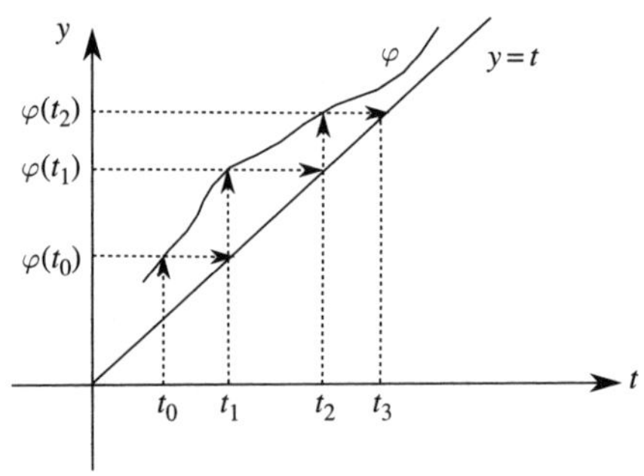

Figure 14.1. Generating variable discrete times via a phi-function.

$$\hat{\varphi}^n f(t) = f(\varphi^{[n]}(t)). \tag{14.27}$$

Given an initial time t_0, the associated temporal $\{t_n\}$ sequence generated by a given phi-function is then given by

$$t_n \equiv \hat{\varphi}^n t\big|_{t=t_0} = \varphi^{[n]}(t_0), \tag{14.28}$$

and this is guaranteed to be a positive-time sequence.

14.4 The phi-derivative

Given a phi-function φ, we define the phi-difference operator

$$\hat{\Delta}_\varphi \equiv \hat{\varphi} - 1 \tag{14.29}$$

and the phi-derivative operator

$$\hat{D}_\varphi \equiv \frac{1}{\hat{\Delta}_\varphi t}\,\hat{\Delta}_\varphi, \qquad t > 0. \tag{14.30}$$

Then the phi-derivative $D_\varphi f$ of a function f is defined by

$$D_\varphi f \equiv \hat{D}_\varphi f \equiv \frac{\hat{\Delta}_\varphi f}{\hat{\Delta}_\varphi t}, \qquad t > 0. \tag{14.31}$$

This is the analogue of the q-derivative (14.1), which we recover by taking the representation $\hat{\varphi} = q$, i.e., the action of the operator is to multiply t by q.

14.4.1 Properties of the phi-derivative

The phi-derivative satisfies the usual linearity property of differentiation:

$$D_\varphi(f + g) = D_\varphi f + D_\varphi g, \tag{14.32}$$

but the standard Leibniz rule

$$\frac{d}{dt}(fg) = \frac{df}{dt}g + f\frac{dg}{dt} \tag{14.33}$$

needs to be modified. As in the case of the q-derivative, there are two useful and equivalent variants of the phi-derivative of a product: we find

$$\begin{aligned}\hat{D}_\varphi(fg) &\equiv \frac{\hat{\varphi}f.\hat{\varphi}g - fg}{\Delta_\varphi t}\\ &= \frac{(\hat{\varphi}f - f)\hat{\varphi}g + f\hat{\varphi}g - fg}{\Delta_\varphi t} = \hat{D}_\varphi f.\hat{\varphi}g + f\ \hat{D}_\varphi g\\ &= \frac{\hat{\varphi}f.(\hat{\varphi}g - g) + \hat{\varphi}fg - fg}{\Delta_\varphi t} = \hat{\varphi}f.\hat{D}_\varphi g + \hat{D}_\varphi f.g. \end{aligned} \tag{14.34}$$

Which one is most useful will depend on context.

Then we find the following results:

$$\hat{D}_\varphi t^p = \sum_{n=0}^{p-1} [\varphi(t)]^n t^{p-n-1}, \qquad p = 0, 1, 2, \ldots, \tag{14.35}$$

$$\hat{D}_\varphi \frac{1}{t} = -\frac{1}{\varphi(t)t}, \qquad t > 0. \tag{14.36}$$

14.5 Phi-integrals

A phi-integral over an interval $[a, b]$, where $b > a$, of a function f with respect to a phi-function φ is denoted by

$$\int_a^b f \, d_\varphi \equiv \int_a^b f(t) d_\varphi(t), \tag{14.37}$$

where d_φ represents a 'phi-measure', which we need not specify at this point (Klimek, 1993). A phi-integral will satisfy the following axiomatic properties.

1. Additivity: for $a < b < c$,

$$\int_a^b f \, d_\varphi + \int_b^c f \, d_\varphi = \int_a^c f \, d_\varphi. \tag{14.38}$$

2. Linearity: for $a \leqslant b$, $\alpha, \beta \in \mathbb{C}$ and arbitrary functions f and g,

$$\int_a^b \{\alpha f + \beta g\} d_\varphi = \alpha \int_a^b f \, d_\varphi + \beta \int_a^b g \, d_\varphi. \tag{14.39}$$

3. Integrability: for $a \leqslant b$ and any function f,

$$\int_a^b \hat{D}_\varphi f . d_\varphi = f(b) - f(a). \tag{14.40}$$

The last property, integrability, justifies the name 'phi-measure' for d_φ, since the choice $f(t) \equiv t$ returns the length of the interval of integration:

$$\int_a^b d_\varphi = \int_a^b \hat{D}_\varphi t . d_\varphi = b - a. \tag{14.41}$$

14.6 The summation formula

We can now derive a most useful expression for a phi-integral for the special case $b = \hat{\varphi}^N t|_{t=a}$, i.e., when the end points of the interval $[a, b]$ are related by an integral number of phi-function compositions. We use the integrability property, (14.40), making the replacement

$$f \to \left\{\varphi^{[n+1]} - \varphi^{[n]}\right\} . \hat{\varphi}^n f. \tag{14.42}$$

Then (14.40) gives

$$\int_a^b \hat{D}_\varphi \left[\left\{\varphi^{[n+1]} - \varphi^{[n]}\right\} \hat{\varphi}^n f\right] d_\varphi = \left[\left\{\varphi^{[n+1]} - \varphi^{[n]}\right\} \hat{\varphi}^n f\right]_a^b. \tag{14.43}$$

On the other hand, by expanding the phi-integral on the left-hand side of (14.43), we find

$$\begin{aligned}\int_a^b \hat{D}_\varphi \left[\left\{\varphi^{[n+1]} - \varphi^{[n]}\right\} \hat{\varphi}^n f\right] d_\varphi &= \int_a^b \frac{1}{\hat{\Delta}_\varphi t}(\hat{\varphi} - 1) \left[\left\{\varphi^{[n+1]} - \varphi^{[n]}\right\} \hat{\varphi}^n f\right] d_\varphi \\ &= \int_a^b \frac{1}{\hat{\Delta}_\varphi t} \left[\left\{\varphi^{[n+2]} - \varphi^{[n+1]}\right\} \hat{\varphi}^{n+1} f \right. \\ &\qquad \left. - \left\{\varphi^{[n+1]} - \varphi^{[n]}\right\}\hat{\varphi}^n f\right] d_\varphi \\ &= \int_a^b \left\{\hat{D}_\varphi \varphi^{[n+1]}.\hat{\varphi}^{n+1} f - \hat{D}_\varphi \varphi^{[n]}.\hat{\varphi}^n f\right\} d_\varphi \\ &= \mathbb{D}_n \int_a^b \hat{D}_\varphi \varphi^{[n]}.\hat{\varphi}^n f \, d_\varphi,\end{aligned} \tag{14.44}$$

where $\mathbb{D}_n$ is the difference operator introduced in Chapter 4. Hence

$$\mathbb{D}_n \int_a^b \hat{D}_\varphi \varphi^{[n]}.\hat{\varphi}^n f \, d_\varphi = \left[\left\{\varphi^{[n+1]} - \varphi^{[n]}\right\} \hat{\varphi}^n f\right]_a^b. \tag{14.45}$$

Now, on summing over n from $n = M$ to $N - 1$, with M, N integers such that $N > M$ and applying the summation theorem (4.21), we find

$$\int_a^b \hat{D}_\varphi \varphi^{[N]}.\hat{\varphi}^N f \, d_\varphi - \int_a^b \hat{D}_\varphi \varphi^{[M]}.\hat{\varphi}^M f \, d_\varphi = \sum_{n=M}^{N-1} \left[\left\{\varphi^{[n+1]} - \varphi^{[n]}\right\} \hat{\varphi}^n f\right]_a^b \tag{14.46}$$

Now take $a = t_0$ and $b = t_N \equiv \varphi^{[N]}(t_0)$ for $M = 0$, whereupon

$$\hat{D}_\varphi \varphi^{[0]} = 1, \qquad \hat{\varphi}^0 f = f, \tag{14.47}$$

and so we deduce

$$-\int_{t_0}^{t_N} f \, d_\varphi = \sum_{n=0}^{N-1} \left[\left\{\varphi^{[n+1]} - \varphi^{[n]}\right\} \hat{\varphi}^n f\right]_a^b. \tag{14.48}$$

But

$$\begin{aligned}\sum_{n=0}^{N-1} \left[\left\{\varphi^{[n+1]} - \varphi^{[n]}\right\} \hat{\varphi}^n f\right]_a^b &= \left[\left\{\varphi^{[1]} - \varphi^{[0]}\right\} f\right]_{t_0}^{t_N} + \left[\left\{\varphi^{[2]} - \varphi^{[1]}\right\} \hat{\varphi}^1 f\right]_{t_0}^{t_N} \\ &\quad + \cdots + \left[\left\{\varphi^{[N]} - \varphi^{[N-1]}\right\} \hat{\varphi}^{N-1} f\right]_{t_0}^{t_N} \\ &= \sum_{n=N}^{2N-1} \mathbb{D}_n t_n.f_n - \sum_{n=0}^{N-1} \mathbb{D}_n t_n.f_n\end{aligned} \tag{14.49}$$

where $f_n \equiv f(\varphi^{[n]}(t_0))$. Hence

$$\int_{t_0}^{t_N} f \, d_\varphi - \sum_{n=0}^{N-1} \mathbb{D}_n t_n f_n = \int_{t_0}^{t_N} \hat{D}_\varphi \varphi^{[N]} \hat{\varphi}^{[N]} f \, d_\varphi - \sum_{n=N}^{2N-1} \mathbb{D}_n t_n f_n. \tag{14.50}$$

This has to be true for all functions f. We note that the left-hand side of (14.50) involves only values of the function f over the interval $[t_0, t_N]$, whilst the right-hand side involves values over the interval $[t_N, t_{2N}]$. Therefore, we deduce that each side must be a constant. By choosing $f = 1$, we see that this constant must be zero, giving the fundamental result

$$\int_{t_0}^{t_N} f \, d_\varphi = \sum_{n=0}^{N-1} \mathbb{D}_n t_n f_n. \tag{14.51}$$

Several comments are in order here.

1. The result (14.51) avoids infinite summation because the limits of integration are taken to be in 'phi-resonance', i.e., we have taken $a \equiv t_0$ and $b \equiv \varphi^{[N]}(t_0)$. If we did not, we would encounter the same sort of infinite summations as found in the definite q-integral (3.26) (Klimek, 1993).
2. We note that the discretization in (14.51) is biased to use the function value f_n at the lower end of the temporal link $[t_n, t_{n+1}]$, which is a consequence of our taking $\varphi(t) > t$. This is in contrast to the analogous result (3.31), which has the function f_{n+1} in place of f_n, which is a consequence of our taking $0 < q < 1$.
3. There is no reasonable interpretation within DT of *not* taking the integration limits a and b to be in 'phi-resonance', simply because, by definition, observations cannot be taken at any other times in DT mechanics. Otherwise, we would be discussing CT mechanics based on difference equations, which is an interesting topic in its own right, but not the focus of DT mechanics per se.

   Therefore, if we want to vary the action integral with respect to φ, we conclude that we may as well replace the phi-calculus in the elegant form given by Klimek by Lee's approach to mechanics, as discussed in Chapter 10, where the temporal links become dynamical variables. The point about that is that then we should automatically find the equivalent of a conserved energy. The downside is that we may encounter solutions that violate causality in the sense that in some cases solutions have $t_{n+1} < t_n$. This is a theoretical possibility in Type-3 discretization that can be avoided in Type-2 discretization, but at the cost of no longer having an automatic FT analogue of conserved energy.

14.7 Conserved currents

A fundamental goal of mechanics is to discover invariants of the motion. The more invariants we can find, the easier it becomes to solve the equations of motion. This was the strategy behind the development of transformation theory in CM, which led to the Hamilton–Jacobi equation (Goldstein *et al.*, 2002; Leech, 1965).

We saw in Chapter 8 that, in the case of DT CM, the Weiss action principle allows us to find conserved quantities, called charges, provided that we can spot symmetries of the system function. Another method of constructing invariants, often referred to as currents, for a certain class of CT differential equations involves the Wronskian, discussed in Appendix B. A DT analogue of this construct is discussed in Section 4.7. In the CT case, suppose that we have a differential equation of the form

$$\frac{d^2}{dt^2}f(t) + p(t)\frac{d}{dt}f(t) + q(t)f(t) \underset{c}{=} 0, \tag{14.52}$$

where $p(t)$ and $q(t)$ are given functions of t. Let f and g be two solutions to this equation, and define the construct

$$J(t) \equiv h(t)\left\{f\frac{d}{dt}g - \frac{d}{dt}f.g\right\}, \tag{14.53}$$

where h is to be determined. We recognize the term in the brackets on the right-hand side of (14.53) as the Wronskian of f and g. Then

$$\frac{d}{dt}J \underset{c}{=} \left\{\frac{d}{dt}h - hp\right\}\left\{f\frac{d}{dt}g - \frac{d}{dt}f.g\right\}. \tag{14.54}$$

Hence J is a constant of the motion if we choose h such that $dh/dt = hp$.

In the DT case, suppose that we have found two solutions $\{f_n\}$ and $\{g_n\}$ to the equation of motion

$$\varphi_{n+1} + \alpha_n\varphi_n + \beta_n\varphi_{n-1} \underset{c}{=} 0, \tag{14.55}$$

where $\{\alpha_n\}$ and $\{\beta_n\}$ are given sequences. Define

$$J_n \equiv h_n\{f_n g_{n-1} - f_{n-1}g_n\}. \tag{14.56}$$

Then

$$J_{n+1} - J_n \underset{c}{=} \{h_{n+1}\beta_n - h_n\}\{f_n g_{n-1} - f_{n-1}g_n\}. \tag{14.57}$$

Hence J_n is an invariant if we ensure that $h_{n+1}\beta_n = h_n$. This is readily arranged, provided that $\beta_n \neq 0$ for any n in the region of interest, since then we can take

$$h_n = \frac{h_0}{\beta_{n-1}\beta_{n-2}\ldots\beta_0} \tag{14.58}$$

with h_0 arbitrary.

A similar concept was exploited and developed by Klimek in the context of phi-evolution (Klimek, 1996). We note, however, that this approach works only for a restricted class of difference equation. The introduction of non-linearity into the linear difference equation of motion (14.55) would make the construction of invariants along these lines problematical.

15
Intermission

This chapter forms a natural divide half-way into the book. In the first half, we discussed the principles of discrete time (DT) classical mechanics (CM). In the second half we focus mainly on quantum principles. This chapter is a good place in which to take stock of what we have done, what we plan to do, and how the two halves of the book are related. The common theme is Lee's approach to DT mechanics, discussed in Chapter 10.

As we implied previously, theorists whose work is relevant to us can be classified into two types: the applied mathematicians and the mathematical physicists, who may also be called fundamentalists. This division is based not on any value judgements but on the motivations and ambitions driving a theorist's work, which are generally easy to identify.

The applied mathematicians fall into two categories. The first consists of those interested in finding better ways of understanding CM and, if necessary, finding ever better approximations to it. They explore DT CM with that in mind and are generally not remotely interested in quantum mechanics (QM). The second category consists of quantum theorists, such as Bender, who develop DT numerical simulation to approximate standard QM. Lattice gauge theorists also fall into the applied group, since their discretization of spacetime is regarded at all times as an approximation to the continuum. The notion of a fundamental chronon is an alien one to all of these applied mathematicians.

On the other hand, the fundamentalists are interested in DT and QM because DT QM may turn out to be a better fundamental picture of the Universe than the CT QM paradigm generally used by fundamentalists to date. We would classify Marsden, Maeda, Moser, Cadzow, Logan and Bender in the applied category, whilst Finkelstein, Lee, Snyder and Caldirola fall into the fundamentalist category.

In this chapter we review the significance of Lee's Type-3 DT mechanics as it relates to modern perspectives of differential geometry. Applied mechanics theorists do not have QM path integrals on their agenda,[1] which was Lee's aim in the beginning, but they are interested in invariants of the motion. In particular, energy conservation is of interest to all parties.

15.1 The continuous time Lagrangian approach

In this section we review CT Lagrangian mechanics from the perspective of differential geometry. A good review of many of the concepts as applied in physics is given by Schutz (1980). We shall mainly follow the terminology of Marsden and West (2001).

In CT CM, a system under observation (SUO) is described by configuration space, an r-dimensional differentiable manifold $\mathcal{C}$. Such manifolds come equipped with useful structure as standard, thus being ideally suited to our purposes in both CT and DT mechanics. We discuss now the geometry of CT CM.

The tangent bundle $T\mathcal{C}$ is a $2r$-dimensional manifold built up from all the tangent spaces associated with all the points of $\mathcal{C}$. The tangent bundle is often referred to as velocity–configuration space: a canonical Lagrangian $L \equiv L(\mathbf{q}, \dot{\mathbf{q}}, t)$ can be interpreted as a real-valued function over $T\mathcal{C}$, with the instantaneous velocity $\dot{\mathbf{q}}$ at a point $\mathbf{q}$ on a trajectory being identified with a vector in the tangent space $T_{\mathbf{q}}$ at the point $\mathbf{q}$ in $\mathcal{C}$.

The cotangent bundle is a $2r$-dimensional manifold built up from all the cotangent spaces associated with all the points of $\mathcal{C}$. The cotangent bundle is the setting for the Hamiltonian formulation of CM and is then referred to as phase space.

A curve Γ_λ is a set of points in $\mathcal{C}$ labelled by some real parameter λ that takes values in some interval $[\lambda_{\mathrm{i}}, \lambda_{\mathrm{f}}]$. The set of image points of a curve in $\mathcal{C}$ is called a trajectory when λ is a time parameter. A differentiable curve is one such that, in a given coordinate patch of $\mathcal{C}$, each of the coordinates $\{q^i(\lambda) : i = 1, 2, \ldots, r\}$ is a differentiable function of λ.

Given a differentiable curve Γ_t, the tangent $\dot{\mathbf{q}}_{\mathrm{P}}$ at any point P on that curve is a vector $\mathbf{v}_{\mathrm{P}}$ in the tangent space T_{P} at P, such that, relative to any coordinate patch containing P, we have

$$v^i_{\mathrm{P}} \equiv \frac{d}{dt} q^i(t_{\mathrm{P}}), \tag{15.1}$$

where $\mathbf{v}_{\mathrm{P}} \equiv v^i_{\mathrm{P}}\, \partial_i$ is a coordinate basis representation of $\mathbf{v}_{\mathrm{P}}$.

A vector field $\mathbf{v}$ is a section of $T\mathcal{C}$: for any point P in $\mathcal{C}$, the vector field assigns an element $\mathbf{v}_{\mathrm{P}}$ in T_{P}. Given a vector field $\mathbf{v}$ over $T\mathcal{C}$, a curve Γ_λ is a flow line or integral curve of $\mathbf{v}$ if, in a patch containing the curve, we can write, for points on the curve,

$$\frac{d}{d\lambda} q^i(\lambda) = v^i(\mathbf{q}), \tag{15.2}$$

where we assume $\mathbf{v} = v^i\, \partial_i$.

[1] Apart from lattice gauge theorists, who use DT path integrals as a method of approximation.

The action integral

Given a canonical Lagrangian L, any two points P and Q in $\mathcal{C}$, and a differentiable curve Γ_t from P to Q, we define the action integral $A_{\mathrm{QP}}[\Gamma]$ by

$$A_{\mathrm{QP}}[\Gamma] \equiv \int_{t_{\mathrm{P}}}^{t_{\mathrm{Q}}} L|_{\Gamma}\, dt. \tag{15.3}$$

As discussed by Marsden and West (2001), the exterior derivative of the action integral may be written formally as

$$\tilde{d}A_{\mathrm{QP}}[\Gamma] = \left[\frac{\partial L}{\partial \dot{q}^i}\, \tilde{d}q^i\right]_{\mathrm{P}}^{\mathrm{Q}} + \int_{t_{\mathrm{P}}}^{t_{\mathrm{Q}}} \left[\left\{\frac{\partial L}{\partial q^i} - \frac{d}{dt}\left[\frac{\partial L}{\partial \dot{q}^i}\right]\right\} \tilde{d}q^i\right] dt. \tag{15.4}$$

We shall call this the action one-form. Usually, the action one-form is encountered via its action on an arbitrary vector field, as in (8.5). The one-form

$$\tilde{\Theta} \equiv \frac{\partial L}{\partial \dot{q}^i}\, \tilde{d}q^i \tag{15.5}$$

occurring at the end points P and Q of the curve is known as the Lagrangian one-form. When the coefficients $\partial L/\partial \dot{q}^i \equiv p_i$ are used to define canonical coordinates $\{p_i\}$ over the relevant patch in cotangent space, the one-form $p_i\, \tilde{d}q^i$ is known as the canonical one-form.

The next step is to restrict Γ to being a solution to the Euler–Lagrange equations of motion. It is generally assumed that a unique trajectory Γ_{c} can always be found. This need not be the case. For example, if we were discussing special relativity and P and Q were relatively spacelike, then there would be no timelike trajectory from P to Q.

Assuming a classical trajectory does exist, denoted by Γ_{c}, we have

$$\frac{\partial L}{\partial q^i} - \frac{d}{dt}\left[\frac{\partial L}{\partial \dot{q}^i}\right] \underset{\mathrm{c}}{=} 0. \tag{15.6}$$

Now the relation between the points P and Q is that Q is the evolution of P along the classical trajectory Γ_{c}, so we may write symbolically $\mathrm{Q} \underset{\mathrm{c}}{=} U_{\mathrm{QP,c}}\mathrm{P}$, where $U_{\mathrm{QP,c}}$ is an evolution operator taking us along the classical trajectory from P to Q. Hence we may write formally (Marsden and West, 2001)

$$\tilde{d}A_{\mathrm{QP}}[\Gamma_{\mathrm{c}}] \underset{\mathrm{c}}{=} U_{\mathrm{QP,c}}\, \tilde{\Theta}\Big|_{\mathrm{P}} - \tilde{\Theta}\Big|_{\mathrm{P}}. \tag{15.7}$$

Since the action one-form is exact, the exterior derivative identity $\tilde{d}\tilde{d} = 0$ means that

$$U_{\mathrm{QP,c}}\, \Omega|_{\mathrm{P}} \underset{\mathrm{c}}{=} \Omega|_{\mathrm{P}}, \tag{15.8}$$

where

$$\Omega \equiv \tilde{d}\tilde{\Theta} = \frac{\partial^2 L}{\partial q^i\, \partial \dot{q}^j}\, \tilde{d}q^i \wedge \tilde{d}q^j + \frac{\partial^2 L}{\partial \dot{q}^i\, \partial \dot{q}^j}\, \tilde{d}\dot{q}^i \wedge dq^j \tag{15.9}$$

is known as the Lagrangian symplectic form. The dynamical equality (15.8) tells us that the Lagrangian symplectic form is conserved modulo the equations of motion.

This discussion can be readily translated into phase-space terms, but we will restrict ourselves to Lagrangians, since, in the long run, the system function concept in our approach to DT mechanics is directly related to the Lagrangian rather than to the Hamiltonian.

15.2 The discrete time Lagrangian approach

In this section we review DT Lagrangian mechanics from the perspective of differential geometry. The problem for us now is that we have to abandon the concept of a continuous classical trajectory.

We start by reorganizing information. Generally, the CT Lagrange approach to CM deals with time-dependent system points, which are points in the tangent space $T\mathcal{C}$. The objective is usually to determine where the system point is at any given time, starting from an initial system point. This means that, if configuration space is r-dimensional, we can pin down the system point at any time by specifying $2r$ real numbers. These are the r configuration-space coordinates and the r velocity components at time t, relative to some convenient coordinate patch.

In DT mechanics, we rearrange this information in the form of a system point in the Cartesian product $\mathcal{C}^{(1)} \times \mathcal{C}^{(2)}$, where $\mathcal{C}^{(i)}$ is a copy of the CT configuration space $\mathcal{C}$. At a given discrete time n, the system point will consist of a point P_n in $\mathcal{C}^{(1)}$ with coordinates denoted $\{\mathbf{q}_n\}$ and a point Q_n in $\mathcal{C}^{(2)}$ with coordinates denoted $\{\mathbf{q}_{n+1}\}$, all relative to suitable coordinate patches. In general, the coordinate patch used to discuss P_n in $\mathcal{C}^{(1)}$ is a copy of that used to discuss Q_n in $\mathcal{C}^{(2)}$.

This way of encoding information does have some issues. First, we should ask what the relevant time interval between $\mathcal{C}^{(1)}$ and $C^{(2)}$ is. In fact, this question is central to the discussion of Lee mechanics and we will discuss this presently. Second, if we think of $\mathcal{C}^{(2)}$ as *later* in some sense than $C^{(1)}$, then it would be more natural to think of the 'state' at time n of the SUO as given by $(\mathbf{q}_{n-1}, \mathbf{q}_n)$, i.e., initial information is usually supplied from the past and used to predict the future. However, since $(\mathbf{q}_{n-1}, \mathbf{q}_n)$ and $(\mathbf{q}_n, \mathbf{q}_{n+1})$ are related by a second-order equation of motion, we assume that they are formally equivalent in terms of their information content.

What we refer to as a system function $F^n \equiv F^n(\mathbf{q}_n, \mathbf{q}_{n+1})$ is related to the discrete Lagrangian $L_{\mathrm{d}}(\mathbf{q}_n, \mathbf{q}_{n+1})$ of Marsden and West (2001), differing by a multiplicative temporal factor equal to the time step between $\mathcal{C}^{(n)}$ and $\mathcal{C}^{(n+1)}$, assuming the architecture of Figure 5.2. This is a trivial difference for Type-1 DT mechanics but not for the other types. We shall use the system function notation, since this covers Type 1, Type 2 and, ultimately, Type 3.

The continuous trajectories Γ discussed in CT CM are replaced by discrete paths $\tilde{\Gamma}$ in DT CM, which are specified by the finite sequences $\{(t_n, \mathbf{q}_n) : n = M, M+1, \ldots, N\}$, where t_n is the time associated with position $\mathbf{q}_n$.

The action integral of CT CM is now replaced by an action sum, defined by

$$A_{NM}[\tilde{\Gamma}] = \sum_{n=M}^{N-1} F^n(\mathbf{q}_n, \mathbf{q}_{n+1}). \tag{15.10}$$

Note that this is valid both for Type-1 and for Type-2 DT CM, since any variability in time-step/chronon can be absorbed into the system function (hence the index n in the system function).

As implied by Marsden and West (2001), the exterior derivative of the action sum may be written formally as

$$\tilde{d}A_{NM}[\tilde{\Gamma}] \underset{c}{=} \mathbb{U}_{N-1,M}\, \tilde{\Theta}_M^{(+)} - \tilde{\Theta}_M^{(-)}, \tag{15.11}$$

modulo the DT equations of motion

$$\frac{\partial}{\partial \mathbf{q}_n}\{F^n + F^{n-1}\} \underset{c}{=} \mathbf{0}, \quad M < n < N. \tag{15.12}$$

In (15.11), the operator $\mathbb{U}_{N-1,M}$ takes us from DT M to *DT* $N-1$.

We shall call (15.11) the discrete action one-form. The novelty for DT CM is that there are two end-point one-forms:

$$\tilde{\Theta}_n^{(+)} \equiv \frac{\partial F^n}{\partial q_{n+1}^i}\, \tilde{d}q_{n+1}^i, \quad \tilde{\Theta}_n^{(-)} \equiv -\frac{\partial F^n}{\partial q_n^i}\, \tilde{d}q_n^i. \tag{15.13}$$

By inspection,

$$\tilde{d}\tilde{\Theta}_n^{(+)} = \frac{\partial^2 F^n}{\partial q_n^i\, \partial q_{n+1}^j}\, \tilde{d}q_n^i \wedge \tilde{d}q_{n+1}^j = \tilde{d}\tilde{\Theta}_n^{(-)}, \tag{15.14}$$

so we may define a unique discrete system function two-form

$$\Omega^n(\mathbf{q}_n, \mathbf{q}_{n+1}) \equiv \frac{\partial^2 F^n}{\partial q_n^i\, \partial q_{n+1}^j}\, \tilde{d}q_n^i \wedge \tilde{d}q_{n+1}^j. \tag{15.15}$$

Then, from (15.11), we deduce that it is conserved modulo the equations of motion, viz.,

$$\mathbb{U}_{N-1,M}\, \Omega^M \underset{c}{=} \Omega^M. \tag{15.16}$$

Hence this form of DT mechanics is ‘discretely symplectic’ (Marsden and West, 2001).

15.3 Extended discrete time mechanics

The above formulation of DT CM has been used extensively to investigate numerous schemes for ever better approximations to CT CM. These fall strictly outside of our agenda, which is a fundamentalist one. However, one line of investigation is of common interest to all parties: the issue of energy conservation.

It is a common experience in any branch of DT CM to discover that, with rare exceptions, any DT CM simulation based on Type-1 or Type-2 temporal discretization fails to conserve the equivalent of energy. This is the case even for discretizations of those CT CM SUOs where energy is conserved. It is also the case for the sort of variationally based discretization discussed above: discretely symplectic equations of motion need not have a DT analogue of conserved energy.

The problem stems from the temporal relationship, i.e, the chronon T_n, between the two copies of configuration space, $\mathcal{C}^{(n)}$ and $\mathcal{C}^{(n+1)}$, used to formulate the system function (or discrete Lagrangian) in the preceding section. A better handle has to be found for fixing the relevant time interval between these two spaces. Certainly, fixing the chronon T_n before the equations of motion have been worked out leads to problems in short order for energy conservation.

Dirac was a pioneer in many fields, but one in particular stands out here: constraint mechanics. In an influential book on the subject (Dirac, 1964), Dirac discussed the possibility of reparametrizing the time parameter in CT CM action integrals. We discussed this in Section 10.3, and considered how it relates to Lee's mechanics. Dirac's approach results in the original time parameter t being treated as a new dynamical variable, with an equation of motion just like the original configuration-space degrees of freedom. Moreover, the conjugate momentum associated with t turns out to be the Hamiltonian, up to a sign.

At this point, alarm bells should ring out, given our discussion of Pauli's theorem in Chapter 1. The matter is resolved by the appearance of a primary constraint, which tells us in effect that the time parameter is not a dynamical variable after all: it can be altered by the equivalent of a change of gauge without any impact on the physics.

The interpretation of Dirac's reparametrization therefore is not that time is now turned into a true dynamical variable, but that we can choose to treat it as a dynamical variable precisely because it can be transformed by a 'gauge transformation' without impact on the physics. This is 'dynamics without dynamics'.

This freedom to reparametrize time has had the most profound consequences for quantum gravitation and quantum cosmology. When theorists attempted to unite general relativity and quantum mechanics, they eventually came across the Wheeler–de Witt equation $0 = \hat{H}|\Psi\rangle$, where $\hat{H}$ is the 'Hamiltonian' for the evolution of the state $|\Psi\rangle$ of the Universe (DeWitt, 1965). The zero on the left-hand side means that the state of the Universe is effectively frozen relative to the parameter conjugate to $\hat{H}$. The physics of the Wheeler–de Witt equation is still being debated vigorously by quantum cosmologists and string theorists (Smolin, 2013).

When Lee suggested that time should be treated as a discrete dynamical variable (Lee, 1983), he was interested in path integrals. We shall discuss these in Chapter 18. His formulation turns out to be analogous to Dirac's reparametrization of time in CT CM. It is also reasonable to think of Lee's variable chronon as a form of Lagrange multiplier.[2] We touched upon this theme in our second relativistic-particle model in Section 9.3. As we know, Lagrange multipliers may be used to enforce constraints in the calculus of variations. When used directly, Lagrange multipliers are treated as dynamical variables in an extended configuration space and are killed off from the physics by the resulting constraints. In Dirac's formulation of constraint mechanics, conventional Lagrange multipliers turn out to be related to the so-called second-class constraints, which are not associated with gauge degrees of freedom. The second-class constraints can in fact be entirely eliminated from the formalism, leading to a reformulation of the extended-space Poisson brackets as Dirac brackets. This is of great importance in the canonical quantization programme, since it is the Dirac brackets that should be used rather than the Poisson brackets.

In Dirac's reparametrization-of-time programme, the time parameter is encoded into an extended configuration space, so that the r configuration-space dynamical variables $\mathbf{q} \equiv \{q^1, q^2, \ldots, q^r\}$ are replaced by the $r+1$ variables $\tilde{\mathbf{q}} \equiv \{t, q^1, q^2, \ldots, q^r\}$. This is the approach taken by Lee and by the various applied mathematicians interested in finding ever better approximations to the equivalent of system functions (they refer to them as discrete Lagrangians).

Therefore, the formalism of the preceding section is taken over more or less wholesale, with the important proviso that now the node times t_n are regarded as dynamical variables. This leads us to work with extended system functions $\tilde{F}^n \equiv \tilde{F}^n(t_n, \mathbf{q}_n, t_{n+1}, \mathbf{q}_{n+1})$, defined over $\tilde{\mathcal{C}}^{(n)} \times \tilde{\mathcal{C}}^{(n+1)}$, where $\tilde{\mathcal{C}}^{(i)}$ is a copy of the extended configuration space $\mathcal{C} \times \mathbb{R}$, with $\mathbb{R}$ representing the newly introduced temporal degree of freedom.

Although it seems as if more information is being introduced into the dynamics, the additional 'equation of motion' for the discrete time variable effectively compensates, so that, in the long run, the physics is the same. However, the variability of the chronon means that successive copies of configuration space are related in such a way as to conserve some quantity that can be identified as a form of energy.

We will not go into the details here: a full account is given in Chapter 4 of Marsden and West (2001). The details are generally similar, if not identical, to those discussed above and in Chapter 10. Suffice it to say that, in the extended DT CM formulation, the equations of motion are the same as those

[2] Not precisely, since Lee's chronon enters into system functions in a non-linear way. The effect is much the same, however.

discussed in Chapter 10 and it is proven that there is a conserved extended discrete Lagrangian two-form.

Only one concern remains. Given a formulation where the chronon is a dynamical variable, we would find it uncomfortable if, during the course of any solution to the classical extended equations of motion, the chronon took on a negative value over some link. Certainly, it seems possible. In Lee's path integral, there is no natural cutoff to ensure positive chronons. We recall at this point those Feynman diagrams where electron lines that run backwards in time have the Feynman–Stueckelberg interpretation of representing antiparticles (Stueckelberg, 1941; Bjorken and Drell, 1964). In the case of Lee's DT path integrals, which would presumably run from infinite negative time to infinite positive time (the standard scenario), there should be no problem arising from negative chronons, since they would be involved in no more than a book-keeping exercise.

For quantum processes occurring over finite times, however, this issue will undoubtedly require some careful attention if causality is to be respected: we cannot expect dynamical effects originating *before* a state was created in a laboratory to affect the forwards dynamical evolution of that state. This concern underlines the point that we have stressed elsewhere in this book: discrete time needs to be discussed in terms of SUOs and observers, not just SUOs alone.

Part III

Discrete time quantum mechanics

16

Discrete time quantum mechanics

16.1 Quantization

We now discuss the quantization of discrete time (DT) classical mechanics (CM). We restrict our attention in this chapter to a system under observation (SUO) consisting of a single point particle of mass m moving in one spatial dimension with Cartesian coordinate x. The generalization to two or more degrees of freedom is as straightforward here as in CT quantum mechanics (QM). The Dirac bra–ket notation will be used for convenience.

The standard principles of QM have proven remarkably successful and consistent over the years (Peres, 1993) and we have no reason to alter them apart from changing from CT to DT. This is a significant change. Quantum mechanics became dominant in physics and chemistry because of the success of the Schrödinger equation, which is a first-order-in-time differential equation. Therefore, we should take care to ensure that discretizing time does not undermine the successes of CT QM.

There are two reasons why discretizing time in QM might be considered. First, the Schrödinger equation is hard, or even impossible, to solve exactly in many situations and temporal discretization might be seen as a step towards numerical simulation by computer. This motivated the work of Bender (Bender *et al.*, 1985a, 1985b, 1993) and others on the DT Schrödinger equation. The second reason is one of principle: we may want to explore the properties of DT QM as a self-consistent theory in its own right rather than as an approximation to CT QM. This motivated Caldirola and the Italian School, as discussed in Section 1.4.

The interpretation of CT QM still generates debate, and the situation is no different in DT QM. Temporal discretization does not answer any of the conceptual questions that arise about observation in QM, such as the origin of the Born probability rule and the superposition principle.

In earlier chapters we argued that architecture, or overall plan, is of fundamental importance in DT CM: we should be clear whether states are evolving in a fixed universe or jumping from one universe to another. The same concern applies to DT QM.

The architecture assumed in this chapter is that of any standard QM experiment: a classical observer describes a prepared state of an SUO as a normalized vector in a fixed Hilbert space $\mathcal{H}$, that vector is then left to evolve unitarily in $\mathcal{H}$ according to the Schrödinger equation and finally the observer tests for the probability that the final state is some selected vector in $\mathcal{H}$. This is the most frequent form of architecture used by physicists in non-relativistic QM and can be discussed either in the Heisenberg picture or in the Schrödinger picture.

The following standard principles of CT QM are used in our approach to DT QM.

1. Pure quantum states of an SUO correspond one-to-one to rays in some predefined separable Hilbert space $\mathcal{H}$. Vectors in $\mathcal{H}$ representing physical states are normalized to unity.

   Our interest here is in modelling particle SUOs moving in physical space, which we shall restrict to one dimension for simplicity, without loss of any content. This means that $\mathcal{H}$ is an infinite-dimensional Hilbert space.
2. At each DT nT, the observer describes states in $\mathcal{H}$ via an improper position basis

$$\boldsymbol{B}^n \equiv \{|x, n\rangle : x \in \mathbb{R}\}, \tag{16.1}$$

   the elements of which satisfy the improper relations

$$\langle x, n|y, n\rangle = \delta(x - y). \tag{16.2}$$

The second principle will allow us to use the Heisenberg picture when required. The resolution of the identity operator $\hat{I}_{\mathcal{H}}$ in $\mathcal{H}$ depends on the basis chosen, which depends on the time at which we make the resolution. Specifically, we have, for each time n,

$$\hat{I}_{\mathcal{H}} = \int |x, n\rangle\langle x, n| dx. \tag{16.3}$$

In the Heisenberg picture, a pure physical state $|\Psi\rangle$ is frozen in time in $\mathcal{H}$ and may be described using any choice of basis. We will choose to describe the prepared state of the SUO at any given DT n by

$$|\Psi\rangle = \int \Psi_n(x)|x, n\rangle dx, \tag{16.4}$$

where $\Psi_n(x)$ is the corresponding Schrödinger-picture wavefunction at time n. Assuming normalization to unity, we must have

$$\int |\Psi_n(x)|^2 \, dx = 1, \quad M \leqslant n \leqslant N. \tag{16.5}$$

In the Heisenberg picture, quantum operators representing observables are generally time-dependent, so it is an advantage to use the time-dependent bases

$\boldsymbol{B}^n$ to describe them as well. An important set of such operators, the *diagonal operators*, is defined as follows.

Definition 16.1 An operator $\hat{A}_n$ is diagonal with respect to basis $\boldsymbol{B}^n$ if it can be written in the form

$$\hat{A}_n = \int |x,n\rangle A_n(x,\partial_x)\langle x,n|dx, \tag{16.6}$$

where the component $A_n(x,\partial_x)$ is some differential operator acting on functions of x to its right.

The action of such an operator on a typical state $|\Psi\rangle$ is given by

$$\hat{A}_n|\Psi\rangle \equiv \int |x,n\rangle A_n(x,\partial_x)\Psi_n(x)dx, \tag{16.7}$$

with matrix elements given by

$$\langle\Phi|\hat{A}_n|\Psi\rangle = \int \Phi_n^*(x)A_n(x,\partial_x)\Psi_n(x)dx. \tag{16.8}$$

We define the Hermitian conjugate or adjoint operator $\hat{A}_n^\dagger$ to have the property that

$$\langle\Phi|\hat{A}_n|\Psi\rangle \equiv \langle\Psi|\hat{A}_n^\dagger|\Phi\rangle^* \tag{16.9}$$

for a dense set of physical states. If $\hat{A}_n$ is diagonal with respect to $\boldsymbol{B}^n$ then writing $\hat{A}_n^\dagger$ in the form

$$\hat{A}_n^\dagger = \int |x,n\rangle \tilde{A}_n(x,\partial_x)\langle x,n|dx \tag{16.10}$$

for some operator component $\tilde{A}_n(x,\partial_x)$ with definition (16.9) gives

$$\int \Phi_n^*(x)A_n(x,\partial_x)\Psi_n(x)dx = \int\left[\tilde{A}_n^*(x,\partial_x)\Phi_n^*(x)\right]\Psi_n(x)dx \tag{16.11}$$

for all normalizable wavefunctions Ψ_n, Φ_n.

We can determine the relationship between $A_n(x,\partial_x)$ and its adjoint counterpart $\tilde{A}_n(x,\partial_x)$ by integrating (16.11) by parts. This will be possible for normalizable wavefunctions, which necessarily fall off to zero as $|x|$ tends to infinity.

If we find $A_n(x,\partial_x) = \tilde{A}_n(x,\partial_x)$ then the operators $\hat{A}_n$ and $\hat{A}_n^\dagger$ are equal and $\hat{A}_n$ is said to be self-adjoint. Self-adjoint operators have real eigenvalues and so can be used to represent observables.

Example 16.2 Given the operator $A_n(x,\partial_x) \equiv z\,\partial_x$, where z is complex, integration by parts gives $\tilde{A}_n(x,\partial_x) = -z^*\,\partial_x$. Choosing $z = -i\hbar$ gives $\tilde{A}_n(x,\partial_x) = A_n(x,\partial_x)$, so the corresponding operator $\hat{A}_n$ is self-adjoint. This particular choice of z gives the observable known as linear momentum.

16.2 Quantum dynamics

In this section we develop a mathematical description of the quantum architecture described above, which is described by Figure 5.1(b). The Hilbert space $\mathcal{H}_n$ used at each time n is the same Hilbert space $\mathcal{H}$, i.e., $\mathcal{H}_n = \mathcal{H}$, for all n running from initial time M to final time N.

In the Schrödinger picture, state vectors are evolved from one node n of DT to the next, $n+1$, by the application of unitary *time-step operators* $\hat{U}_n$, i.e., according to the rule

$$|\Psi, n+1\rangle = \hat{U}_n |\Psi, n\rangle, \tag{16.12}$$

where both $|\Psi, n\rangle$ and $|\Psi, n+1\rangle$ are vectors in $\mathcal{H}$. The quantum time-step operator $\hat{U}_n$ should not be confused with the classical step operator $\mathbb{U}_n$ defined previously.

Because we are using the DT equivalent of the Heisenberg picture, the transformation rule relating basis $\boldsymbol{B}^n$ to basis $\boldsymbol{B}^{n+1}$ is

$$|x, n+1\rangle = \hat{U}_n^\dagger |x, n\rangle, \qquad x \in \mathbb{R}. \tag{16.13}$$

From this we deduce the relations

$$\langle x, n+1| = \langle x, n|\hat{U}_n, \quad |x, n\rangle = \hat{U}_n |x, n+1\rangle, \quad \langle x, n| = \langle x, n+1|\hat{U}_n^\dagger. \tag{16.14}$$

The operator $\hat{U}_n$ provides an isometry between $\boldsymbol{B}^n$ to $\boldsymbol{B}^{n+1}$, that is

$$\langle x, n+1|y, n+1\rangle = \langle x, n|y, n\rangle = \delta(x-y). \tag{16.15}$$

Using the resolution of the identity (16.3), we may represent the time-step operators in the non-diagonal form

$$\hat{U}_n = \int dx |x, n\rangle\langle x, n+1|, \quad \hat{U}_n^\dagger = \int dx |x, n+1\rangle\langle x, n|. \tag{16.16}$$

We now have the structure in place to define the fundamental evolution functions for DT QM with the given architecture, i.e., the *system amplitudes* $U_n(x, y)$. These are defined by

$$\begin{aligned} U_n(x,y) &\equiv \langle x, n+1|y, n\rangle = \langle x, n|\hat{U}_n|y, n\rangle, \\ U_n^*(x,y) &\equiv \langle y, n|x, n+1\rangle = \langle y, n|\hat{U}_n^\dagger|x, n\rangle. \end{aligned} \tag{16.17}$$

With these we may construct the time-step operators in non-diagonal form, viz.,

$$\begin{aligned} \hat{U}_n &= \int dx\, dy |x, n\rangle U_n(x,y)\langle y, n|, \\ \hat{U}_n^\dagger &= \int dx\, dy |x, n\rangle U_n^*(y,x)\langle y, n|. \end{aligned} \tag{16.18}$$

The condition that the time-step operators are unitary, viz., $\hat{U}_n\hat{U}_n^\dagger = \hat{I}_{\mathcal{H}}$, then leads to the *closure condition*

$$\int dy\, U_n(x,y)U_n^*(z,y) = \delta(x-z) \tag{16.19}$$

on the system amplitudes.

There are two classes of system amplitude of interest to us.

(i) *Autonomous quantum systems.* These are SUOs for which the system amplitudes are independent of time, i.e., $U_n(x, y) = U(x, y)$ for all n.
(ii) Time-reversible *quantum systems.* These are SUOs for which the system amplitudes $U_n(x, y)$ have the symmetry $U_n(x, y) = U_n(y, x)$.

Most system amplitudes of interest to us will be autonomous and time-reversal-invariant. This occurs whenever we construct system amplitudes from system functions obtained from conventional time-translation-invariant and time-reversal-invariant Lagrangians using the virtual-path approach.

16.3 The Schrödinger picture

Given a Heisenberg-picture state $|\Psi\rangle$ we may write

$$|\Psi\rangle = \int \Psi_{n+1}(x)|x, n+1\rangle dx = \int \Psi_n(x)|x, n\rangle dx, \tag{16.20}$$

from which we deduce

$$\Psi_{n+1}(x) = \int U_n(x, y)\Psi_n(y)dy. \tag{16.21}$$

From this we see that the system amplitudes play a role analogous to that of the Feynman kernel in CT QM (Feynman and Hibbs, 1965), as discussed in Chapter 18.

We may set up a Schrödinger-picture description as follows. Given a Heisenberg-picture state $|\Psi\rangle$ and knowledge of the component functions $\Psi_n(x)$, define the sequence of states

$$|\Psi_{n,M}\rangle \equiv \int dx\, \Psi_n(x)|x, M\rangle, \tag{16.22}$$

for some chosen initial time M. Then, if we take $|\Psi_{M,M}\rangle \equiv |\Psi\rangle$, we find

$$\hat{U}_M|\Psi_{M,M}\rangle = |\Psi_{M+1,M}\rangle. \tag{16.23}$$

It is straightforward to extend this to jumps over more than one time interval. This establishes the Schrödinger picture in this theory.

16.4 Position eigenstates

Given an improper basis $\boldsymbol{B}^n \equiv \{|x, n\rangle : x \in \mathbb{R}\}$ at time n, the Heisenberg-picture position operator $\hat{x}_n$ at that time is given by

$$\hat{x}_n \equiv \int |x, n\rangle x \langle x, n| dx, \tag{16.24}$$

which has the property

$$\hat{x}_n|x, n\rangle = x|x, n\rangle. \tag{16.25}$$

Any one of the position operators $\hat{x}_n$ is diagonal with respect to basis $\boldsymbol{B}^n$ but not necessarily with respect to any other basis. From the Heisenberg-picture operator evolution rule

$$\hat{x}_{n+1} = \hat{U}_n^\dagger \hat{x}_n \hat{U}_n \tag{16.26}$$

we find

$$\hat{x}_{n+1} = \int dx\, dy\, dz |x, n\rangle U_n^*(y, x) y U_n(y, z) \langle z, n|, \tag{16.27}$$

which is self-adjoint but not necessarily diagonal with respect to $\boldsymbol{B}^n$. Whether $\hat{x}_{n+1}$ is diagonal with respect to $\boldsymbol{B}^n$ depends on the details of the system amplitudes.

In CT QM, Dirac's canonical quantization principle

$$[\hat{p}_i, \hat{q}^j] = i\hbar \{p_i, q^j\}_{\text{PB}} = -i\hbar\delta_{ij} \tag{16.28}$$

relates equal-time commutation relations between conjugate observables to classical Poisson brackets (PBs) of conjugate pairs of coordinates (Dirac, 1925). In DT QM, we have to take into account the different geometrical structure of DT CM. The phase space of CT CM is identified with the cotangent bundle $T^*\mathcal{C}$, where $\mathcal{C}$ is the configuration-space manifold. On the other hand, the corresponding geometrical structure in DT CM is $\mathcal{C} \times \mathcal{C}$ (Marsden and West, 2001).

One possibility is to consider the node momenta which we have already encountered, viz.,

$$\mathbf{p}_n^{(-)} \equiv -\frac{\partial F^n}{\partial \mathbf{q}_n}, \quad \mathbf{p}_{n-1}^{(+)} \equiv \frac{\partial F^{n-1}}{\partial \mathbf{q}_n}. \tag{16.29}$$

Then the equations of motion are equivalent to

$$\mathbf{p}_n^{(-)} \underset{c}{=} \mathbf{p}_{n-1}^{(+)}, \tag{16.30}$$

so we could define our momentum $\mathbf{p}_n$ at node n to be either $\mathbf{p}_n^{(-)}$ or $\mathbf{p}_{n-1}^{(+)}$. We note that (16.30) holds over classical trajectories only and not off them, whereas in CT CM, equal-time PBs are independent of the equations of motion, so the canonical quantization principle (16.28) is independent of the solutions. Indeed, canonical equal-time quantization is of value precisely because it is independent of the operator equations of motion and is generally used to generate *them* via dynamical commutator algebras.

There are several routes we could take in the proposed quantization.

1. In the quantum theory, we could interpret (16.30) as an operator equation of motion, impose the commutator

$$[\hat{p}_{n,i}^{(-)}, \hat{q}_n^j] = [\hat{p}_{n-1,i}^{(+)}, \hat{q}_n^j] = -i\hbar\delta_{ij} \tag{16.31}$$

for both sides of that equation of motion and check for consistency.

2. We might consider looking at *unequal-time* PBs, which we shall refer to as Peierls–Poisson brackets (PPBs) after Peierls, who gave a formulation of PPBs that is based on the so-called *equation of small disturbances* (Peierls, 1952). This approach can be used to discuss unequal-time quantum field commutation relations in relativistic quantum field theory (DeWitt, 1965).

   The PPBs are related to Schwinger functions, which can be used to discuss the evolution of classical field information from an initial spacelike hypersurface in relativistic spacetime to some final spacelike hypersurface (Roman, 1969).
3. We may use a DT version of the Schwinger action principle (Schwinger, 1959), which is itself the quantized version of the Weiss action principle (Weiss, 1936), to derive the equivalent of n-point (Green) functions, from which we can infer quantum commutators. This is most conveniently done using Schwinger's source-function technique (Schwinger, 1969), adapted to DT QM (Jaroszkiewicz and Norton, 1997a).

The reader unfamiliar with the PPB will readily understand its power and its relevance to DT CM by considering the CT CM one-dimensional free-particle systems with Lagrangian $L = \frac{1}{2}m\dot{x}^2$. Hamilton's equations of motion are readily found to be

$$\dot{p}(t) \underset{c}{=} 0, \quad \dot{x}(t) \underset{c}{=} p(t)/m, \tag{16.32}$$

from which we easily find the general solution

$$x(t) = x(0) + \frac{p(0)}{m}t, \tag{16.33}$$

where $x(0)$ and $p(0)$ are the initial position and linear momentum, respectively. We may now work out the unequal-time Poisson bracket, i.e., the Peierls–Poisson bracket, $\{x(t), x(0)\}_{\text{PPB}}$ from knowing that the equal-time PB is given by $\{p(0), x(0)\}_{\text{PB}} = -1$. We deduce from this condition and the solution (16.33) the PPB

$$\{x(t), x(0)\}_{\text{PPB}} = -\frac{t}{m}. \tag{16.34}$$

The significant feature of the PPB (16.34) is that it seems ready made for DT: we simply set $t = T$, the value of the chronon. Moreover, the PPB is not a phase-space object per se: it lives in the Cartesian product $\mathcal{C} \times \mathcal{C}$, where $\mathcal{C}$ is configuration space, making it suitable for DT quantization.

If now we extend Dirac's canonical quantization principle (16.28) to PPBs and apply it to (16.34) we arrive at the unequal-time commutator

$$[\hat{x}(t), \hat{x}(0)] = -\frac{i\hbar}{m}t. \tag{16.35}$$

This is consistent with the canonical commutation relation $[\hat{p}(0), \hat{x}(0)] = -i\hbar$ if we recall that, in the classical theory, $p(t) = m\dot{x}(t)$. Note that we have to differentiate (16.35) with respect to t *before* taking the equal-time limit $t \to 0$.

Using (16.27), we can work out a formal expression for the unequal DT position operators' commutators, viz.,

$$[\hat{x}_{n+1}, \hat{x}_n] = \int dx\ dy\ dz |x, n\rangle U_n^*(y, x) y(z - x) U_n(y, z) \langle z, n|. \tag{16.36}$$

16.5 Normal-coordinate systems

In this section we discuss the quantization of systems described by *normal* coordinates, which we encountered in Section 13.9. If such coordinates can be found, then the right-hand side of the commutator (16.36) is a multiple of the identity operator for each value of n. The DT oscillator is an example of a system with normal coordinates.

A class of system amplitudes with normal coordinates may be constructed from autonomous, time-reversal-invariant system functions of the form

$$F(x_n, x_{n+1}) = -\beta x_n x_{n+1} + \frac{1}{2} W(x_n) + \frac{1}{2} W(x_{n+1}), \tag{16.37}$$

where β is a non-zero constant and $W(x)$ is a differentiable function of x. The DT equation of motion for such a system is

$$x_{n+1} \underset{c}{=} \beta^{-1} W'(x_n) - x_{n-1}, \tag{16.38}$$

which has the merit of giving x_{n+1} explicitly in terms of x_n and x_{n-1}.

Quantization amounts to finding a system amplitude for this system. We take as our guide Feynman's path-integral approach, which we will discuss in greater detail in Chapter 18. In Feynman's original approach, the propagation kernel $K(a, b)$ for a one-dimensional system to go from (x_a, t_a) to (x_b, t_b) is a path integral of the form (Feynman and Hibbs, 1965)

$$K(a, b) = \lim_{\varepsilon \to 0} \frac{1}{A} \int \dots \int \int e^{(i/\hbar) S[b,a]} \frac{dx_1}{A} \frac{dx_2}{A} \dots \frac{dx_{N-1}}{A}, \tag{16.39}$$

where A is a normalization constant,

$$S[b, a] \equiv \int_{t_a}^{t_b} L(\dot{x}, x, t) dt \tag{16.40}$$

and the temporal interval $[t_a, t_b]$ is discretized into $N - 1$ links. By inspection of this path integral, we equate $S[b, a]$ with an action sum, leading to the ansatz

$$U(x, y) = k e^{iF(x,y)/\hbar}, \tag{16.41}$$

where k is some constant to be determined and $F(x, y)$ is a normal system function. The magnitude of the constant k can be determined from the unitarity condition (16.19). We find

$$|k|^2 = \frac{\beta}{2\pi\hbar}. \tag{16.42}$$

From (16.36) we find

$$[\hat{x}_{n+1}, \hat{x}_n] = -\frac{i\hbar}{\beta}\hat{I}_{\mathcal{H}}, \tag{16.43}$$

so that the above coordinates for this system are normal. Moreover, with the momentum p_n conjugate to x_n defined by (8.27), we recover the conventional commutator

$$[\hat{p}_n, \hat{x}_n] = -i\hbar\hat{I}_{\mathcal{H}}. \tag{16.44}$$

This result is not expected to hold for systems that are not normal.

The operator equations of motion are found to be

$$\hat{x}_{n+1} = \beta^{-1}\,\hat{W}' - \hat{x}_{n-1}, \tag{16.45}$$

where $\hat{W}'$ is the diagonal operator

$$\hat{W}' \equiv \int dx|x, n\rangle \left\{\frac{dW(x)}{dx}\right\} \langle x, n|. \tag{16.46}$$

From this we obtain the DT version of Ehrenfest's theorem:

$$\langle\hat{x}_{n+1}\rangle = \langle\beta^{-1}\,\hat{W}'\rangle - \langle\hat{x}_{n-1}\rangle \tag{16.47}$$

for expectation values over a physical state.

Example 16.3 The free Newtonian particle

In this example, we show how the exact free-particle system amplitude can be read off from the Feynman kernel, and we verify the commutation properties.

Given the Hamiltonian $H = p^2/(2m)$, the Feynman kernel is given by (Feynman and Hibbs, 1965)

$$K(x, t; y, t_0) = \sqrt{\frac{m}{i2\pi\hbar(t - t_0)}} \exp\left\{i\frac{(x - y)^2 m}{2(t - t_0)\hbar}\right\}. \tag{16.48}$$

If now we imagine discretizing time, we make the replacements $t_0 \to Tn$, and $t \to T(n+1)$, $y \to x_n$, and $x \to x_{n+1}$ in (16.48), giving the system amplitude

$$U_n(x_{n+1}, x_n) = \sqrt{\frac{m}{i2\pi\hbar T}} \exp\left\{i\frac{(x_{n+1} - x_n)^2 m}{2T\hbar}\right\}, \tag{16.49}$$

where $T \equiv (t - t_0)$ is the chronon. This system function satisfies the closure condition (16.19) as required.

From (16.41) we read off the system function:

$$F^n(x_{n+1}, x_n) = \frac{m}{2T}(x_{n+1} - x_n)^2, \tag{16.50}$$

which means that it is an example of a normal system. Comparison with (16.37) gives

$$\beta = \frac{m}{T}, \quad W(x_n) = \frac{m x_n^2}{T} \tag{16.51}$$

and from (16.45) we find

$$\hat{x}_{n+1} \underset{c}{=} 2\hat{x}_n - \hat{x}_{n-1} \tag{16.52}$$

as expected. Finally, the important commutator is given by

$$[\hat{x}_{n+1}, \hat{x}_n] = -i\frac{\hbar T}{m}\hat{I}_{\mathcal{H}}, \tag{16.53}$$

which is consistent with (16.35)

16.6 Compatible operators

In our approach to DR QM, the quantum dynamics is completely determined by the system amplitudes. Suppose now that the system is autonomous and time-reversal-invariant. This means that for each time n we may write $U_n(x,y) = U(x,y)$, where $U(x,y)$ is independent of n. We expect that there should be constants of the motion comparable to the DT Noether invariants discussed in the classical theory in Chapter 8.

Consider an operator $\hat{C}$ that is diagonal with respect to $\boldsymbol{B}^n$, viz.,

$$\hat{C} \equiv \int dx |x,n\rangle C(x,\partial_x)\langle x,n|, \tag{16.54}$$

where $C(x,\partial_x)$ is some differential operator. The matrix elements of the commutator of $\hat{C}$ with $\hat{U}$ are given by

$$\langle\Phi| \left[\hat{C},\hat{U}\right] |\Psi\rangle = \int dx\, dy\, \Phi_n^*(x)\{C(x,\partial_x)U(x,y) - \tilde{C}^*(y,\partial_y)U(x,y)\}\Psi_n(y), \tag{16.55}$$

where $|\Phi\rangle$ and $|\Psi\rangle$ are arbitrary physical states. From this we arrive at the following result.

Theorem 16.4 *A diagonal operator commutes with the time-step operator of an autonomous system if*

$$C(x,\partial_x)U(x,y) = \tilde{C}^*(y,\partial_y)U(x,y). \tag{16.56}$$

A diagonal operator that commutes with the time-step operator of an autonomous system will be said to be *compatible* (with the time-step operator). It is not necessary for a diagonal operator $\hat{C}$ to be self-adjoint for it to be compatible with the time-step operator.

We can show that eigenvalues of compatible operators are invariants of the motion; i.e., given a state $|\Psi\rangle$ that is an eigenstate of the diagonal operator $\hat{C}$ with eigenvalue c, then $\hat{C}|\Psi\rangle = c|\Psi\rangle$. Then we find

$$C(x, \partial_x)\Psi_n(x) = c\Psi_n(x). \qquad (16.57)$$

Moreover, we can show that

$$C(x, \partial_x)\Psi_{n+1}(x) = c\Psi_{n+1}(x), \qquad (16.58)$$

which demonstrates explicitly that eigenvalues of compatible diagonal operators are constants of the motion.

17

The quantized discrete time oscillator

17.1 Introduction

The quantized harmonic oscillator has proven to be perhaps the most important system in continuous time (CT) quantum mechanics (QM). It was the atomic oscillators assumed to line the containing walls of a black-body radiation field that Planck referred to when he postulated the quantization of energy in 1900 (Planck, 1900, 1901), thereby triggering the development of QM. Subsequently, the quantized harmonic oscillator resurfaced in many contexts, most notably in quantum optics and relativistic quantum field theory, including string theory.

There are two reasons for the importance of the quantized oscillator: (i) it can be solved completely; and (ii) it has ladder operators that generate Planck's quantized energy levels in particle QM and particle-like states in relativistic quantum field theory.

Temporal discretization does not destroy the properties of the oscillator. Rather, it enhances them. We shall show that the parabolic barrier encountered in the classical theory of the discrete time (DT) oscillator, which was discussed in the previous chapter, provides a natural upper bound for important physical quantities such as energy and momentum. In other words, the quantized DT oscillator has a natural cutoff. This may turn out to be of fundamental importance in the regularization (renormalization) of relativistic quantum field theories, where there exist immense problems related to the fact that the CT oscillator has no upper bound to its energy spectrum.

Canonical quantization is the term used to describe the traditional process of finding a quantum analogue theory corresponding to a given classical theory: given a canonical Lagrangian, find the Hamiltonian, replace the phase-space classical variables by operators, and apply the resulting Hamiltonian operator to a wavefunction, thereby arriving at the Schrödinger equation or its equivalent. In the case of DT mechanics we do not have a Hamiltonian per se, simply because a Hamiltonian is a generator of translations in CT, and time is not continuous insofar as we are concerned here. What we do have, however, are jumps in time, as discussed in the previous chapter, and we can follow the approach to quantization discussed there.

17.2 Canonical quantization

17.2.1 The quantization of the stroboscopic oscillator

In CT QM, given the Hamiltonian

$$H = \frac{p^2}{2m} + \frac{m\omega^2}{2}x^2, \tag{17.1}$$

the Feynman kernel is found via path integration to be given by (Feynman and Hibbs, 1965)

$$\begin{aligned} K(x,t;y,t_0) = &\sqrt{\frac{m\omega}{i2\pi\hbar\sin(\omega t - \omega t_0)}} \\ &\times \exp\left\{\frac{im\omega[(x^2+y^2)\cos(\omega t - \omega t_0) - 2xy]}{2\hbar\sin(\omega t - \omega t_0)}\right\}. \end{aligned} \tag{17.2}$$

This is an exact result. On making the DT replacements $t_0 \to Tn$, $t \to T(n+1)$, $y \to x_n$, and $x \to x_{n+1}$, we can immediately read off the stroboscopic system function as in the previous chapter for the free particle. We find

$$\begin{aligned} U_n(x_{n+1}, x_n) = &\sqrt{\frac{m\omega}{i2\pi\hbar\sin(\omega T)}} \\ &\times \exp\left\{\frac{im\omega\left[(x^2_{n+1} + x^2_{n+1})\cos(\omega T) - 2x_n x_{n+1}\right]}{2\hbar\sin(\omega T)}\right\}. \end{aligned} \tag{17.3}$$

This satisfies the closure condition (16.19).

From (17.3) we read off the system function:

$$F^n(x_{n+1}, x_n) = \frac{m\omega\left[(x^2_{n+1} + x^2_{n+1})\cos(\omega T) - 2x_n x_{n+1}\right]}{2\sin(\omega T)}, \tag{17.4}$$

which is of the standard DT oscillator form (13.11) when

$$\alpha = \frac{m\omega\cos(\omega T)}{\sin(\omega T)}, \quad \beta = \frac{m\omega}{\sin(\omega T)}. \tag{17.5}$$

At this stage, everything is exact. If we perform a Laurent expansion about $T = 0$, we find

$$\alpha \approx \frac{m}{T}(6 - 2\omega^2T^2 + O(T^4)), \quad \beta \approx \frac{m}{6T}(6 + \omega^2T^2 + O(T^4)), \tag{17.6}$$

with agreement to order $O(T^2)$ with (13.12), which were obtained by our basic virtual-path approach.

This is the place to comment on the virtual-path approach to temporal discretization. It has been noted by various authors that how discretization is carried out affects the outcome (Klimek, 1996; Farias and Recami, 2010). We emphasize that it is not enough to discuss only the merits of forwards, backwards or symmetric discretization of velocities when it comes to finding good discretizations of system functions. Potentials must be treated carefully as well. Our virtual-path approach to the Newtonian oscillator gives the agreement noted above precisely because we applied the virtual-path methodology to the quadratic potential as well as to the kinetic-energy term.

17.2.2 Quantizing the generic discrete time oscillator

In this section we apply the quantization principles discussed in the previous chapter to the generic DT oscillator, the system function for which is given by (13.11), where α and β are taken to be positive constants. From comparison with (16.37), we see that (13.11) is an example of a normal system. Hence from (16.41) the system amplitude is given by

$$U_n(x,y) = \sqrt{\frac{\beta}{2\pi\hbar}}\exp(iF(x,y)/\hbar), \tag{17.7}$$

up to an arbitrary phase. This satisfies the unitarity condition (16.19).

At this point we are working in the Heisenberg picture, with successive position operators being related dynamically by equation (16.26). From (16.36) we find (16.43), which confirms that the above system coordinates are indeed normal. The operator equations of motion for the DT oscillator are

$$\hat{x}_{n+1} \underset{c}{=} 2\eta\hat{x}_n - \hat{x}_{n-1}, \qquad \eta \equiv \frac{\alpha}{\beta}, \quad \beta \neq 0, \tag{17.8}$$

with commutators given by (16.43).

A Logan invariant for the classical DT oscillator corresponding to energy is given by (13.25) and is quantized according to the rule

$$\hat{C}_n = \frac{1}{2}\beta(\hat{x}_n\hat{x}_n + \hat{x}_{n+1}\hat{x}_{n+1}) - \frac{1}{2}\alpha(\hat{x}_n\hat{x}_{n+1} + \hat{x}_{n+1}\hat{x}_n). \tag{17.9}$$

Using the operator equation of motion (17.8) we readily find $\hat{C}_n \underset{c}{=} \hat{C}_{n-1}$ as expected.

This operator is symmetrized for two reasons. First, the canonical quantization of products of variables in standard QM leads to the following question: given a term such as pq in a classical Hamiltonian, is its quantum analogue operator given by $\hat{p}\hat{q}$, $\hat{q}\hat{p}$, or some combination of these? It turns out that, where such answers make sense, the most suitable operator is usually the *symmetrized* operator $\frac{1}{2}\{\hat{p}\hat{q} + \hat{q}\hat{p}\}$. The second reason is more convincing: the system function

(13.11) is time-reversal-invariant, which in our context means that it is form-invariant with respect to the interchange $x_n \leftrightarrow x_{n+1}$. Symmetrization of (17.9) is consistent with that interchange.

We can find a suitable coordinate representation of the position operators by first choosing a particular initial time $n = M$, constructing a coordinate representation for the position operator $\hat{x}_M$, and then expressing all other operators in those terms. For example, we consider the representation

$$\hat{x}_M \mapsto x, \quad \hat{x}_{M+1} \mapsto \eta x - i\frac{\hbar}{\beta}\,\partial_x, \tag{17.10}$$

which is consistent with (16.43). Using these expressions directly in the operator equation of motion (17.8) gives the rule $\hat{x}_{M-1} \mapsto \eta x + i(\hbar/\beta)\partial_x$. In this coordinate representation, (17.9) is represented by the self-adjoint diagonal operator with operator component

$$\overrightarrow{C_x} \equiv \frac{1}{2}\beta^{-1}[-\hbar^2\,\partial_x^2 + (\beta^2 - \alpha^2)x^2]. \tag{17.11}$$

This is readily proved to be compatible with the system amplitude, in the sense of definition (16.6).

By inspection of the potential term in the differential operator (17.11) and comparison with standard Schrödinger wave mechanics for the quantized harmonic oscillator, we deduce that a complete set of normalizable physical states can be found as eigenstates of the operator (17.11), provided that $\beta^2 > \alpha^2$. This corresponds precisely to the elliptic region in the classical theory discussed in Chapter 13.

Assuming that the constants α and β do indeed satisfy the elliptic condition $\beta^2 > \alpha^2$, we may construct annihilation and creation operators for the system. These are diagonal with respect to any of the bases $\boldsymbol{B}_n$ and are given by

$$\begin{aligned}
\hat{a}_n &\equiv ie^{in\theta}\left[\hat{x}_{n+1} - e^{i\theta}\hat{x}_n\right] = e^{in\theta}\int dx|x,n\rangle\left\{\sqrt{1-\eta^2}x + \frac{\hbar}{\beta}\,\partial_x\right\}\langle x,n|,\\
\hat{a}_n^\dagger &\equiv -ie^{-i\theta}[\hat{x}_{n+1} - e^{-i\theta}\hat{x}_n] = e^{-i\theta}\int dx|x,n\rangle\left\{\sqrt{1-\eta^2}x - \frac{\hbar}{\beta}\,\partial_x\right\}\langle x,n|.
\end{aligned} \tag{17.12}$$

Here θ is the *harmonic angle* which satisfies the condition $\cos\theta = \eta \equiv \alpha/\beta$. These operators satisfy the commutation relation

$$\left[\hat{a}_n, \hat{a}_n^\dagger\right] = \frac{2\hbar\sqrt{1-\eta^2}}{\beta}. \tag{17.13}$$

Clearly, these operators become physically problematical at the parabolic barrier $\eta^2 = 1$ and in the hyperbolic regime $\eta^2 > 1$. This is associated with the concept of a stationary state, which we shall encounter in more detail in our chapters on DT field theory. Essentially, we shall see that the concept of stable particle states is meaningful only in the elliptic regime, $\eta^2 < 1$. This provides a fundamental cutoff mechanism in our approach to DT QM.

For the rest of this chapter we shall assume that we are in the elliptic regime. Using the evolution relation (16.26) and the DT oscillator operator equation of motion (17.8) we find

$$\hat{a}_{n+1} \underset{c}{=} \hat{a}_n, \tag{17.14}$$

but this does not mean that this operator is conserved. A conserved operator according to our definition must be compatible with the time-step operator $\hat{U}_n$. We find that the creation and annihilation operators satisfy the relations

$$\hat{U}_n \hat{a}_n - e^{i\theta} \hat{a}_n \hat{U}_n = 0, \quad \hat{U}_n \hat{a}_n^\dagger - e^{-i\theta} \hat{a}_n^\dagger \hat{U}_n = 0, \tag{17.15}$$

which is reminiscent of various deformed commutators encountered in q-mechanics.

The above ladder operators demonstrate the technical difference between an *invariant* operator and a *conserved* operator. According to (17.14) the operators $\hat{a}_n$ are invariant with respect to the action of the classical temporal displacement operator $\mathbb{U}_n$, i.e., $\mathbb{U}_n \hat{a}_n = \hat{a}_{n+1}$, but according to (17.15) they are not conserved because of the appearance of the phase factors $e^{\pm i\theta}$.

Two comments are relevant here: (i) in classical phase space, a function $f(\mathbf{p}, \mathbf{q}, t)$ has two kinds of time derivative, namely a *dynamical one*, given by the Poisson bracket $\{f, H\}_{\mathrm{PB}}$ with the Hamiltonian, and a *kinematical one*, given by $\partial_t f$; and (ii) a choice of time $n = m$ at which we represent our operators underpins all of the above, although this choice is ultimately arbitrary.

The operator (17.9) corresponding to the invariant $C^n = \frac{1}{2}\beta a_n^* a_n$, however, is compatible with the quantum time-step operator $\hat{U}_n$ and is therefore an invariant of the motion and is conserved. We find

$$\begin{aligned} \hat{C} &= \frac{1}{4}\beta\{\hat{a}^\dagger \hat{a} + \hat{a}\hat{a}^\dagger\} = \frac{1}{2}\beta \hat{a}^\dagger \hat{a} + \frac{1}{2}\sqrt{1-\eta^2}\hbar \\ &= \int dx |x, n\rangle \left\{ -\frac{\hbar^2}{2\beta}\frac{\partial^2}{\partial x^2} + \frac{1}{2}\beta(1-\eta^2) \right\} x^2 \langle x, n| \\ &= \int dx |x, n\rangle \overrightarrow{C_x} \langle x, n|. \end{aligned} \tag{17.16}$$

This invariant is the nearest analogue of the oscillator Hamiltonian in CT mechanics which we can find here, and its eigenstates follow the same pattern as the eigenstates of that Hamiltonian. For example, there is a ground state $|\Psi_0\rangle$ satisfying the relation

$$\hat{a}_n |\Psi_0\rangle = 0 \tag{17.17}$$

with normalizable wavefunction

$$\Psi_0(x) = \Psi_0(0) \exp\left\{ -\frac{1}{2}\beta\sqrt{1-\eta^2}x^2/\hbar \right\}. \tag{17.18}$$

This wavefunction is also an eigenstate of the invariant operator, with

$$\overrightarrow{C}\Psi_0(x) = \frac{1}{2}\sqrt{1-\eta^2}\hbar\Psi_0(x). \tag{17.19}$$

These results hold only for $|\eta| < 1$, i.e., in the elliptic regime. The appearance of the factor $\sqrt{1-\eta^2}$ in many DT oscillator equations serves as a reminder that the DT oscillator has a richer space of solutions than the CT oscillator, and, indeed, in a significant sector of those solutions (the hyperbolic regime), the solutions cannot be described as oscillating.

17.3 The inhomogeneous oscillator

We now turn to the inhomogeneous harmonic oscillator, which serves as a prototype for the application of our quantization principles to field theories. Adding an inhomogeneous term to the equation of motion (17.8) allows us to use the source-functional techniques of Schwinger (1965) to obtain the ground-state functional and various n-point functions of interest. Because the Schwinger method deals with time-ordered products, we should expect the discretization of the time parameter to involve some changes in the details of the calculations. In particular, we need to specify what is meant by the discrete analogues of the Heaviside theta and the Dirac delta. These are discussed in Section 4.4.

Schwinger's method adapted to DT QM proceeds as follows. First, given a system function $F^n \equiv F(x_n, x_{n+1})$, we are free to introduce an external source j_n in any convenient way, since ultimately this source term will be set to zero. The basic idea is to gently 'poke' the SUO with a long stick represented by j_n and monitor the response. Our choice is to define the system function $F^n[j]$ in the presence of the external source as

$$F^n[j] \equiv F^n + \frac{1}{2}Tj_{n+1}x_{n+1} + \frac{1}{2}Tj_nx_n. \tag{17.20}$$

This choice satisfies time-reversal symmetry if we include the interchange $j_n \leftrightarrow j_{n+1}$, and allows the construction of DT ordered product expectation values directly.

The second step is to calculate the new classical equation of motion. In the presence of the external source, the action sum from time MT to time NT $(N > M)$ now becomes

$$A^{NM}[j] \equiv A^{NM} + \frac{1}{2}Tj_Mx_M + \frac{1}{2}Tj_Nx_N + T\sum_{n=M+1}^{N-1} j_nx_n, \quad M < N. \tag{17.21}$$

Using the Weiss action principle as before gives the DT equations of motion

$$\frac{\partial}{\partial x_n}\{F^n + F^{n-1}\} + j_n \underset{c}{=} 0, \quad M < n < N. \tag{17.22}$$

The third step is to invoke the Schwinger action principle modified for DT. In CT QM, this principle is the quantum analogue of the Weiss action principle discussed in Chapter 8. Schwinger's approach is powerful because it deals with quantum matrix elements between different states rather than with quantum operators directly.

We start with the amplitude $\langle\phi, N|\psi, M\rangle^j$ for an initial state ψ at time M to be found in state ϕ at a later time $N > M$, assuming that the system function has a classical source term involving j. Suppose the source j_n is given a small change δ_{j_n} at DT$_n$. We postulate that, for an infinitesimal variation $\delta\hat{A}^{NM}[j]$ of the action operator,

$$\delta\langle\phi, N|\psi, M\rangle^j = \frac{i}{\hbar}\langle\phi, N|\delta\hat{A}^{NM}[j]|\psi, M\rangle^j, \quad M < N, \tag{17.23}$$

for any states $|\phi, N\rangle$ and $|\psi, M\rangle$ at times NT and MT, respectively, with evolution in the presence of the source.

The idea behind this principle is intuitively reasonable. Consider a rocket on its way from the Earth to the Moon. If during the motion the rocket is subjected to a slight deflection arising from a passing asteroid and no course correction is made, then the rocket may well miss its target by a significant amount.

Independent variation of the j_n for $M \leq n \leq N$ then leads to the equations

$$\begin{aligned}
-\frac{i\hbar}{T}\frac{\partial}{\partial j_M}\langle\phi, N|\psi, M\rangle^j &= \frac{1}{2}\langle\phi, N|\hat{x}_M|\psi, M\rangle^j, \\
-\frac{i\hbar}{T}\frac{\partial}{\partial j_n}\langle\phi, N|\psi, M\rangle^j &= \langle\phi, N|\hat{x}_n|\psi, M\rangle^j, \\
-\frac{i\hbar}{T}\frac{\partial}{\partial j_N}\langle\phi, N|\psi, M\rangle^j &= \frac{1}{2}\langle\phi, N|\hat{x}_N|\psi, M\rangle^j,
\end{aligned} \tag{17.24}$$

for $M < n < N$. Further application of the principle leads to expectation values of time-ordered products of operators, such as

$$\left(-\frac{i\hbar}{T}\right)^2 \frac{\partial^2}{\partial j_m\,\partial j_n}\langle\phi, N|\psi, M\rangle^j = \langle\phi, N|\widetilde{T}\,\hat{x}_m\hat{x}_n|\psi, M\rangle^j, \tag{17.25}$$

with $M < m$ and $n < N$, where the symbol $\widetilde{T}$ denotes DT ordering. For example,

$$\begin{aligned}
\widetilde{T}\,\hat{x}_m\hat{x}_n &= \left(\Theta_{m-n} + \frac{1}{2}\delta_{m-n}\right)\hat{x}_m\hat{x}_n + \left(\Theta_{n-m} + \frac{1}{2}\delta_{m-n}\right)\hat{x}_n\hat{x}_m \\
&= \Theta_{m-n}\hat{x}_m\hat{x}_n + \delta_{m-n}\hat{x}_n\hat{x}_n + \Theta_{n-m}\hat{x}_n\hat{x}_m,
\end{aligned} \tag{17.26}$$

where Θ_n and δ_n are the discrete Heaviside and Dirac functions discussed in Chapter 4.

Given the harmonic oscillator system function as obtained from virtual paths applied to the standard oscillator Lagrangian

$$F^n = \frac{m(x_{n+1} - x_n)^2}{2T} - \frac{Tm\omega^2}{6}\left(x_{n+1}^2 + x_{n+1}x_n + x_n^2\right), \tag{17.27}$$

the classical DT oscillator in the presence of the external source j_n satisfies the equation

$$x_{n+1} \underset{c}{=} 2\eta x_n - x_{n-1} + \frac{T}{\beta} j_n, \tag{17.28}$$

where $\eta = \alpha/\beta$, with

$$\alpha = \frac{m(1-2T^2\omega^2)}{6T}, \quad \beta = \frac{m(6+T^2\omega^2)}{6T}. \tag{17.29}$$

As discussed previously, elliptic (oscillatory) solutions occur for $\eta^2 < 1$, whereas hyperbolic solutions occur for $\eta^2 > 1$. We will now discuss these possibilities separately.

17.4 The elliptic regime

The importance of the elliptic regime $\eta^2 < 1$ stems from the fact that in field theory this corresponds to physical particle configurations of the fields, i.e., solutions that can be normalized.

The DT inhomogeneous oscillator classical equation of motion (17.28) may be written in the form

$$(\mathbb{U}_n - 2\eta + \mathbb{U}_n^{-1})x_n \underset{c}{=} \frac{T}{\beta} j_n, \tag{17.30}$$

where $\mathbb{U}_n$ is the classical jump operator. To help us solve this equation for the elliptic case we introduce the following notation: since the elliptic regime means $\eta^2 < 1$, we write $c \equiv \cos\,\theta = \eta$ and $s \equiv \sin\,\theta = +\sqrt{1-\eta^2} > 0$, taking $0 < \theta < \pi$. If we define $s_p \equiv \sin(p\theta)$, where p is an integer, then a useful identity is

$$s_p s_{q-r} + s_q s_{r-p} + s_{rc} s_{p-q} = 0, \tag{17.31}$$

where q and r are integers. From this we deduce

$$s_{p+1} + s_{p-1} = 2cs_p, \tag{17.32}$$

which is equivalent to

$$\left(\mathbb{U}_p - 2\eta + \mathbb{U}_p^{-1}\right) s_p = 0. \tag{17.33}$$

Next, noting that $s > 0$, we define the matrices

$$\Lambda^n = \frac{1}{s}\begin{bmatrix} s_{1+n} & -s_n \\ s_n & s_{1-n} \end{bmatrix}, \quad n = 0, 1, 2, \ldots, \tag{17.34}$$

where we define $\Lambda \equiv \Lambda^1$. We can use (17.31) to prove

$$\Lambda^p \Lambda^q = \Lambda^{p+q}. \tag{17.35}$$

If now we write

$$\mathbf{X}_n \equiv \begin{bmatrix} x_{n+1} \\ x_n \end{bmatrix}, \quad \mathbf{J}_n \equiv \begin{bmatrix} (T/\beta)j_n \\ 0 \end{bmatrix} \tag{17.36}$$

then (17.28) may be written in the form

$$\mathbf{X}_n = \Lambda \mathbf{X}_{n-1} + \mathbf{J}_n. \tag{17.37}$$

This equation may be readily solved using the properties of the s_p functions and by diagonalizing the matrix Λ. We choose Feynman boundary conditions, specifying the particle to be at position x_M in the past (at time MT) and at position x_N in the future (at time NT). We find

$$\begin{aligned} ss_{N-M}x_n \underset{c}{=} ss_{N-n}x_M + ss_{n-M}x_N \\ + \frac{T}{\beta}\left\{\sum_{m=M}^{n-1} s_{N-n}s_{M-m}j_m + s_{N-n}s_{M-n}j_n \right. \\ \left. + \sum_{m=1+n}^{N} s_{M-n}s_{N-m}j_m \right\}, \end{aligned} \tag{17.38}$$

which is valid only for $M < n < N$. This can be tidied up into the form

$$x_n = \frac{s_{N-n}x_M}{s_{N-M}} + \frac{s_{M-n}x_N}{s_{M-N}} - T\sum_{m=M}^{N} G_{NM}^{nm} j_m, \tag{17.39}$$

where

$$G_{NM}^{nm} = \begin{cases} -\dfrac{s_{N-n}s_{M-m}}{\beta ss_{N-M}}, & M \leq m < n < N, \\ -\dfrac{s_{N-n}s_{M-m}}{\beta ss_{N-M}}, & M < m = n < N, \\ -\dfrac{s_{M-n}s_{N-m}}{\beta ss_{N-M}}, & M < n < m \leq N. \end{cases} \tag{17.40}$$

Then G_{NM}^{nm} satisfies the inhomogeneous equation

$$\beta(\mathbb{U}_n - 2\eta + \mathbb{U}_n^{-1})G_{NM}^{nm} = -\delta_{n-m}, \tag{17.41}$$

for $M < n < N$.

Up to this point we have taken $N > M$ with both finite, but normally we will be interested in the scattering limit $N \to +\infty$, $M \to -\infty$. Also, we have appeared to have overlooked the possibility that s_{N-M} vanishes in the denominator of the propagator (17.40) for some values of N and M. We shall address both of these issues directly now.

Our method of avoiding possible singularities is to extend the Feynman $-i\epsilon$ prescription to the θ parameter. By inspection of the equation

$$\eta \equiv \cos\theta = \frac{6 - 2T^2\omega^2}{6 + T^2\omega^2} \tag{17.42}$$

we deduce that Feynman's prescription corresponds to the substitutions

$$\omega^2 \to \omega^2 - i\epsilon \to \theta \to \theta - i\epsilon, \quad \eta \to \eta + i\epsilon. \tag{17.43}$$

With this deformation of the θ parameter and the taking of the limit $N \to +\infty$, $M \to -\infty$, we find

$$x_n = \tilde{x}_n - T \sum_{m=-\infty}^{\infty} G_{\mathrm{F}}^{n-m} j_m, \tag{17.44}$$

where $\tilde{x}_n$ satisfies the homogeneous equation

$$(\mathbb{U}_n - 2\eta + \mathbb{U}_n^{-1})\tilde{x}_n = 0 \tag{17.45}$$

and

$$G_{\mathrm{F}}^{n-m} \equiv \frac{1}{2\beta is}(e^{i(m-n)\theta}\Theta_{n-m} + \delta_{m-n} + e^{i(n-m)\theta}\Theta_{m-n}). \tag{17.46}$$

This is the DT analogue of the harmonic oscillator Feynman propagator and reduces to it in the continuum limit $T \to 0$, $nT \to t$. A direct application of the DT Schwinger action principle to the operator equation of motion

$$(\mathbb{U}_n - 2\eta + \mathbb{U}_n^{-1})\hat{x}_n = \frac{T}{\beta} j_n \tag{17.47}$$

then gives the ground-state vacuum functional

$$Z[j] \equiv Z[0] \exp\left\{-\frac{iT^2}{2\hbar} \sum_{n,m=-\infty}^{\infty} j_n G_{\mathrm{F}}^{n-m} j_m\right\}, \tag{17.48}$$

essentially solving the quantum problem. For example, using this expression for $Z[j]$ and (17.24), we find that, in the limit $j \to 0$,

$$\langle 0|\hat{x}_n \hat{x}_n|0\rangle = \frac{\hbar}{2\beta s}, \tag{17.49}$$

$$\langle 0|\hat{x}_{n+1} \hat{x}_n|0\rangle = \frac{\hbar}{2\beta s} e^{-i\theta}. \tag{17.50}$$

From this we deduce the ground-state expectation values of the DT commutation relations to be

$$\langle 0|[\hat{x}_{n+1}, \hat{x}_n]|0\rangle = -\frac{i\hbar}{\beta}, \tag{17.51}$$

which agrees exactly with the oscillator commutation relation

$$[\hat{x}_{n+1}, \hat{x}_n] = -\frac{i\hbar}{\beta} \hat{I}_{\mathcal{H}} \tag{17.52}$$

found previously.

Further ground-state expectation values of commutators may be obtained by using the result

$$\langle 0|\tilde{T}\hat{x}_m \hat{x}_n|0\rangle = i\hbar G_{\mathrm{F}}^{n-m} = \frac{\hbar}{2\beta \sin\theta} e^{-i|n-m|\theta}. \tag{17.53}$$

For example, we find

$$\langle 0|[\hat{x}_{n+2}, \hat{x}_n]|0\rangle = -\frac{2i\hbar\eta}{\beta}, \tag{17.54}$$

which agrees with the commutator

$$[\hat{x}_{n+2}, \hat{x}_n] = -\frac{2i\hbar\eta}{\beta}\hat{I}_{\mathcal{H}} \tag{17.55}$$

obtained from the operator equation of motion (17.8) and the commutator (17.52).

Another verification of the consistency of these methods is that (17.50) may be used directly to find the ground-state expectation value of the invariant (17.9) for the DT oscillator in the elliptic regime. We find

$$\langle 0|\hat{C}^n|0\rangle = \frac{1}{2}\hbar\sqrt{1-\eta^2}, \tag{17.56}$$

which agrees exactly with previous results.

17.5 The hyperbolic regime

Because $\kappa \equiv T\omega$ is real and positive, the controlling parameter η as given by (17.42) takes values only in the following regions:

$$\begin{aligned} \text{elliptic:} &\quad -1 < \eta < 1, & 0 < \kappa < 2\sqrt{3}, \\ \text{parabolic:} &\quad \eta = -1, & \kappa = 2\sqrt{3}, \\ \text{hyperbolic:} &\quad -2 < \eta < -1, & \kappa > 2\sqrt{3}. \end{aligned} \tag{17.57}$$

If we parametrize η by the rule $\eta = \cos\ z$, where z is complex, then, if we take

$$\eta = \begin{cases} \cos\ \theta, & 0 < \kappa < 2\sqrt{3}, \\ -\cosh\ \lambda, & \kappa > 2\sqrt{3}, \end{cases} \tag{17.58}$$

then the range of possibilities (17.57) corresponds to a contour Γ in the complex $z = \theta - i\lambda$ plane that runs just below the real axis from the origin to π and then runs from π to $\pi - i\infty$. The elliptic regime corresponds to values of z on the first part of the contour, for which $\lambda = 0 + \epsilon$, where ϵ is infinitesimal and positive, corresponding to the Feynman $-i\epsilon$ prescription.

The hyperbolic region corresponds to the part of the contour given by $z = \pi - i\lambda, \lambda > 0$. For this region analytic continuation of the s_n functions leads to

$$s_n \to i(-1)^{n+1}\tilde{s}_n, \tag{17.59}$$

where $\tilde{s}_n \equiv \sinh(n\lambda)$. From this the analytic continuation of the finite-interval propagator (17.40) gives

$$\tilde{G}_{NM}^{nm} \equiv \frac{(-1)^{n-m+1}}{\beta\tilde{s}\tilde{s}_{N-M}}\{\tilde{s}_{N-n}\tilde{s}_{M-m}\Theta_{n-m} + \tilde{s}_{N-m}\tilde{s}_{M-m}\delta_{n-m} + \Theta_{m-n}\tilde{s}_{M-n}\tilde{s}_{N-m}\}, \tag{17.60}$$

for $M < n$ and $m < N$, which satisfies the equation

$$\beta\{\mathbb{U}_n - 2\eta + \mathbb{U}_n^{-1}\}\tilde{G}_{NM}^{nm} = -\delta_{n-m}. \tag{17.61}$$

Taking the limit $N = -M \to \infty$ gives the infinite-interval propagator

$$\tilde{G}_{\mathrm{F}}^{n-m} \equiv \frac{(-1)^{1+n-m}}{2\beta\tilde{s}}\{e^{(m-n)\lambda}\Theta_{n-m} + \delta_{n-m} + e^{(n-m)\lambda}\Theta_{m-n}\}, \tag{17.62}$$

which satisfies the equation

$$\beta\{\mathbb{U}_n - 2\eta + \mathbb{U}_n^{-1}\}\tilde{G}_{\mathrm{F}}^{n-m} = -\delta_{n-m}. \tag{17.63}$$

The elliptic and hyperbolic Feynman boundary condition propagators can be summarized in the analytic form

$$\begin{aligned}\Delta_{\Gamma}^{n}(\eta) &= \frac{T(2+\cos z)}{6m\imath\sin z}e^{-i|n|z} \\ &= \frac{T(2+\cos z)}{6mi\sin z}\left(e^{-inz}\Theta_n + \delta_n + e^{inz}\Theta_{-n}\right),\end{aligned} \tag{17.64}$$

where $\eta = \cos z$ and z lies somewhere on the contour Γ discussed above. Then $\Delta_{\mathrm{F}}^{n}(\eta)$ satisfies the equation

$$\beta\{\mathbb{U}_n - 2\eta + \mathbb{U}_n^{-1}\}\Delta_{\mathrm{F}}^{n}(\eta) = -\delta_n. \tag{17.65}$$

Physical states correspond only to points in the elliptic regime, since the wave functions will not remain normalizable in time otherwise. For example, we showed previously that the ground-state wavefunction for the DT quantized oscillator in the elliptic regime is given by

$$\Psi_0(x) = \Psi_0(0)\exp\left(-\frac{1}{2}\beta\sqrt{1-\eta^2}x^2/\hbar\right), \tag{17.66}$$

demonstrating that analytic continuation to the hyperbolic regime will not give a normalizable ground-state wavefunction.

17.6 The time-dependent oscillator

In this section we discretize the CT time-dependent oscillator reviewed in Appendix B. The generic system function for a time-dependent oscillator is defined here by

$$F^n \equiv \frac{1}{2}\alpha_n\left(x_n^2 + x_{n+1}^2\right) - \beta_n x_n x_{n+1}, \tag{17.67}$$

where all quantities are real and $\beta_n \neq 0$ for all n. This system function serves as a template for our discussion of DT quantized scalar fields in Robertson–Walker-type spacetimes in Chapter 28, so we bypass discussing classical solutions and move directly to the quantum theory. Fortunately, the system function (17.67) is simple enough to permit quantization directly.

The system function gives for the upper and lower momenta

$$\begin{aligned} p_n^{(-)} &\equiv -\frac{\partial}{\partial x_n} F^n = -\alpha_n x_n + \beta_n x_{n+1}, \\ p_n^{(+)} &\equiv \frac{\partial}{\partial x_n} F^{n-1} = \alpha_{n-1} x_n - \beta_{n-1} x_{n-1}. \end{aligned} \tag{17.68}$$

The DT equation of motion (8.26) equates these momenta at each temporal node, so we find

$$x_{n+1} \underset{c}{=} 2\eta_n x_n - \gamma_n x_{n-1}, \tag{17.69}$$

where

$$2\eta_n \equiv \frac{\alpha_n + \alpha_{n-1}}{\beta_n}, \quad \gamma_n = \frac{\beta_{n-1}}{\beta_n}. \tag{17.70}$$

We note that time-reversal invariance is lost at each node n where $\gamma_n \neq 1$.

Definition 17.1 A sequence $\{\Phi_n\}$ of elements that satisfies (17.69) for given sequences $\{\eta_n\}$ and $\{\gamma_n\}$ will be called an *oscillator sequence*, or O-sequence. The Φ_n need not be coordinates.

By inspection, the system function (17.67) is a *normal-coordinate system* according to the definition given in Section 16.5. Therefore we may impose the quantization condition

$$\left[\hat{p}_n^{(-)}, \hat{x}_n\right] = -i, \tag{17.71}$$

taking $\hbar = 1$. Then we find

$$[\hat{x}_{n+1}, \hat{x}_n] = -\frac{i}{\beta_n}. \tag{17.72}$$

We note that, by inspection of the system function, β_n is associated with the *link* $[n, n+1]$, so (17.72) is a consistent link equation.

The $\hat{x}_n$ operators are in the Heisenberg picture and the commutators (17.72) are required to be compatible with the Heisenberg operator equations of motion

$$\hat{x}_{n+1} \underset{H}{=} 2\eta_n \hat{x}_n - \gamma_n \hat{x}_{n-1}. \tag{17.73}$$

We may use this requirement to generate all relevant commutators, as follows. First, define the commutators

$$C_{n,m} = -C_{m,n} \equiv [\hat{x}_n, \hat{x}_m] \tag{17.74}$$

for arbitrary times n, m. Then, from the operator equation of motion (17.73), we deduce the relations

$$\begin{aligned} C_{n+1,m} &= 2\eta_n C_{n,m} - \gamma_n C_{n-1,m}, \\ C_{n,m+1} &= 2\eta_m C_{n,m} - \gamma_m C_{n,m-1}. \end{aligned} \tag{17.75}$$

By inspection, we see that the C_{nm} are elements of O-sequences. These commutators are also related to the DT Green function of the system.

If now we take the conditions

$$C_{n,n} = 0, \quad C_{n+1,n} = -\frac{i}{\beta_n}, \tag{17.76}$$

then we can build up a table of commutators. See Table 17.1.

We note that, because the C_{nm} are c-numbers, the Jacobi identity is automatically satisfied.

To construct linear invariant operators analogous to particle creation and annihilation operators, we define

$$\begin{aligned} a_n &\equiv i\beta_n\left(z_n^* \hat{x}_{n+1} - z_{n+1}^* \hat{x}_n\right), \\ a_n^\dagger &\equiv -i\beta_n\left(z_n \hat{x}_{n+1} - z_{n+1} \hat{x}_n\right), \end{aligned} \tag{17.77}$$

where the z_n are complex-valued elements of an O-sequence. Using the Heisenberg operator equations (17.73), we readily find

$$a_n \underset{c}{=} a_{n-1}, \quad a_n^\dagger \underset{c}{=} a_{n-1}^\dagger, \tag{17.78}$$

i.e., these are invariants of the motion. Hence we may define $a \equiv a_n$ and $a^\dagger \equiv a_n^\dagger$.

An important classical invariant of the motion is given by

$$f_n\left(z_n\right) \equiv i\beta_n\left(z_n^* z_{n+1} - z_{n+1}^* z_n\right). \tag{17.79}$$

As discussed in Section 4.7 and with reference to Appendix B, this invariant is related to a DT version of Abel's identity (Abel, 1829) applied to the DT Wronskian for the difference equation (17.69). We find

$$\left[a, a^\dagger\right] = f_n(z_n), \tag{17.80}$$

so that, if we set $f = 1$, we have

$$\left[a, a^\dagger\right] = 1. \tag{17.81}$$

We may invert the relations (17.77) to find

$$\hat{x}_n = z_n a + z_n^* a^\dagger \tag{17.82}$$

and hence

$$C_{n,m} = z_n z_m^* - z_n^* z_m. \tag{17.83}$$

If we write $z_n = u_n + iv_n$ then we find

$$u_n v_m - v_n u_m = \frac{1}{2} i C_{mn}. \tag{17.84}$$

Table 17.1 *Some of the commutators derived from the Heisenberg equations of motion*

$C_{n,m} : m \rightarrow$	n	$n+1$	$n+2$	$n+3$
n	0	i/β	$i(2\eta_{n+1}/\beta_n)$	$i(4\eta_{n+1}\eta_{n+2} - \gamma_{n+2})/\beta_n$
$n+1$	$-i/\beta_n$	0	i/β_{n+1}	$i(2\eta_{n+2}/\beta_{n+1})$
$n+2$	$-i(2\eta_{n+1}/\beta_n)$	$-i/\beta_{n+1}$	0	i/β_{n+2}
$n+3$	$-i(4\eta_{n+1}\eta_{n+2} - \gamma_{n+2})/\beta_n$	$-i\,(2\eta_{n+2}/\beta_{n+1})$	$-i/\beta_{n+2}$	0

To consider further the interpretation of these results, consider the expectation value of position relative to some given initial state Ψ. We find

$$\begin{aligned}\langle \hat{x}_n \rangle_\Psi &= z_n \langle a \rangle_\Psi + z_n^* \langle a^\dagger \rangle_\Psi \\ &= 2\ \mathrm{Re}\{\mathrm{z_n} \langle \mathrm{a} \rangle_\Psi\}. \end{aligned} \tag{17.85}$$

Since the expectation value $\langle a \rangle_\Psi$ will be independent of time, we see that the temporal behaviour of $\langle \hat{x}_n \rangle_\Psi$ depends solely on the behaviour of the O-sequence (z_n). Now consider a region where the α_n and β_n are relatively slowly varying functions of n, so that we may write

$$\beta_n \sim \beta, \quad \eta_n \sim \eta, \quad \gamma_n \sim 1. \tag{17.86}$$

In such a regime we may apply the results on the DT harmonic oscillator discussed by Jaroszkiewicz and Norton (1997a). Given the equation of motion for u_n,

$$u_{n+1} \sim 2\eta u_n - u_{n-1}, \tag{17.87}$$

we identify three distinct dynamical regimes, which are referred to as the elliptic, parabolic and hyperbolic regions, respectively. For the elliptic regime, we have $\eta^2 < 1$, and in this case the motion is bounded. If we define $\eta \equiv \cos\theta$ then an explicit solution is

$$u_n \underset{c}{=} \frac{\sin(n\theta)u_1 - \sin((n-1)\theta)u_0}{\sin\theta}, \quad n = 0, 1, 2, \ldots, \tag{17.88}$$

which explicitly shows that the motion is bounded. In the limit $\eta^2 \to 1$ we find the parabolic solution

$$u_n \underset{c}{=} n(u_1 - u_0) + u_0, \quad n = 0, 1, 2, \ldots. \tag{17.89}$$

The hyperbolic region occurs when $\eta^2 > 1$. If we write $\eta = \cosh\chi$, taking without loss of generality positive η, then we find

$$u_n = \frac{\sinh(n\chi)u_1 - \sinh((n-1)\chi)u_0}{\sinh\theta}. \tag{17.90}$$

For large n we see that in general

$$u_n \sim \frac{u_1 - e^{-\chi}u_0}{2\sinh\chi} e^{n\chi}, \tag{17.91}$$

so that the motion is unbounded. In the special case $u_1 = e^{-\chi}u_0$ the hyperbolic solution tends to zero. The same argument applies to the imaginary parts v_n.

The general conclusion is that, for weakly varying coefficients α_n and β_n, there will be the same split, approximately, into elliptic, parabolic and hyperbolic motion, depending on the ratio $\eta_n \sim \alpha_n/\beta_n$. In the hyperbolic case, the expectation value $\langle x_n \rangle_\psi$ will therefore in general diverge.

Another way of seeing the problem in the case of the hyperbolic regime is via the sequence of basis sets $\boldsymbol{B}_n = \{|x_n\rangle, x \in \mathcal{R}\}$ discussed by Jaroszkiewicz and Norton (1997a). In the hyperbolic regime the ground-state wavefunction for the DT oscillator, when expressed in terms of any of these bases, is clearly not normalizable. In the field theory case, this will correspond to a regime where the definition of a localized particle state will lose its meaning.

18
Path integrals

18.1 Introduction

The Feynman path-integral (PI) approach to quantum mechanics (QM) has proven to be an invaluable technical tool with applicability far beyond the non-relativistic QM from which it sprang (Feynman and Hibbs, 1965). The PI approach is based on a fundamental quantum principle, which is that nothing can or should be said with certainty about a system under observation (SUO) when it is not actually being observed. This is in complete contrast to classical mechanics (CM), which asserts that an SUO exists in a well-defined state at all times, regardless of whether it is being observed or not. It is a remarkable fact that what appears to be a metaphysical point actually underpins the principles of quantum physics.

The quantum principle that we cannot say precisely in classical terms what an SUO is doing when we are not observing it has a powerful corollary: if we cannot say what the SUO is doing when we are not observing it, then we have to allow for the possibility that it could be doing anything, or at least anything that would be consistent with required conservation laws, behind our backs.

Following a lead from Dirac (1933), Feynman developed and popularized a mathematical way of expressing this idea: the path integral. We shall discuss his approach in the case of an SUO consisting of a Newtonian point particle moving in one spatial dimension. Discrete time (DT) comes into the picture right at the start of the formulation.

In the next section we shall review the continuous time (CT) Feynman PI for a non-relativistic quantized free particle moving in one spatial dimension. Then we shall discuss Lee's extension of the PI to encompass Type-3 DT QM.

18.2 Feynman's path integrals

One of the main objectives in quantum theory is to calculate the QM transition amplitude $\mathcal{A}(\Phi, t_f|\Psi, t_i)$ for the detection of a QM state Φ at a final time t_f, given that the SUO was prepared in state Ψ at initial time $t_\mathrm{i} < t_\mathrm{f}$. The related conditional outcome probability $P(\Phi, t_\mathrm{f}|\Psi, t_\mathrm{i})$ is given by the Born rule (Born, 1926)

$$P(\Phi, t_\mathrm{f}|\Psi, t_\mathrm{i}) \sim |\mathcal{A}(\Phi, t_\mathrm{f}|\Psi, t_\mathrm{i})|^2, \tag{18.1}$$

up to various normalization constants and subject to technical issues to do with densities of states, etc., which we can ignore here.

The focus in path integration is generally on the amplitude $\mathcal{A}(\Phi, t_\mathrm{f}|\Psi, t_\mathrm{i})$ rather than the more physical probability $P(\Phi, t_\mathrm{f}|\Psi, t_\mathrm{i})$: once the former is known, the physical quantities of interest such as cross-sections and decay lifetimes can be calculated.

The original and indeed standard formulation of the PI is to use a single Hilbert space $\mathcal{H}$ and follow the evolution of states in it (Feynman and Hibbs, 1965). Therefore, the architecture of Figure 5.1 is the one employed, where the single 'universe' U is replaced by the fixed Hilbert space $\mathcal{H}$. In fact, both Figure 5.1(a) and Figure 5.1(b) are relevant here: the former represents what it is imagined happens in CT, whilst the latter represents how the PI is formulated via temporal discretization.

Our choice here is to use the architecture illustrated in Figure 5.2, where, in the case of the PI, each 'universe' U_n is a copy $\mathcal{H}_n$ of the original one-particle Hilbert space $\mathcal{H}$. If we denote $t_M \equiv t_i$ then the initial Hilbert space is $\mathcal{H}_M$, successive Hilbert spaces are denoted $\mathcal{H}_n$ and, taking $t_N \equiv t_\mathrm{f}$, the final Hilbert space is $\mathcal{H}_N$. We shall assume that N and M are integers with $N \geqslant M$.

The point about doing things in the multi-Hilbert-space approach is that it opens the door to having time-dependent Hilbert spaces, representing changes in the observer's apparatus. We shall discuss this idea towards the end of this book, in Chapter 29.

In our chosen architecture, the dynamics requires us to map a state from one Hilbert space, $\mathcal{H}_n$, to the next one, $\mathcal{H}_{n+1}$, the two spaces being separated in time by a chronon. The original approach of Feynman was to slice up the time interval $[t_\mathrm{i}, t_\mathrm{f}]$ into equal time slices, intervals of duration $T \equiv (t_\mathrm{f} - t_\mathrm{i})/(N - M)$, work out a discretized approximation to the amplitude $\mathcal{A}(\Phi, t_\mathrm{f}|\Psi, t_\mathrm{i})$ and then take the CT limit $N - M \to \infty$.

Our interest will be not so much in the CT limit but in the discretization details prior to the taking of that limit, because, if there is a fundamental chronon in reality, then this limit is not taken. Although it should be possible to use Type-2 discretization, Type 1 is invariably chosen for convenience, with $T_n \equiv t_{n+1} - t_n = (t_\mathrm{f} - t_\mathrm{i})/(M - N)$. Historically, there was an attempt to use Type-3 chronons by Lee (1983), which we will discuss in the next section.

In practice, path integration does not focus on the complete amplitude $\mathcal{A}(\Phi, t_\mathrm{f}|\Psi, t_\mathrm{i})$ but on a related quantity known as the Feynman kernel, denoted

$K(x, t_f; y, t_i)$, where x and y are coordinates in the case of a one-particle system in one dimension. The formalism runs as follows. First, for each discrete time $t_n \equiv t_M, t_{M+1}, t_{M+2}, \ldots, t_N$ we set up a position basis set $\boldsymbol{B}_n \equiv \{|x, n\rangle : x \in \mathbb{R}\}$ for $\mathcal{H}_n$. The improper normalization rule for this basis is

$$\langle x, n|y, n\rangle = \delta(x - y), \quad M \leqslant n \leqslant N, \tag{18.2}$$

where $\delta(x)$ is the Dirac delta. Then, for each universe $\mathcal{H}_n$, the resolution of the identity operator $\mathbb{I}_n$ for that space is given by

$$\mathbb{I}_n \equiv \int_{-\infty}^{\infty} |x, n\rangle\langle x, n|dx. \tag{18.3}$$

Completeness means that the state $|\Psi, n\rangle$ of an SUO at time n can be written in the form

$$|\Psi, n\rangle = \int_{-\infty}^{\infty} \Psi_n(x)|x, n\rangle dx, \tag{18.4}$$

where $\Psi_n(x)$ is a complex function of x.

Quantum dynamics enters the discussion in the way that states in $\mathcal{H}_n$ are mapped into $\mathcal{H}_{n+1}$. Throughout this chapter we shall be interested in pure states only, which means that statistical mixtures involving density matrices are not discussed. Therefore, we will assume that any given vector $|\Psi, n\rangle$ in $\mathcal{H}_n$ is mapped into a unique vector $|\Psi, n+1\rangle$ in $\mathcal{H}_{n+1}$ under dynamical evolution. We shall represent this by the evolution rule

$$|\Psi, n\rangle \to |\Psi, n+1\rangle \equiv \mathsf{U}_n|\Psi, n\rangle, \tag{18.5}$$

where U_n is a jump operator from $\mathcal{H}_n$ to $\mathcal{H}_{n+1}$.

A jump operator has to have certain features if the theory is to represent QM as we know it. An important property is that evolution is *linear*, i.e., for any two states $|\Psi, n\rangle$ and $|\Phi, n\rangle$ in $\mathcal{H}_n$, and for arbitrary complex numbers α and β, we have

$$\alpha|\Psi, n\rangle + \beta|\Phi, n\rangle \to \alpha|\Psi, n+1\rangle + \beta|\Phi, n+1\rangle. \tag{18.6}$$

In this expression, addition of vectors on the left-hand side of the arrow is in $\mathcal{H}_n$, whilst addition on the right-hand side is in $\mathcal{H}_{n+1}$. In terms of the jump operators, the linearity rule (18.6) becomes

$$\begin{aligned}\alpha|\Psi, n\rangle + \beta|\Phi, n\rangle &\to \mathsf{U}_n\{\alpha|\Psi, n\rangle + \beta|\Phi, n\rangle\} \\ &= \alpha\, \mathsf{U}_n|\Psi, n\rangle + \beta\, \mathsf{U}_n|\Phi, n\rangle \\ &= \alpha|\Psi, n+1\rangle + \beta|\Phi, n+1\rangle.\end{aligned} \tag{18.7}$$

Another feature generally incorporated into quantum dynamics is that the evolution rule (18.5) preserves inner products, i.e., for any vectors $|\Psi, n\rangle$ and $|\Phi, n\rangle$ in $\mathcal{H}_n$ we require that

$$\langle \Phi, n+1|\Psi, n+1\rangle = \langle \Phi, n|\Psi, n\rangle. \tag{18.8}$$

This will be satisfied if $\mathsf{U}_n^\dagger \mathsf{U}_n = \mathbb{I}_n$. This condition guarantees that the dynamical evolution conserves probability.

We are now in a position to develop the PI formalism. We consider an initial state $|\Psi, M\rangle$ of the SUO that is set up at initial time M and then allowed to evolve undisturbed to time $N > M$ by a number $(N - M)$ of DT steps. The final state $|\Psi, N\rangle$ is therefore given by iteration of (18.5):

$$|\Psi, N\rangle = \mathsf{U}_{N-1}\mathsf{U}_{N-2}\dots\mathsf{U}_M|\Psi, M\rangle. \tag{18.9}$$

The object of interest is the amplitude $\mathcal{A}(\Phi, N|\Psi, M)$ to catch the state (18.9) registering in detectors at time N as some predefined final state $|\Phi, N\rangle$. The Born interpretation gives

$$\begin{aligned}\mathcal{A}(\Phi, N|\Psi, M) &= \langle \Phi, N|\Psi, N\rangle \\ &= \langle \Phi, N|\mathsf{U}_{N-1}\mathsf{U}_{N-2}\dots\mathsf{U}_M|\Psi, M\rangle.\end{aligned} \tag{18.10}$$

We now insert identity operators on either side of each evolution operator, which does not change the amplitude:

$$\mathcal{A}(\Phi, N|\Psi, M) = \langle \Phi, N|\mathbb{I}_N\mathsf{U}_{N-1}\mathbb{I}_{N-1}\mathsf{U}_{N-2}\mathbb{I}_{N-2}\dots\mathbb{I}_{M+1}\mathsf{U}_M\mathbb{I}_M|\Psi, M\rangle. \tag{18.11}$$

Next, we use a resolution of each identity in terms of the position basis for the relevant Hilbert space, as in (18.3), giving

$$\begin{aligned}\mathcal{A}(\Phi, N|\Psi, M) = \langle \Phi, N| &\left\{\int_{-\infty}^{\infty} |x_N, N\rangle\langle x_N, N|dx_N\right\}\mathsf{U}_{N-1} \\ &\times\left\{\int_{-\infty}^{\infty} |x_{N-1}, N-1\rangle\langle x_{N-1}, N-1|dx_{N-1}\right\}\mathsf{U}_{N-2}\dots\mathsf{U}_M \\ &\times\left\{\int_{-\infty}^{\infty} |x_M, M\rangle\langle x_M, M|dx_M\right\}|\Psi, M\rangle \\ = \int_{-\infty}^{\infty} &\Phi_N^*(x_N)K(x_N, t_N; x_M, t_M)\Psi_M(x_M)dx_N\, dx_M,\end{aligned} \tag{18.12}$$

where the Feynman kernel $K(x_N, t_N; x_M, t_M)$ is given by

$$K(x_N, t_N; x_M, t_M) \equiv \left\{\prod_{m=M+1}^{N-1}\int dx_m\right\}\left\{\prod_{n=M}^{N-1}\langle x_{n+1}, n+1|\mathsf{U}_n|x_n, n\rangle\right\}. \tag{18.13}$$

No approximations have been made up to this point.

The strategy employed by Feynman in his formulation of the PI was to discretize the temporal interval $[t_\mathrm{i}, t_\mathrm{f}]$ into $N - M$ equal-duration links, devise a suitable approximation for the single time-step kernel

$K(x_{n+1}, t_{n+1}; x_n, t_n) \equiv \langle x_{n+1}, n+1|\mathsf{U}_n|x_n, n\rangle$, evaluate the $N-1-M$ integrals in (18.13) and take the limit $N \to \infty$, with t_i and t_f fixed.

The story behind how Feynman was led to his development of the PI has been recounted by L. M. Brown in his review of Feynman's doctoral thesis (Brown, 2005) and goes as follows. In an article on the Lagrangian in QM (Dirac, 1933), which was at that time an unusual thing to discuss in QM, Dirac had related the quantum-mechanical step kernel $\langle x_{n+1}, n+1|\mathsf{U}_n|x_n, n\rangle$ to the exponential of the classical action. Specifically, he suggested

$$\langle x_{n+1}, n+1|\mathsf{U}_n|x_n, n\rangle \approx \exp\left\{\frac{i}{\hbar}\int_{t_n}^{t_{n+1}} L\, dt\right\}. \tag{18.14}$$

Now, one of the hallmarks of Dirac's work was his thorough familiarity with the principles of CM, including transformation theory. It turns out that Hamilton's principal function is a central element in transformation theory and the Hamilton–Jacobi equation (Goldstein *et al.*, 2002; Leech, 1965). Hamilton's principal function is the integral of the Lagrangian over Γ_c, the true or classical trajectory.

Dirac used the correspondence (18.14) between the step kernel and the exponential of the action to give an argument based on the cancellation of phases as to why the classical trajectory Γ_c would appear significant from a quantum-mechanical perspective. Dirac, in other words, was giving an argument for the appearance of the classical world as a first approximation to a quantum world.

The story goes that a visitor to Princeton, where Feynman was based, informed Feynman of Dirac's article. This immediately stimulated Feynman to develop his particular approach to the path integral. Specifically, for $t_{n+1} - t_n \equiv T$ very small (in an appropriate sense), Feynman wrote (18.14) in the approximate form

$$\langle x_{n+1}, n+1|\mathsf{U}_n|x_n, n\rangle \approx \exp\left\{\frac{i}{\hbar}\int_{t_n}^{t_{n+1}} L\, dt\right\} \approx \exp\left\{\frac{iT}{\hbar}L_n\right\}, \tag{18.15}$$

where L_n is some discretization assumed valid on the temporal interval $[t_n, t_n + T]$.

For sufficiently small T, Feynman discretized standard Lagrangians of the form

$$L(\dot{x}, x) = \frac{\dot{x}^2}{2m} - V(x) \tag{18.16}$$

according to the rule

$$L(\dot{x}, x) \to L_n \equiv L\left(\frac{x_{n+1} - x_n}{T}, x_{n+1}\right). \tag{18.17}$$

Then (18.13) is replaced by

$$K(x_N, t_N; x_M, t_M) \approx \left\{ \prod_{m=M+1}^{N-1} \int dx_m \right\} \left\{ \prod_{n=M}^{N-1} \exp\left\{ \frac{iT}{\hbar} L_n \right\} \right\}$$
$$= \left\{ \prod_{m=M+1}^{N-1} \int dx_m \right\} \exp\left\{ \frac{iT}{\hbar} \sum_{n=M}^{N-1} L_n \right\} \qquad (18.18)$$

and then the CT limit $T \to 0$, $N - M \to \infty$, with $(N - M)T = t_{\mathrm{f}} - t_{\mathrm{i}}$ fixed, was taken, to give the complete result.

We make the following observations.

1. The product TL_n in (18.18) corresponds to what we have termed a system function. Specifically, in equation (8.19) we wrote

$$F^n \sim \int_{t_n}^{t_{n+1}} L\, dt, \qquad (18.19)$$

   with integration over the classical trajectory if we know it, or else over some virtual path if we do not.
2. The argument of the exponential in (18.18) is an action sum multiplied by $i/\hbar$.
3. In the CT limit, there are infinitely many integrals to evaluate in (18.18): this is the 'famous sum over all paths', which encodes the principle we touched upon at the start of this chapter, namely that, if we are not monitoring an SUO, we have to take all possible paths into account in the quantum calculation.
4. Depending on the Lagrangian, there will be either a trivial overall constant in (18.18) or a 'Jacobian' factor that needs to be carefully evaluated (Abers and Lee, 1973).
5. The difference between a PI in CT QM and the corresponding expression in DT QM is that the CT limit is not taken in the latter.
6. In his analysis of Feynman's thesis, Brown emphasizes two points: first, that standard QM can be obtained from an SUO described by a Lagrangian from which a Hamiltonian can be obtained; and second, that Dirac's relationship (18.14) is exact for vanishingly small time steps, up to an overall normalization constant.

Feynman's approach works well for Lagrangians that are at most quadratic in the velocities. This is analogous in DT to working with system functions that are in normal form, i.e., given by

$$F^n = -\beta x_{n+1} x_n + \frac{1}{2} W(x_{n+1}) + \frac{1}{2} W(x_n). \qquad (18.20)$$

For such system functions, we find

$$\langle x_{n+1}, n+1|\mathsf{U}_n|x_n, n\rangle \approx \exp\{iF^n/\hbar\}, \tag{18.21}$$

up to a normalization constant.

18.3 Lee's path integral

In Chapter 10 we discussed Lee's DT CM Lagrangians, which are based on Type-3 temporal intervals, i.e., the time t_n at node n is now regarded as a dynamical variable with an equation of motion. Lee's motivation for turning the node times into dynamical variables is explained in his paper. We quote:

> Each measurement determines the physical field as well as the space-time location of the measurement itself. In this sense, this new approach [i.e., his Type-3 mechanics] seems to be nearer to our actual experimental experience. Our usual concept that all physical fields should be embedded in a continuous space-time manifold may well be only an approximation, far from fundamental.
>
> In this new formulation, any finite number of observations may be made very close to each other in space and in time. However, within an infinitesimal space-time volume it is not possible to perform an infinite number of measurements.
>
> (Lee, 1983)

Lee took Feynman's PI as formulated above and modified it to incorporate the new temporal degrees of freedom. Taking the Lagrangian (18.16), Lee discretized the action integral, turning it into an action sum given by

$$\int_{t_0}^{t_f} L(\dot{x}, x)dt \rightarrow \sum_{n=M}^{N-1} (t_{n+1} - t_n) L_n^{\text{Lee}}, \tag{18.22}$$

where

$$L_n^{\text{Lee}} \equiv \frac{(x_{n+1} - x_n)^2}{2m(t_{n+1} - t_n)^2} - \frac{1}{2}[V(x_{n+1}) + V(x_n)], \tag{18.23}$$

assuming $t_{n+1} > t_n$. Then the path integral (18.18) was replaced by

$$K(x_N, t_N; x_M, t_M) \approx \left\{ \prod_{m=M+1}^{N-1} \int dx_m \, dt_m \right\} J \exp\left\{ \frac{iT}{\hbar} \sum_{n=M}^{N-1} L_n^{\text{Lee}} \right\}, \tag{18.24}$$

where J is a suitable Jacobian.

We make the following observations.

1. Lee did not insist on the CT limit, so his basic ambition matches ours. Indeed, he ends his paper with the following remark: '... one may also regard the fundamental length [read chronon in the case of particle Lagrangians] l of the

discrete mechanics as a covariant regularization parameter for the elimination of ultraviolet divergence'.

2. There will be contributions to the Lee path integral (18.24) for which time ordering is reversed, i.e., for which $t_{n+1} < t_n$. However, preliminary analysis of the action sum in (18.22) indicates that such reversal will not occur over most reasonable classical trajectories, and phase cancellations should kill off contributions that are well away from the classical trajectory.

19
Quantum encoding

19.1 Introduction

In this chapter we discuss an approach to discrete time (DT) equations of motion that is based on the quantum theory of coherent states (Kowalski, 1994). These are reviewed in Appendix A.

Coherent states of a single oscillator degree of freedom are elements of a Hilbert space $\mathcal{H}$ that are indexed by a complex parameter z. We will use the following notation. Coherent states with standard normalization will be denoted with angular Dirac brackets, viz., $|z\rangle$, $z \in \mathbb{C}$. Then for any complex numbers z, w we have the inner product rule $\langle z|w\rangle = e^{-|z-w|^2}$.

However, it is more convenient to adopt a simplified normalization. We denote our simplified coherent states by $|z)$. If a and a^+ are harmonic oscillator ladder operators satisfying the commutation relation

$$[a, a^+] = 1 \tag{19.1}$$

then we define

$$|z) \equiv \exp\{za^+\}|0), \tag{19.2}$$

where $|0)$ is the normalized-to-unity ground state. The ground state is annihilated by the a operator, i.e., $a|0) = 0$. Our inner product rule is

$$(z|w) = e^{z^* w}, \quad z, w \in \mathbb{C}, \tag{19.3}$$

which means that our normalization convention is

$$(z|z) = e^{|z|^2}, \quad z \in \mathbb{C}, \tag{19.4}$$

which is never zero. Our simplified normalization convention does not affect the defining property of coherent states, which is that they are eigenstates of the lowering ladder operator a, satisfying the eigenvalue condition

$$a|z) = z|z), \quad z \in \mathbb{C}. \tag{19.5}$$

Quantum coherent states can be used to discuss classical difference equations. We shall first discuss first-order autonomous difference equations.

19.2 First-order quantum encoding

Our aim in this section is to encode difference equations of the form

$$z_{n+1} \underset{\mathrm{c}}{=} f(z_n), \quad n \in \mathbb{N}, \tag{19.6}$$

into a coherent state representation and investigate possible methods of solution. Here the jump function f is assumed to be analytic. The formal solution is

$$z_n = f^{[n]}(z_0), \quad n \in \mathbb{N}, \tag{19.7}$$

where $f^{[n]}$ denotes the nth composition of the function f, satisfying the rule

$$f^{[n+1]} \equiv f \circ f^{[n]}. \tag{19.8}$$

The first step in the encoding is to define the jump operator M_f:

$$\mathsf{M}_f \equiv \sum_{k=0}^{\infty} \frac{a^{+k}}{k!}(f(a) - a)^k. \tag{19.9}$$

Then, with the above properties, we readily find

$$\mathsf{M}_f|z) = |f(z)). \tag{19.10}$$

The difficulty in solving (19.6) is that it involves multiple compositions of the function f. The quantum encoding method transforms the process of function composition to one of operator multiplication. To see this, consider two jump functions f and g. Then

$$\mathsf{M}_f\mathsf{M}_g|z) = |f(g(z)) = \mathsf{M}_{fog}|z). \tag{19.11}$$

Using the overcompleteness property discussed in Appendix A, we deduce

$$\mathsf{M}_f\mathsf{M}_g = \mathsf{M}_{fog}. \tag{19.12}$$

Hence we deduce that a formal solution may be obtained by the rule

$$|z_n) = \mathsf{M}_f^n|z_0). \tag{19.13}$$

19.2.1 Example (i): first-order linear inhomogeneous transformations

To illustrate the encoding method, consider the jump equation

$$z_{n+1} \underset{\mathrm{c}}{=} \alpha z_n + \beta, \tag{19.14}$$

where α and β are constants. Then we see that $f(z) = \alpha z + \beta$.

To work out $f^{[n]}$, we first define two auxiliary operators.

1. **The scaling operator**, S_α, is defined by

$$\mathsf{S}_\alpha|z) \equiv |\alpha z). \tag{19.15}$$

2. **The displacement operator**, D_β, is defined by

$$\mathsf{D}_\beta|z) \equiv |z+\beta). \tag{19.16}$$

These operators have the following properties:

$$\begin{aligned} \mathsf{S}_\alpha \mathsf{S}_\lambda &= \mathsf{S}_{\alpha\lambda}, \\ \mathsf{D}_\beta \mathsf{D}_\gamma &= \mathsf{D}_{\beta+\gamma}. \end{aligned} \tag{19.17}$$

We deduce that S_α has an inverse, given by $\mathsf{S}_\alpha^{-1} = \mathsf{S}_{\alpha^{-1}}$, provided that $\alpha \neq 0$, and D_β always has an inverse, given by $\mathsf{D}_\beta^{-1} = \mathsf{D}_{-\beta}$.

To solve (19.14), we use the product rule

$$\mathsf{S}_\alpha \mathsf{D}_\beta = \mathsf{D}_{\alpha\beta} \mathsf{S}_\alpha, \tag{19.18}$$

which is readily proved.

By inspection of (19.14), we take

$$\mathsf{M}_f = \mathsf{D}_\beta \mathsf{S}_\alpha, \tag{19.19}$$

which gives $\mathsf{M}_f|z) = |\alpha z + \beta)$. Now, using the above operator properties, we find

$$\begin{aligned} \mathsf{M}_f \mathsf{M}_f &= \mathsf{D}_\beta \mathsf{S}_\alpha \mathsf{D}_\beta \mathsf{S}_\alpha \\ &= \mathsf{D}_\beta \mathsf{D}_{\alpha\beta} \mathsf{S}_\alpha \mathsf{S}_\alpha \\ &= \mathsf{D}_{\beta(1+\alpha)} \mathsf{S}_{\alpha^2}. \end{aligned} \tag{19.20}$$

By inspection, we deduce

$$\mathsf{M}_f^{\,n} = \mathsf{D}_{\beta(1+\alpha+\alpha^2 \cdots + \alpha^{n-1})} \mathsf{S}_{\alpha^n}, \tag{19.21}$$

which is readily proved by induction.

Hence we deduce that $|z_n) \equiv \mathsf{M}_f^{\,n}|z_0)$ is given by

$$|z_n) = |\beta(1+\alpha+\alpha^2+\cdots+\alpha^{n-1}) + \alpha^n z_0), \tag{19.22}$$

giving the solution

$$z_n = \beta \sum_{k=0}^{n-1} \alpha^k + \alpha^n z_0, \tag{19.23}$$

which is in agreement with the Green-function method and the Laplace-transform method discussed in Chapter 6.

19.2.2 Example (ii): the specialized logistic equation

The logistic equation is a non-linear first-order difference equation given by

$$z_{n+1} \underset{c}{=} \mu(z_n - z_n^2). \tag{19.24}$$

We shall solve it for the special case $\mu = 2$ using quantum encoding.

First, define the quadratic operator Q by its action on a coherent state:

$$\mathsf{Q}|y) \equiv |y^2). \tag{19.25}$$

Then, by inspection,

$$\mathsf{Q}^n|y) \equiv |y^{2^n}), \quad n \in \mathbb{N}. \tag{19.26}$$

Now consider the difference equation

$$y_{n+1} \underset{c}{=} y_n^2. \tag{19.27}$$

This is encoded as $|y_{n+1}) = \mathsf{Q}|y_n)$. Then, by inspection,

$$|y_n) = \mathsf{Q}^n|y_0) = |y_0^{2^n}). \tag{19.28}$$

There is a relation between the solutions to (19.24) when $\mu = 2$ and (19.27): We can transform the latter into the former by the transformation $y_n = 1 - 2z_n$. Hence (19.28) becomes

$$\mathsf{D}_1\mathsf{S}_{-2}|z_n) = |1 - 2z_n) = |y_n) = |y_0^{2^n}) = |(1 - 2z_0)^{2^n}), \tag{19.29}$$

which gives

$$\begin{aligned} |z_n) &= \mathsf{S}_{-\frac{1}{2}}\mathsf{D}_{-1}|(1 - 2z_0)^{2^n}) \\ &= \mathsf{S}_{-\frac{1}{2}}|(1 - 2z_0)^{2^n} - 1) \\ &= |\frac{1}{2} - \frac{1}{2}(1 - 2z_0)^{2^n}). \end{aligned} \tag{19.30}$$

The solution to (19.24) for $\mu = 2$ is therefore

$$z_n = \frac{1}{2} - \frac{1}{2}(1 - 2z_0)^{2^n}. \tag{19.31}$$

19.3 Second-order quantum encoding

We may extend the above first-order formalism to second-order difference equations of the explicit form

$$x_{n+1} \underset{c}{=} f(x_n, x_{n-1}), \qquad n \in \mathbb{N}. \tag{19.32}$$

The DT harmonic oscillator and anharmonic oscillators are of this type.

The trick here is to rewrite the second-order equation (19.32) as a pair of first-order equations (Jaroszkiewicz, 1995). We define $y_n \equiv x_{n-1}$ and write

$$x_{n+1} \underset{c}{=} f(x_n, y_n), \quad y_{n+1} \underset{c}{=} x_n. \tag{19.33}$$

Next, we introduce the ladder operators a, a^+ associated with the x_n degrees of freedom and the ladder operators b, b^+ associated with the y_n degrees of freedom. The non-zero commutation relations are

$$[a, a^+] = [b, b^+] = 1. \tag{19.34}$$

Bi-coherent states are states $|z, w)$ in the tensor product $\mathcal{H} \otimes \mathcal{H}$ defined by

$$|z, w) \equiv \exp\{za^+ + wb^+\}|0, 0), \tag{19.35}$$

where $|0, 0) \equiv |0) \otimes |0)$.

The jump operator M_f is defined by

$$\mathsf{M}_f \equiv \sum_{r=0}^{\infty}\sum_{s=0}^{\infty} \frac{(a^+)^r}{r!}\frac{(b^+)^s}{s!}[f(a,b) - a]^r[a - b]^s. \tag{19.36}$$

Then, given a starting state $|x_0, x_{-1})$, we find

$$\mathsf{M}_f|x_0, x_{-1}) = |f(x_0, x_{-1}), x_0) \equiv |x_1, x_0). \tag{19.37}$$

Similarly, we find

$$\mathsf{M}_f|x_n, x_{n-1}) = |x_{n+1}, x_n). \tag{19.38}$$

The formal solution is therefore

$$|x_n, x_{n-1}) = \mathsf{M}_f^n|x_0, x_{-1}). \tag{19.39}$$

19.4 Invariants of the motion

Invariants of the second-order equations of motion (19.32) will typically be first-order functions, by which we mean functions $C_f(x_n, x_{n-1})$ of the pair $\{x_n, x_{n-1}\}$ such that

$$C_f(x_{n+1}, x_n) \underset{c}{=} C_f(x_n, x_{n-1}). \tag{19.40}$$

Given such a function, we associate with it the operator $\mathsf{C}_f \equiv \mathsf{C}_f(a, b)$. Then we must have

$$\mathsf{C}_f|x_n, x_{n-1}) = C_f(x_n, x_{n-1})|x_n, x_{n-1}). \tag{19.41}$$

To investigate the relationship between such an invariant and the step generator M_f we first apply M_f to both sides of (19.41), which gives

$$\mathsf{M}_f \mathsf{C}_f |x_n, x_{n-1}) = C_f(x_n, x_{n-1}) |x_{n+1}, x_n). \tag{19.42}$$

On the other hand, applying C_f to both sides of (19.38) gives

$$\mathsf{C}_f \mathsf{M}_f |x_n, x_{n-1}) = C_f(x_{n+1}, x_n) |x_{n+1}, x_n). \tag{19.43}$$

Subtracting (19.42) from (19.43) then gives

$$[\mathsf{C}_f, \mathsf{M}_f] |x_n, x_{n-1}) = \{C_f(x_{n+1}, x_n) - C_f(x_n, x_{n-1})\} |x_{n+1}, x_n). \tag{19.44}$$

If $C_f(x_n, x_{n-1})$ is an invariant, the right-hand side of (19.44) vanishes. Then, using overcompleteness, we deduce that the commutator $[\mathsf{C}_f, \mathsf{M}_f]$ must be zero. This is a necessary and sufficient condition for any operator $\mathsf{C}_f(a, b)$ to represent a conserved first-order quantity for such a system.

19.4.1 Example (iii): the DT oscillator

As an example of a second-order system for which the encoding approach works, we discuss the DT oscillator, with equation of motion

$$x_{n+1} \underset{c}{=} 2\eta x_n - x_{n-1}, \tag{19.45}$$

where $\eta \equiv \alpha/\beta$ and α and β are constants. Then

$$f(a, b) = 2\eta a - b, \tag{19.46}$$

so from (19.36) we have

$$\mathsf{M}_f \equiv \sum_{r=0}^{\infty} \sum_{s=0}^{\infty} \frac{(a^+)^r}{r!} \frac{(b^+)^s}{s!} [2\eta a - a - b]^r [a - b]^s. \tag{19.47}$$

Now, by inspection, this second order system has first-order invariant

$$C_f(x_n, x_{n+1}) = \frac{1}{2}\beta(x_n^2 + x_{n+1}^2) - \alpha x_n x_{n+1}, \tag{19.48}$$

so the associated operator is

$$\mathsf{C}_f = \frac{1}{2}\beta(a^2 + b^2) - \alpha ab. \tag{19.49}$$

To verify that the commutator of C_f and M_f vanishes, we quote two results:

$$\begin{aligned} a\mathsf{M}_f &= \mathsf{M}_f f(a) = \mathsf{M}_f(2\eta a - b), \\ b\mathsf{M}_f &= \mathsf{M}_f a. \end{aligned} \tag{19.50}$$

These are readily proved using the commutators $[a, a^{+r}] = ra^{+(r-1)}$ and $[b, b^{+r}] = rb^{+(r-1)}$, $r \in \mathbb{N}$, and definition (19.36).

Then, using (19.50), we find

$$\begin{aligned}
\mathsf{C}_f\mathsf{M}_f &= \left\{\frac{1}{2}\beta(a^2+b^2) - \alpha ab\right\}\mathsf{M}_f \\
&= \frac{1}{2}\beta a\mathsf{M}_f(2\eta a - b) + \frac{1}{2}\beta b\mathsf{M}_f a - \alpha a\mathsf{M}_f a \\
&= \mathsf{M}_f\left\{\frac{1}{2}\beta(2\eta a - b)^2 + \frac{1}{2}\beta a^2 - \alpha(2\eta a - b)a\right\} \\
&= \mathsf{M}_f\left\{\frac{1}{2}\beta(a^2+b^2) - \alpha ab\right\} = \mathsf{M}_f\mathsf{C}_f,
\end{aligned} \tag{19.51}$$

i.e., $[\mathsf{C}_f, \mathsf{M}_f] = 0$, which establishes that (19.48) is conserved.

Part IV

Discrete time classical field theory

20

Discrete time classical field equations

20.1 Introduction

We have now reached the point where we can start discussing classical and quantum discrete time (DT) field theories. Such theories are fundamentally different in character from the theories we have discussed so far, in the following specific ways.

(i) We can no longer maintain a particulate view of the world, by which we mean that we can no longer view matter as consisting solely of localized point-like particles running along worldlines in spacetime.
(ii) We now have to develop techniques for dealing with mechanical systems involving continuously many degrees of freedom.
(iii) Some degrees of freedom, such as quark fields, might not correspond to directly observable degrees of freedom.
(iv) Special relativity emphasizes certain spacetime symmetries between inertial frames, but simultaneity is not one of those symmetries. The problem we face in DT mechanics is that the formalism explicitly breaks Lorentz symmetry, raising the question of the compatibility of DT field theory with relativity.

The first three points in this list are encountered in continuous time (CT) field theories but the fourth is specific to our subject: temporal discretization is done in our approach in a given inertial frame, the preferred frame, and this breaks Lorentz covariance explicitly. We shall address this fundamental point later in this book. We shall develop the equations of DT classical and quantum field theory (QFT) as seen from the preferred frame in this and the next few chapters.

20.2 System functions for discrete time field theories

In this section we extend the principles of DT mechanics as discussed in previous chapters to encompass systems described by classical fields $\varphi^{\alpha}(t, \mathbf{x})$. These are dynamical variables parametrized by one time dimension t and labelled by three spatial dimensions $\mathbf{x} = \{x, y, z\}$. We assume that these coordinates refer to the preferred inertial frame, relative to which temporal discretization is being considered.

The field superscript α will denote whatever spin degrees of freedom and other properties, such as electric charge, are associated with the field φ^α. Natural units will be used for convenience, so we have $c = \hbar = 1$. The summation convention will apply to repeated small Greek indices except where stated otherwise.

In CT field theory, the analogues of particle Lagrangians L are spatial integrals of Lagrange densities $\mathcal{L}$, i.e., $L = \int \mathcal{L}\, d^3\mathbf{x}$. Because we shall follow standard CT field theory principles as closely as possible, all Lagrange densities $\mathcal{L}$ discussed in this book will be canonical, i.e., assumed to be analytic functions of the fields and their first space and time derivatives only. We write

$$\mathcal{L} = \mathcal{L}(\varphi^\alpha, \partial_t \varphi^\alpha, \nabla \varphi^\alpha). \tag{20.1}$$

We shall restrict our attention to the DT analogue of flat spacetime, i.e., Minkowski spacetime, in most of the following chapters. The development of DT field theories over curved spacetimes undoubtedly touches upon interesting and fundamental problems, some of which have not been solved yet in CT. For instance, some general relativistic spacetimes such as Gödel's famous spacetime (Gödel, 1949) contain closed timelike curves. The existence of such curves appears to be inconsistent with the probabilistic concepts of conventional quantum mechanics. It is not surprising therefore that conventional CT QFT has not been extended properly to such situations, assuming such an idea can make sense. Our intuition is that a better understanding of the observer concept would be needed in order to resolve this conundrum, most probably by modifying the rules of general relativity, which after all, came down to us from a pre-quantum view of the Universe.

Another deep issue and unsolved problem in curved spacetime is the particle concept itself. Relativistic QFT provides a good framework for high-energy particle physics in Minkowski spacetime, but only because the issue of locality, or the need to localize particle detection, can be swept under the theoretical rug of the asymptotics of the LSZ reduction formalism (Lehmann *et al.*, 1955; Bjorken and Drell, 1965). The extension of such concepts to curved spacetimes awaits further development (Colosi and Rovelli, 2009).

20.3 System functions for node variables

We shall construct system function densities $\mathcal{F}^n$ from CT Lagrange densities using the virtual-path principles outlined in earlier chapters. The choice of virtual path is based on the assumption that we start from Minkowski spacetime and are discretizing in a preferred inertial frame.

We shall first derive the DT equations of motion for *node* variables, i.e., for dynamical degrees of freedom associated with single values of time t_n. We shall deal with *link variables*, which are associated with *two* successive times, later on in this chapter. Link-variable fields are naturally associated with gauge bosonic degrees of freedom, whereas node-variable fields are naturally associated with matter fields, such as leptons. Indeed, one of the interesting and unexpected consequences of temporal discretization is that such a distinction between matter and gauge fields arises relatively naturally.

As we discussed in Chapter 14, it is possible to use Type-2 temporal lattices, but we shall restrict our attention to Type 1, i.e., the regular case $T_n \equiv \mathbb{D}_n t_n = T$, where T is a constant. The reason for this choice is that Type-1 discretization treats time homogeneously, which is how it is treated in standard inertial frames, and we have no reason at this stage to impose Type-2 discretization.

Given that $\varphi^\alpha(t, \mathbf{x})$ is a dynamical variable depending on CT, we define a virtual path $\varphi^\alpha_{n\lambda}$ from $\varphi^\alpha_n(\mathbf{x}) \equiv \varphi^\alpha(t_n, \mathbf{x})$ to $\varphi^\alpha_{n+1}(\mathbf{x}) \equiv \varphi^\alpha(t_{n+1}, \mathbf{x})$ in φ-value space by the linear rule

$$\varphi^\alpha_{n\lambda} \equiv \lambda \varphi^\alpha_{n+1}(\mathbf{x}) + \bar{\lambda} \varphi^\alpha_n(\mathbf{x}), \tag{20.2}$$

where λ runs from 0 to 1 and $\bar{\lambda} \equiv 1 - \lambda$. This choice of virtual path is based on the metric structure of Minkowski spacetime and the use of standard Cartesian coordinates.

We shall find that the above prescription is modified when we include gauge invariance, discussed in Chapter 24. In that situation, the field index α changes as we move from time t_n to t_{n+1} along a virtual path, in such a way as to preserve the gauge invariance of the theory.

With the above virtual-path prescription, time derivatives are naturally turned into finite differences, i.e.,

$$\frac{\partial}{\partial t} \varphi \to \frac{1}{T} \frac{\partial}{\partial \lambda} \varphi^\alpha_{n\lambda} \equiv \frac{\mathbb{D}_n \varphi^\alpha_n}{T}. \tag{20.3}$$

The next step is to define the virtual-path Lagrange density $\mathcal{L}_{n\lambda}$ by the replacements

$$\varphi^\alpha \to \varphi^\alpha_{n\lambda}, \quad \partial_t \varphi^\alpha \to T^{-1}\, \partial_\lambda \varphi^\alpha_{n\lambda}, \quad \nabla \varphi^\alpha \to \nabla \varphi^\alpha_{n\lambda} \tag{20.4}$$

in the CT Lagrange density, i.e.,

$$\mathcal{L}(\varphi^\alpha, \partial_t \varphi^\alpha, \nabla \varphi^\alpha) \to \mathcal{L}_{n\lambda} \equiv \mathcal{L}(\varphi^\alpha_{n\lambda}, T^{-1}\, \partial_\lambda \varphi^\alpha_{n\lambda}, \nabla \varphi^\alpha_{n\lambda}). \tag{20.5}$$

Then the system function density is the integral

$$\mathcal{F}^n \equiv T \langle \mathcal{L}_{n\lambda} \rangle = T \int_0^1 \mathcal{L}_{n\lambda}\, d\lambda. \tag{20.6}$$

The system function $F^n[\boldsymbol{\varphi}]$ is the spatial integral of the system function density and is a functional of the field values $\varphi^\alpha_n(\mathbf{x})$, i.e.,

$$F^n[\varphi] = \int \mathcal{F}^n \, d^3\mathbf{x}. \tag{20.7}$$

As in particle mechanics, the action sum $A_{MN}[\varphi]$ is a functional of the dynamical degrees of freedom over a given path in configuration space, given by the sum

$$A_{MN}[\varphi] = \sum_{n=M}^{N-1} F^n[\varphi]. \tag{20.8}$$

20.4 Equations of motion for node variables

The equations of motion for node variables are obtained from (20.8) using the Weiss action principle (Weiss, 1936) suitably modified for DT field theory. Given a path $[\varphi]$ in field configuration space from time M to time $N > M$, consider a variation of path of the form

$$\varphi_n^\alpha \to \varphi_n^\alpha + \varepsilon u_n^\alpha, \quad M \leqslant n \leqslant N, \tag{20.9}$$

where ε is a positive parameter that will be taken to zero eventually and $[u] \equiv \{u_n^\alpha : M \leqslant n \leqslant N\}$ is a sequence of well-behaved field variations that need not vanish at either of the end times M and N.

Now define the variation $\delta_u A_{MN}[\varphi]$ of the action sum with respect to $[u]$ by

$$\delta_u A_{MN}[\varphi] \equiv A_{MN}[\varphi + \varepsilon u] - A_{MN}[\varphi]. \tag{20.10}$$

Because the original CT Lagrange density $\mathcal{L}$ is assumed analytic in the fields and their spatial derivatives, the same is true for the system function, so we can make a Taylor expansion in the parameter ε, giving

$$\begin{aligned}\delta_{\varepsilon u} A_{MN}[\varphi] = \varepsilon \sum_{n=M}^{N-1} \int \left\{ u_n^\alpha(\mathbf{y}) \cdot \frac{\delta}{\delta\varphi_n^\alpha(\mathbf{y})} + u_{n+1}^\alpha(\mathbf{y}) \cdot \frac{\delta}{\delta\varphi_{n+1}^\alpha(\mathbf{y})} \right\} F^n[\varphi] d^3\mathbf{y} \\ + O(\varepsilon^2),\end{aligned} \tag{20.11}$$

where $\delta/\delta\varphi_n$ denotes functional differentiation with respect to φ_n. The directional functional derivative $\mathcal{D}_u A_{MN}[\varphi]$ at field configuration $[\varphi]$ and in the direction of $[u]$ is then defined as the limit

$$\mathcal{D}_u A_{MN}[\varphi] = \lim_{\varepsilon \to 0} \frac{\delta_{\varepsilon u} A_{MN}[\varphi]}{\varepsilon}, \tag{20.12}$$

which is assumed to exist. With some rearrangement, we find

$$\begin{aligned}\mathcal{D}_u A_{MN}[\varphi] = \int d^3\mathbf{x} \left\{ u_M^\alpha(\mathbf{x}) \cdot \frac{\delta}{\delta\varphi_M^\alpha(\mathbf{x})} F^M[\varphi] + u_N^\alpha(\mathbf{x}) \cdot \frac{\delta}{\delta\varphi_N^\alpha(\mathbf{x})} F^{N-1}[\varphi] \right\} \\ + \int d^3\mathbf{x} \sum_{n=M+1}^{N-1} u_n^\alpha(\mathbf{x}) \cdot \frac{\delta}{\delta\varphi_n^\alpha(\mathbf{x})} \{F^n + F^{n-1}\}.\end{aligned} \tag{20.13}$$

Now the Weiss action principle asserts that, for classical trajectories, i.e., those satisfying the equations of motion, the above directional functional derivative (20.13) depends only on the end-point contributions, for any choice of variation $[u]$, i.e.,

$$\mathcal{D}_u A_{MN}[\varphi] \underset{c}{=} \int d^3\mathbf{x} \left\{ u_M^\alpha(\mathbf{x}) \cdot \frac{\delta}{\delta\varphi_M^\alpha(\mathbf{x})} F^M[\varphi] + u_N^\alpha(\mathbf{x}) \cdot \frac{\delta}{\delta\varphi_N^\alpha(\mathbf{x})} F^{N-1}[\varphi] \right\}. \tag{20.14}$$

This means that

$$\int d^3\mathbf{x} \sum_{n=M+1}^{N-1} u_n^\alpha(\mathbf{x}) \cdot \frac{\delta}{\delta\varphi_n^\alpha(\mathbf{x})} \{F^n + F^{n-1}\} \underset{c}{=} 0 \tag{20.15}$$

for any $[u]$, which can be true only if

$$\frac{\delta}{\delta\varphi_n^\alpha(\mathbf{x})} \{F^n + F^{n-1}\} \underset{c}{=} 0, \qquad M < n < N. \tag{20.16}$$

This is the formal field theory analogue of Cadzow's equation of motion in point mechanics (8.26) for node variables.

We can go further because we know that the system function is related to the system function density by (20.7). We find

$$\begin{aligned} \frac{\delta}{\delta\varphi_n^\alpha(\mathbf{x})} \{F^n + F^{n-1}\} &= \int d^3\mathbf{y} \left\{ \frac{\delta\varphi^\beta(\mathbf{y})}{\delta\varphi_n^\alpha(\mathbf{x})} \frac{\partial}{\partial\varphi^\beta(\mathbf{y})} \right. \\ &\qquad \left. + \nabla_\mathbf{y} \left[\frac{\delta\varphi^\beta(\mathbf{y})}{\delta\varphi_n^\alpha(\mathbf{x})} \right] \cdot \frac{\partial}{\partial\nabla\varphi^\beta(\mathbf{y})} \right\} \\ &\quad \times \{\mathcal{F}^n + \mathcal{F}^{n-1}\} \\ &= \left\{ \frac{\partial}{\partial\varphi^\alpha(\mathbf{x})} - \nabla_\mathbf{x} \cdot \frac{\partial}{\partial\nabla\varphi^\alpha(\mathbf{x})} \right\} \{\mathcal{F}^n + \mathcal{F}^{n-1}\}, \end{aligned} \tag{20.17}$$

integrating by parts and discarding the surface terms at spatial infinity. Hence, using the above two results, we arrive at the most useful expression for our DT field equations of motion:

$$\frac{\partial}{\partial\varphi_n^\alpha} \{\mathcal{F}^n + \mathcal{F}^{n-1}\} \underset{c}{=} \nabla \cdot \frac{\partial}{\partial\nabla\varphi_n^\alpha} \{\mathcal{F}^n + \mathcal{F}^{n-1}\}, \quad M < n < N, \tag{20.18}$$

where the field index α ranges over all node field variables involved.

20.5 Exact and near symmetry invariants

The construction of constants of the motion associated with various symmetries is straightforward and follows the pattern used in CT field theory. Recall that there are two classes of transformations normally considered in mechanics: *active* and *passive*. Active transformations consist of actual changes in the dynamical degrees of freedom at fixed spacetime events or changes in dynamical degrees induced by real displacements of apparatus in spacetime, whereas passive transformations generally involve a relabelling of dynamical variables and/or the coordinate frame coordinates. Active transformations involve what physicists might actually do in a laboratory and have empirical content, whereas passive transformations are usually carried out for mathematical convenience or purposes of analysis, do not involve any actual change in the physical Universe, and therefore cannot have any empirical content.

Provided that we are clear about the nature of the transformation involved, we can regard an infinitesimal active or passive transformation T as a change in the dynamical field variables at a given event $(t_n, \mathbf{x})$ of the form

$$\varphi_n^\alpha(\mathbf{x}) \underset{T}{\rightarrow} \tilde{\varphi}_n^\alpha(\mathbf{x}) \equiv \varphi_n^\alpha(\mathbf{x}) + \varepsilon\, \delta\varphi_n^\alpha(\mathbf{x}), \tag{20.19}$$

where, as before, ε is an infinitesimal parameter that will be eventually taken to zero and $\delta\varphi_n^\alpha(\mathbf{x})$ is a function specific to the transformation T.

Given the above infinitesimal transformation (20.19), one of two things must happen: either the system function F^n is invariant with respect to the transformation or it is not. If it is invariant, then that is an example of an *exact symmetry*, i.e.,

$$\delta_{\varepsilon\delta\boldsymbol{\varphi}} A_{MN}[\boldsymbol{\varphi}] \equiv A_{MN}[\tilde{\boldsymbol{\varphi}}] - A_{MN}[\boldsymbol{\varphi}] = 0. \tag{20.20}$$

If it is not invariant, then suppose the transformation is a *near symmetry*, which is a transformation for which we may write

$$\delta_{\varepsilon\delta\boldsymbol{\varphi}} A_{MN}[\boldsymbol{\varphi}] \equiv A_{MN}[\tilde{\boldsymbol{\varphi}}] - A_{MN}[\boldsymbol{\varphi}] + O(\varepsilon^2). \tag{20.21}$$

Then, in the case of an exact symmetry or a near symmetry, we find

$$\mathcal{D}_{\delta\boldsymbol{\varphi}} A_{MN}[\boldsymbol{\varphi}] = \lim_{\varepsilon\rightarrow 0} \frac{\delta_{\varepsilon\mathbf{u}} A_{MN}[\boldsymbol{\varphi}]}{\varepsilon} = 0. \tag{20.22}$$

This result holds regardless of whether $[\boldsymbol{\varphi}]$ satisfies the equations of motion (20.18), the only input being that the transformation T is an exact or near symmetry of the system function.

If we now apply the Weiss action principle to the case of an exact or near symmetry transformation, we conclude from (20.22), (20.13) and (20.18) that the construction

$$C^n[\boldsymbol{\varphi}_c] \equiv \sum_\alpha \int d^3\mathbf{x} \left\{ \frac{\partial \mathcal{F}^n}{\partial \varphi_n^\alpha} - \nabla \cdot \frac{\partial \mathcal{F}^n}{\partial \nabla \varphi_n^\alpha} \right\} \delta\varphi_n^\alpha \tag{20.23}$$

is conserved over dynamical field trajectories, i.e.,

$$C^{n+1}[\boldsymbol{\varphi}_c] \underset{c}{=} C^n[\boldsymbol{\varphi}_c], \qquad M < n < N. \tag{20.24}$$

Here $\boldsymbol{\varphi}_c$ satisfies the classical equation of motion (20.18). The analogue of this result in CT mechanics is known as *Noether's theorem*, which states that there is a conserved quantity for every differentiable symmetry of the action integral of a physical system (Noether, 1918).

In the following sections we discuss two important exact symmetries associated with inertial frames: spatial translation invariance and spatial rotation invariance.

20.6 Linear momentum

Consider a transformation of the fields by an infinitesimal displacement in spatial position in the preferred frame, i.e., we imagine an extended-in-space field configuration being displaced without distortion by an infinitesimal amount $\varepsilon\,\delta\mathbf{a}$ from every point. Then this is equivalent to a change at a fixed point $\mathbf{x}$ of the field given by

$$\varphi_n(\mathbf{x}) \to \tilde{\varphi}_n(\mathbf{x}) \equiv \varphi_n(\mathbf{x}-\varepsilon\,\delta\mathbf{a}) = \varphi_n(\mathbf{x}) - \varepsilon\,\delta\mathbf{a}\cdot\nabla\varphi_n(\mathbf{x}) + O(\varepsilon^2). \qquad (20.25)$$

Suppose now that, in the analogue CT theory, the Lagrange density is not explicitly dependent on position. Then the same will be true of the system function density, and hence of the system function. Hence

$$\begin{aligned} F^n[\tilde{\varphi}] &\equiv \int d^3\mathbf{x}\; \mathcal{F}(\tilde{\varphi}_n(\mathbf{x}), \tilde{\varphi}_{n+1}(\mathbf{x}), \nabla\tilde{\varphi}_n(\mathbf{x}), \nabla\tilde{\varphi}_{n+1}(\mathbf{x})) \\ &= \int d^3\mathbf{x}\; \mathcal{F}(\varphi_n(\mathbf{x}-\varepsilon\,\delta\mathbf{a}), \varphi_{n+1}(\mathbf{x}-\varepsilon\,\delta\mathbf{a}), \nabla\varphi_n(\mathbf{x}-\varepsilon\,\delta\mathbf{a}), \\ &\qquad\qquad \nabla\varphi_{n+1}(\mathbf{x}-\varepsilon\,\delta\mathbf{a})) \\ &= \int d^3\mathbf{x}\; \mathcal{F}(\varphi_n(\mathbf{x}), \varphi_{n+1}(\mathbf{x}), \nabla\varphi_n(\mathbf{x}), \nabla\varphi_{n+1}(\mathbf{x})) \\ &= F^n[\varphi], \end{aligned} \qquad (20.26)$$

after a passive transformation of integration variables. Hence the above transformation is an exact symmetry of the action sum. The conserved quantity is known as the *linear momentum* $\mathbf{P}^n$ and given by the integral

$$\mathbf{P}^n \underset{\alpha}{\equiv} \int d^3\mathbf{x}(\nabla\varphi_n^\alpha)\left\{\frac{\partial\mathcal{F}^n}{\partial\varphi_n^\alpha} - \nabla\cdot\left(\frac{\partial\mathcal{F}^n}{\partial\nabla\varphi_n^\alpha}\right)\right\}. \qquad (20.27)$$

It satisfies the equation

$$\mathbf{P}^{n+1} \underset{c}{=} \mathbf{P}^n, \qquad M < n < N. \qquad (20.28)$$

The translation invariance discussed above is valid for Minkowski spacetime, which is flat, but it is believed that real spacetime will be curved. Therefore, the above symmetry arguments are valid only over some neighbourhood of the laboratory at the centre of the discussion, and then only when gravitational curvature is negligible.

20.7 Orbital angular momentum

Another fundamental symmetry is rotational invariance. Consider an infinitesimal rotation R of spatial coordinate axes about some axis, such that the coordinates change according to the rule

$$\mathbf{x} \underset{R}{\rightarrow} \mathbf{x}' \equiv \mathbf{x} - \varepsilon \boldsymbol{\omega} \times \mathbf{x}, \tag{20.29}$$

where ε is infinitesimal and $\boldsymbol{\omega}$ is an arbitrary vector parallel to the chosen axis of rotation. Then the dynamical fields will appear to transform according to the rule

$$\varphi_n^\alpha(\mathbf{x}) \rightarrow \tilde{\varphi}_n^\alpha(\mathbf{x}) \equiv \varphi_n^\alpha(\mathbf{x}) + \varepsilon(\boldsymbol{\omega} \times x) \cdot \nabla \varphi_n^\alpha(\mathbf{x}) + \varepsilon \boldsymbol{\omega} \cdot \mathbf{s}^{\alpha\beta} \varphi_n^\beta(\mathbf{x}) + O(\varepsilon^2), \tag{20.30}$$

where $\mathbf{s}^{\alpha\beta}$ is the transformation matrix involved in changes in any spin indices that the field may have. The assumption that the system is isotropic then leads to the conserved angular momentum

$$\mathbf{L}^n \equiv \sum_\alpha \int d^3\mathbf{x} \left(\boldsymbol{x} \times \nabla \varphi_n^\alpha + \mathbf{s}^{\alpha\beta} \varphi_n^\beta \right) \left\{ \frac{\partial \mathcal{F}^n}{\partial \varphi_n^\alpha} - \nabla \cdot \left(\frac{\partial \mathcal{F}^n}{\partial \nabla \varphi_n^\alpha} \right) \right\}. \tag{20.31}$$

Other conserved quantities such as electric charge are just as readily obtained by the same method.

20.8 Link variables

The above discussion has focussed on node variables, but, as we shall see in the chapter on Maxwell's equations, it is possible to encounter dynamical link variables, i.e., variables associated with the temporal link between times n and $n+1$, rather than the nodes or end points of the link.

In such circumstances, the CT Lagrangian densities involved do not have any time derivatives of the corresponding CT fields. This happens in electrodynamics when we use the four-vector electromagnetic potential A^μ. Because the spacetime derivatives of this potential occur only in the special combination given by the components of the Faraday tensor, i.e.,

$$F^{\mu\nu} \equiv \partial^\mu A^\nu - \partial^\nu A^\mu, \tag{20.32}$$

it is found that the time derivative of A^0 is absent from the Lagrange density. This is the case relative to any standard inertial frame and requires the use of the Dirac–Bargmann theory of constraints in order to develop a Hamiltonian formulation of the theory (Dirac, 1964).

Denoting link variables generically by ϕ^A, the Lagrange densities involved have the form

$$\mathcal{L} \equiv \mathcal{L}(\phi^A, \nabla \phi^A, \ldots) \tag{20.33}$$

and then the system functions are of the form

$$\mathcal{F}^n = T\langle \mathcal{L}(\phi_n^A, \nabla\phi_n^A, \ldots)\rangle. \tag{20.34}$$

Following the same approach to variations as with the node variables, we arrive at the corresponding DT equations of motion for link field variables:

$$\frac{\partial}{\partial \phi_n^A}\mathcal{F}^n \underset{c}{=} \nabla \cdot \frac{\partial}{\partial \nabla \phi_n^A}\mathcal{F}^n. \tag{20.35}$$

21
The discrete time Schrödinger equation

21.1 Introduction

In this chapter we apply the discrete time (DT) field theory formalism discussed in the previous chapter to non-relativistic Schrödinger wave mechanics. Temporal discretization is fully compatible with and can be readily applied to non-relativistic partial differential equations, because in the non-relativistic domain there are no issues involving simultaneity or the choice of preferred inertial frame.

The Schrödinger wave equation for a single-particle non-relativistic wavefunction $\Psi(t, \mathbf{x})$ is a classical partial differential field equation that can be derived from a classical field-theoretic approach using a suitable Lagrangian density that is based on Ψ and its complex conjugate Ψ^* treated as dynamical field variables. To show this, consider the continuous time (CT) Lagrangian density

$$\mathcal{L} = \frac{1}{2} i\hbar(\Psi^* \overrightarrow{\partial_t} \Psi - \Psi^* \overleftarrow{\partial_t} \Psi) - \frac{\hbar^2}{2m} \Psi^* \overleftarrow{\nabla} \cdot \overrightarrow{\nabla} \Psi - V(\mathbf{x}, t) \Psi^* \Psi, \tag{21.1}$$

where $V(\mathbf{x}, t)$ is some external potential, possibly of electromagnetic origin, but not necessarily so. We shall see in a later chapter how to construct a gauge-invariant Schrödinger equation for a non-relativistic charged particle interacting with external electromagnetic fields.

The Euler–Lagrange field equation of motion

$$\partial_t \frac{\partial \mathcal{L}}{\partial \partial_t \Psi^*} + \nabla \cdot \frac{\partial \mathcal{L}}{\partial \nabla \Psi^*} \underset{c}{=} \frac{\partial \mathcal{L}}{\partial \Psi^*} \tag{21.2}$$

then gives the time-dependent Schrödinger equation

$$i\hbar \, \partial_t \Psi \underset{c}{=} \overrightarrow{H} \Psi, \tag{21.3}$$

where $\overrightarrow{H} \equiv -(\hbar^2 \nabla^2/(2m)) + V(\mathbf{x}, t)$, and similarly for the complex-conjugate wavefunction.

We may construct a system function density from the above Lagrange density using the virtual-path method described in the previous chapter. The Schrödinger field is taken to be a node variable because it is associated with electrons, which are associated with matter fields, and these tend to be node variables. We first define the following virtual paths in wavefunction space:

$$\begin{aligned} \Psi(t,\mathbf{x}) &\rightarrow \Psi_{n\lambda}(\mathbf{x}) \equiv \lambda\Psi_{n+1}(\mathbf{x}) + \bar{\lambda}\Psi_n(\mathbf{x}), \\ \Psi(t,\mathbf{x})^* &\rightarrow \Psi^*_{n\lambda}(\mathbf{x}) \equiv \lambda\Psi^*_{n+1}(\mathbf{x}) + \bar{\lambda}\Psi^*_n(\mathbf{x}), \\ V(t,\mathbf{x}) &\rightarrow V_{n\lambda}(\mathbf{x}) \equiv V_n(\mathbf{x}), \end{aligned} \tag{21.4}$$

where $\Psi_n(\mathbf{x}) \equiv \Psi(t_n, \mathbf{x})$, $t_{n+1} = t_n + T$, $0 \leq \lambda \leq 1$ and we assume for convenience here that the potential V is a link variable. This assumption is consistent with the potential being of electromagnetic origin, because when we come to discuss Maxwell's equations in Chapter 24 we shall find that the electromagnetic scalar potential is a link variable rather than a node variable.

We commented on the question of a virtual path for interaction terms in Section 17.2.1, and this issue surfaces here. If we took the Schrödinger equation as fundamental, i.e., not derived from a Lagrangian as if the wavefunction were a dynamical degree of freedom, then the potential term in (21.3) should properly be discretized according to the rule $V(\mathbf{x},t) \rightarrow V(\mathbf{x}_{n\lambda}, t_{n\lambda})$. This is because it is then the position $\mathbf{x}$ which is regarded as a quantized degree of freedom. On the other hand, if the wavefunction is regarded as a dynamical field derived from field equations of motion, as in this chapter, then the potential term should be discretized according to whether it is a node variable, a link variable, or some variant, such as an external potential. In this latter scenario, the coordinates $\mathbf{x}$ are merely field labels and have no dynamical role per se.[1]

The system function density is given by (Jaroszkiewicz and Norton, 1997b) $\mathcal{F}^n \equiv T\langle \mathcal{L}_{n\lambda}\rangle$, where

$$\mathcal{L}_{n\lambda} \equiv \mathcal{L}(\Psi_{n\lambda}, \Psi^*_{n\lambda}, \nabla\Psi_{n\lambda}, \nabla\Psi^*_{n\lambda}, T^{-1}\,\mathbb{D}_n\Psi_n, T^{-1}\,\mathbb{D}_n\Psi^*_n, V_n) \tag{21.5}$$

and the angular brackets denote integration over λ from 0 to 1, i.e.,

$$\langle \mathcal{L}_{n\lambda}\rangle \equiv \int_0^1 \mathcal{L}_{n\lambda}\, d\lambda. \tag{21.6}$$

It is frequently advantageous not to work out all the virtual-path integrals until they are explicitly needed. The system function density for this system is therefore best given in the form

$$\begin{aligned} \mathcal{F}^n = \frac{1}{2} i\hbar \langle \Psi^*_{n\lambda} \overrightarrow{\mathbb{D}}_n \Psi_n - \Psi^*_n \overleftarrow{\mathbb{D}}_n \Psi_{n\lambda}\rangle - \frac{\hbar^2 T}{2m}\langle \Psi^*_{n\lambda} \overleftarrow{\nabla}\cdot\overrightarrow{\nabla}\Psi_{n\lambda}\rangle \\ - T\langle V_n \Psi^*_{n\lambda}\Psi_{n\lambda}\rangle. \end{aligned} \tag{21.7}$$

[1] Unless we were digging deeper into the structure of space and time, along the lines taken by Snyder (1947a, 1947b), in which case all bets are off and spacetime coordinates are related to operators. This is discussed in Chapter 28.

To illustrate the complexity residing in such system functions, we can evaluate the λ integrals explicitly, making use of (12.33). This gives the system function density

$$\begin{aligned}\mathcal{F}^n =&\frac{i\hbar}{2}\{\Psi_n^*\Psi_{n+1} - \Psi_{n+1}^*\Psi_n\}\\ &-\frac{\hbar^2 T}{12m}\left\{2|\nabla\Psi_{n+1}|^2 + \Psi_{n+1}^*\overleftarrow{\nabla}\cdot\overrightarrow{\nabla}\Psi_n + \Psi_n^*\overleftarrow{\nabla}\cdot\overrightarrow{\nabla}\Psi_{n+1} + 2|\nabla\Psi_n|^2\right\}\\ &-\frac{T}{6}\left\{2|\Psi_{n+1}|^2 + \Psi_{n+1}^*\Psi_n + \Psi_n^*\Psi_{n+1} + 2|\Psi_n|^2\right\}V_n, \end{aligned} \tag{21.8}$$

which shows the relative complexity of DT field theory compared with its CT analogue. Part of the art of dealing with such equations is to know how to organize and hide the complex structures via an economical notation.

We shall restrict our attention in the remainder of this chapter to time-independent potentials. Using (20.18), the equation of motion for $\Psi_n(\mathbf{x})$ is then found to be

$$i\hbar\frac{\Psi_{n+1} - \Psi_{n-1}}{2T} \underset{c}{=} \overrightarrow{H}\frac{\Psi_{n+1} + 4\Psi_n + \Psi_{n-1}}{6}, \tag{21.9}$$

and similarly for the complex conjugate. Here $\overrightarrow{H} \equiv -(\hbar^2\,\nabla^2/(2m)) + V(\mathbf{x})$.

Unlike the non-relativistic CT Schrödinger equation, which is a first-order-in-time differential equation, the DT Schrödinger equation (21.9) is a *second-order*-in-time difference-differential equation. Unlike some discretizations that are based on forwards or backwards differences, which can lead to manifestly non-unitary evolution, as discussed for instance by Farias and Recami (2010), our approach automatically leads to symmetric discretization schemes.

An important point related to this is the question of conservation of probability. In our approach, there is an analogue of charge conservation, and this is what should be focussed on. What turns out to be important to the physical interpretation of the theory is how the taking of 'inner products' is defined. This is seen clearly in the CT Klein–Gordon equation, where the positive definite density $\rho \equiv \varphi^*\varphi$, assumed naively to be a 'probability' density, does not satisfy a continuity equation.

Our experience with the quantized field version of the DT Schrödinger equation and the others which we will discuss in the next few chapters tells us that the best way to discuss physics is in terms of particle state creation operators and their commutators, rather than trying to extract physics from the classical discretized wave equations.

Nevertheless, there is important mileage to be gained by exploring the properties of classical discretized wave equations, which is what we do for the Schrödinger equation in the next section.

21.2 Stationary states

We define a stationary state to be one for which the magnitude of its DT wavefunction at any point in physical space is constant over discrete time. To investigate such states, consider the ansatz

$$\Psi_n(\mathbf{x}) \equiv e^{-i\varpi n}\psi_\varpi(\mathbf{x}), \qquad \varpi \in \mathbb{R}, \tag{21.10}$$

where ϖ is called the *vibrancy*. This is a dimensionless real parameter that is assumed to be independent of time n and position $\mathbf{x}$. The reason for not calling it a frequency will emerge presently. The properties of vibrancy are contingent on the discretization procedure employed.

Given (21.10), the equation of motion (21.9) becomes

$$\frac{3\hbar}{T}\frac{\sin\varpi}{\cos\varpi + 2}\psi_\varpi(\mathbf{x}) = \overrightarrow{H}\psi_\varpi(\mathbf{x}). \tag{21.11}$$

Because this equation has the form of an eigenvalue equation for the operator $\overrightarrow{H}$, we *define* the DT energy $E(\varpi)$ to be the coefficient on the left-hand side of (21.11), i.e.,

$$E(\varpi) \equiv \frac{3\hbar}{T}\frac{\sin\varpi}{\cos\varpi + 2}. \tag{21.12}$$

To relate the DT energy to more familiar concepts, consider the CT limit $T \to 0$, $\varpi \to \omega T$, where ω is fixed. Then $E(\varpi) \to \hbar\omega \equiv \varepsilon(\omega)$, which is interpreted in CT quantum mechanics (QM) as the energy $\varepsilon(\omega)$ of a quantum associated with angular frequency ω. Taking the limit $n \to \infty$, $nT \to t$, where t is fixed, then turns the ansatz (21.43) into the familiar energy eigenfunction statement

$$\Psi(t, \mathbf{x}) = e^{-i\varepsilon(\omega)t/\hbar}\psi_\omega(\mathbf{x}). \tag{21.13}$$

We shall find it convenient to work with the dimensionless *reduced energy* $\kappa(\varpi)$, defined by

$$\kappa(\varpi) \equiv \frac{E(\varpi)T}{\hbar} = \frac{3\sin\varpi}{\cos\,\varpi + 2}. \tag{21.14}$$

A plot of reduced energy versus vibrancy, Figure 21.1, shows that the DT energy $E(\varpi)$ is positive only for vibrancy in the range $0 < \varpi < \pi$, which therefore defines the *physical region* for vibrancy for such stationary solutions when we deal with positive Hamiltonians, such as in the case of free particles and the harmonic oscillator. Moreover, we do not need to consider vibrancies that differ from the physical values by multiples of 2π, since these represent no new physics. The same view is taken in standard Schrödinger wave mechanics, where it is recognized that the important quantity is an equivalence class of functions.

From Figure 21.1 we see that there is a maximum of the reduced energy at a value of the vibrancy ϖ_0 such that $\cos\,\varpi_0 = -\frac{1}{2}$, and then $\kappa(\varpi_0) = \sqrt{3}$. Hence the maximum possible DT energy for such a system is given by $E(\varpi_0) = \sqrt{3}\hbar/T$. We saw in our discussion of the physics of temporal discreteness in Chapter 2

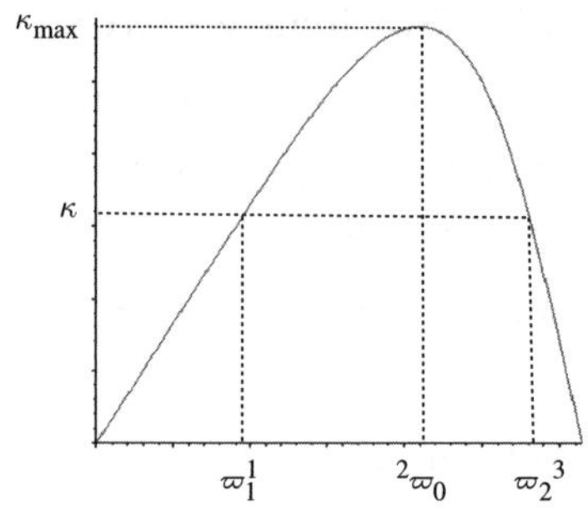

Figure 21.1. The reduced energy–vibrancy diagram.

that the chronon, T, if it exists, must be at least on the grand unified theory scale of time, of the order of 10^{-36} s or less. Therefore, the above upper bound for the DT energy is far beyond anything that could be accessed in non-relativistic physics.

A significant feature evident from Figure 21.1 is that, in the physical vibrancy region, a given value of energy corresponds to *two* distinct vibrancies, ϖ_1 and ϖ_2. The smaller value, ϖ_1, corresponds to *ordinary* particle solutions, whilst the greater value, ϖ_2, corresponds to *oscillon* solutions. The latter solutions, the oscillons, are best described by the statement that they change sign more or less once per chronon, with oscillons possessing the lowest reduced energy demonstrating this behaviour the most. At higher energies, as the reduced energy approaches what will turn out to be the *parabolic barrier*, the ordinary and oscillon vibrancies approach ϖ_0 from either direction, at which point the distinction between ordinary particles and oscillons becomes blurred.

An interesting question is whether every eigenfunction of the CT energy eigenstate equation

$$E\Psi(\mathbf{x}) = \overrightarrow{H}\Psi(\mathbf{x}) \tag{21.15}$$

represents a stationary DT state of the form (21.10). The answer is no, if we require the vibrancy ϖ to be real. For energy eigenvalues greater than $\sqrt{3}\hbar/T$, the associated vibrancy develops an imaginary component, giving DT wavefunctions that grow or decay exponentially, and such wavefunctions cannot be regarded as being stationary. An analogous cutoff phenomenon will be seen when we discuss the Klein–Gordon equation in Chapter 22.

21.2.1 Momentum eigenstates

We now consider the free-particle system, for which the external potential V is zero. For momentum eigenstate solutions, which are stationary, we write

$$\Psi_{n,\mathbf{p}}(\mathbf{x}) \equiv e^{-i\varpi_{\mathbf{p}}n} e^{i\mathbf{p}\cdot\mathbf{x}/\hbar}, \tag{21.16}$$

which gives the DT energy

$$E(\varpi_{\mathbf{p}}) = \frac{\mathbf{p}\cdot\mathbf{p}}{2m}. \tag{21.17}$$

From this and the upper bound $\sqrt{3}\hbar/T$ for the DT energy, we deduce that the maximum magnitude $|\mathbf{p}|_{\max}$ of the momentum for particle states with sharp momentum is given by

$$|\mathbf{p}|^2_{\max} = 2mE_{\max} = \frac{2\sqrt{3}m\hbar}{T}. \tag{21.18}$$

Although this is far above any accessible momentum in the laboratory, this result means that there is a natural cutoff in both energy and momentum in the version of DT mechanics being discussed here.

21.2.2 Relation with the discrete time oscillator

Another way of obtaining the above results is to take the spatial Fourier transform of the wavefunction, defining

$$\tilde{\Psi}_n(\mathbf{p}) \equiv \int d^3\mathbf{x}\, e^{i\mathbf{p}\cdot\mathbf{x}/\hbar}\Psi_n(\mathbf{x}). \tag{21.19}$$

This gives the free-particle equation of motion in momentum space

$$\frac{i\hbar}{2T}(\tilde{\Psi}_{n+1} - \tilde{\Psi}_{n-1}) \underset{c}{=} \frac{\mathbf{p}^2}{2m}\frac{\tilde{\Psi}_{n+1} + 4\tilde{\Psi}_n + \tilde{\Psi}_{n-1}}{6}, \tag{21.20}$$

which reduces to

$$(3 + i\kappa)\tilde{\Psi}_{n+1} \underset{c}{=} -i4\kappa\tilde{\Psi}_n + (3 - i\kappa)\tilde{\Psi}_{n-1} = 0, \tag{21.21}$$

where $\kappa \equiv \mathbf{p}^2 T/(2m\hbar)$. If now we define the *skew angle* λ by

$$\sin\lambda \equiv \frac{\kappa}{\sqrt{9+\kappa^2}}, \quad \cos\lambda \equiv \frac{3}{\sqrt{9+\kappa^2}} \tag{21.22}$$

then the equation of motion becomes

$$e^{i\lambda}\tilde{\Psi}_{n+1} \underset{c}{=} -\frac{i4\kappa}{\sqrt{9+\kappa^2}}\tilde{\Psi}_n + e^{-i\lambda}\tilde{\Psi}_{n-1}. \tag{21.23}$$

The appearance of the skew angle is a by-product, or artefact, of our discretization procedure when applied to first-order-in-time differential field equations. It does not appear with the DT Klein–Gordon equation but reappears in the DT Dirac equation.

We may reduce the equation of motion (21.23) to the DT oscillator by the ansatz

$$\tilde{\Psi}_n(\mathbf{p}) \equiv (-i)^n e^{-in\lambda}\phi_n(\mathbf{p}). \tag{21.24}$$

Then (21.23) becomes

$$\phi_{n+1} \underset{c}{=} \frac{4\kappa}{\sqrt{9+\kappa^2}}\phi_n - \phi_{n-1}, \tag{21.25}$$

which is precisely of the form of a DT harmonic oscillator discussed in Chapter 13, with the oscillator parameter η given by

$$\eta \equiv \cos\theta = \frac{2\kappa}{\sqrt{9+\kappa^2}} = 2\sin\lambda. \tag{21.26}$$

The parabolic barrier $\eta = 1$ is reached when κ reaches its maximum permitted physical value, $\kappa_{\max} = \sqrt{3}$, in agreement with the results of the previous section.

21.3 Vibrancy relations

We may invert (21.14) to give the vibrancies for a given value κ of the reduced energy. The ordinary solution vibrancy ϖ_1 is given by

$$\cos\varpi_1 = \frac{-2\kappa^2 + 3\sqrt{9-3\kappa^2}}{9+\kappa^2}, \qquad \sin\varpi_1 = \frac{k(6+\sqrt{9-3\kappa^2})}{9+\kappa^2}, \tag{21.27}$$

whilst for the oscillon vibrancy ϖ_2 we find

$$\cos\varpi_2 = \frac{-2\kappa^2 - 3\sqrt{9-3\kappa^2}}{9+\kappa^2}, \qquad \sin\varpi_2 = \frac{k(6-\sqrt{9-3\kappa^2})}{9+\kappa^2}. \tag{21.28}$$

The oscillator parameter η is related to the oscillator angle θ by $\eta = \cos\theta$, so we find

$$\cos\theta = \frac{2\kappa}{\sqrt{9+\kappa^2}}, \quad \sin\theta = \frac{\sqrt{9-3\kappa^2}}{\sqrt{9+\kappa^2}}. \tag{21.29}$$

On taking (21.22), (21.27), (21.28) and (21.29) we find that the ordinary and oscillon vibrancies are related to the oscillator and skew angles by

$$\varpi_1 = \frac{\pi}{2} + \lambda - \theta, \quad \varpi_2 = \frac{\pi}{2} + \lambda + \theta. \tag{21.30}$$

21.4 Linear independence and inner products

We have seen that, for a given reduced energy κ, there are two possible vibrancies in general. Consider the wavefunctions

$$\Psi_{n\mathbf{p}}(\mathbf{x}) \equiv e^{-i\varpi_1(\mathbf{p})n + i\mathbf{p}\cdot\mathbf{x}/\hbar}, \quad \Phi_{n\mathbf{p}}(\mathbf{x}) \equiv e^{-i\varpi_2(\mathbf{p})n + i\mathbf{p}\cdot\mathbf{x}/\hbar}, \tag{21.31}$$

where Ψ denotes an ordinary wavefunction and Φ denotes an oscillon wavefunction, and these two wavefunctions have the same reduced energy and momentum. These wavefunctions are improperly normalized wavefunctions, just as in CT Schrödinger wave mechanics. The question is, what is the relation between Ψ and Φ?

To examine this further, consider the standard equal-time 'inner product' $(\Phi_{n\mathbf{q}}, \Psi_{n\mathbf{p}})$ of Φ_n with Ψ_n constructed in the usual way by integrating $\Phi^*_{n\mathbf{q}}(\mathbf{x})\Psi_{n\mathbf{p}}(\mathbf{x})$ over physical space. We find

$$(\Phi_{n\mathbf{q}}, \Psi_{n\mathbf{p}}) \equiv \int d^3\mathbf{x}\ \Phi^*_{n\mathbf{p}}(\mathbf{x})\Psi_{n\mathbf{p}}(\mathbf{x}) = (2\pi\hbar)^3\delta^3(\mathbf{p}-\mathbf{q})e^{i2\theta(\mathbf{p})n}, \tag{21.32}$$

using (21.30). For low reduced energy, we find

$$(\Phi_{n\mathbf{q}}, \Psi_{n\mathbf{p}}) \approx (-1)^n(2\pi\hbar)^3\delta^3(\mathbf{p}-\mathbf{q}) + O(\kappa). \tag{21.33}$$

This demonstrates the relationship between ordinary solutions and oscillon solutions: the latter change sign once per chronon relative to the former, hence the nomenclature 'oscillon'.

It would be misleading to suggest that oscillon solutions vanish in the CT limit: they simply do not have a CT limit and are outside of the normal framework of the solutions to CT differential equations. They are a manifestation of the fact that DT mechanics is inherently richer than CT mechanics, and therefore the former should not be regarded as no more than an approximation to the latter. This in no way impinges on the question of which form of mechanics is most suited to describe the real world.

What this result implies is not only that the ordinary and oscillon solutions are linearly independent, but also that they are not manifestly orthogonal according to the above 'inner product'. Therefore, a more suitable 'inner product' should be considered. This is reminiscent of the situation encountered with the Klein–Gordon equation, where positive- and negative-energy solutions with the same linear momentum occur. In that situation, we define the 'inner product' (f, g) of two solutions f and g by

$$(f, g) \equiv i\int d^3\mathbf{x}\ f^*(x)\ \overleftrightarrow{\partial_0}\, g(x) \tag{21.34}$$

in standard relativistic notation (Gasiorowicz, 1967). Then, if that f is a positive-energy free-particle solution and g is a negative-energy free-particle solution, then their inner product as defined by (21.34) vanishes, because the sum of their energies vanishes.

We can devise an analogous construction in the DT Schrödinger case. The wave equation (21.9) can be written

$$\begin{aligned} \overrightarrow{\Omega}\Psi_{n+1} &\underset{c}{=} -4iT\ \overrightarrow{H}\Psi_n + \overrightarrow{\Omega}^\dagger\Psi_{n-1}, \\ \overrightarrow{\Omega}^\dagger\Psi^*_{n+1} &\underset{c}{=} 4iT\ \overrightarrow{H}\Psi^*_n + \overrightarrow{\Omega}\Psi^*_{n-1}, \end{aligned} \tag{21.35}$$

where $\overrightarrow{\Omega} \equiv 3\hbar + iT\ \overrightarrow{H}$. Now take any two solutions Ψ_n and Φ_n and consider the 'inner product'

$$\langle\Psi_n, \Phi_n\rangle \equiv \int d^3\mathbf{x}\left\{\Psi^*_n\ \overleftarrow{\Omega}^\dagger\Phi_{n-1} + \Psi^*_{n-1}\ \overrightarrow{\Omega}\Psi_n\right\}. \tag{21.36}$$

Then we find

$$\begin{aligned}\mathbb{U}_n\langle\Psi_n,\Phi_n\rangle &\equiv \int d^3\mathbf{x}\left\{\Psi^*_{n+1}\,\overleftarrow{\Omega}^\dagger\Phi_n+\Psi^*_n\,\overrightarrow{\Omega}\Psi_{n+1}\right\}\\ &=\int d^3\mathbf{x}\left\{\left[4iT\,\overrightarrow{H}\Psi^*_n+\overrightarrow{\Omega}\Psi^*_{n-1}\right]\Phi_n\right.\\ &\qquad\left.+\Psi^*_n\left[-4iT\,\overrightarrow{H}\Psi_n+\overrightarrow{\Omega}^\dagger\Psi_{n-1}\right]\right\}\\ &=\int d^3\mathbf{x}\left\{\Psi^*_{n-1}\,\overrightarrow{\Omega}\Phi_n+\overrightarrow{\Omega}^\dagger\Psi^*_n.\Psi_{n-1}\right\}=\langle\Psi_n,\Phi_n\rangle,\end{aligned} \tag{21.37}$$

so this inner product is an invariant of the dynamics.

Now take two free-particle solutions

$$\Psi_{n\mathbf{p}}(\mathbf{x})\equiv e^{-i\varpi_{\mathbf{p}}n+i\mathbf{p}\cdot\mathbf{x}/\hbar},\quad \Phi_{n\mathbf{q}}(\mathbf{x})\equiv e^{-i\varpi_{\mathbf{q}}n+i\mathbf{q}\cdot\mathbf{x}/\hbar}, \tag{21.38}$$

where the vibrancies $\varpi_{\mathbf{p}}$ and $\varpi_{\mathbf{q}}$ are either normal or oscillon. This means that we take $\overrightarrow{H}\equiv-\hbar^2\,\nabla^2/(2m)$.

Then we find the following cases.

1. For $\varpi_{\mathbf{p}}=\varpi_1(\mathbf{p})$ and $\varpi_{\mathbf{q}}=\varpi_1(\mathbf{q})$,

$$\langle\Psi_{n\mathbf{p}},\Phi_{n\mathbf{q}}\rangle=(2\pi\hbar)^3\delta^3(\mathbf{p}-\mathbf{q})2\sqrt{9-3\kappa^2_{\mathbf{p}}}. \tag{21.39}$$

2. For $\varpi_{\mathbf{p}}=\varpi_1(\mathbf{p})$ and $\varpi_{\mathbf{q}}=\varpi_2(\mathbf{q})$,

$$\langle\Psi_{n\mathbf{p}},\Phi_{n\mathbf{q}}\rangle=0. \tag{21.40}$$

3. For $\varpi_{\mathbf{p}}=\varpi_2(\mathbf{p})$ and $\varpi_{\mathbf{q}}=\varpi_2(\mathbf{q})$,

$$\langle\Psi_{n\mathbf{p}},\Phi_{n\mathbf{q}}\rangle=-(2\pi\hbar)^3\delta^3(\mathbf{p}-\mathbf{q})2\sqrt{9-3\kappa^2_{\mathbf{p}}}. \tag{21.41}$$

In (21.39) and (21.41) $\kappa_{\mathbf{p}}$ is the reduced energy associated with momentum $\mathbf{p}$. These 'inner products' suggest that normal and oscillon solutions are orthogonal relative to this inner product and have equal and opposite charge, an interpretation taken from the inner product (21.34).

21.5 Conservation of charge

To find a conserved charge we consider the global U(1) gauge transformation

$$\Psi_n\to e^{i\theta}\Psi_n,\quad \Psi^*_n\to e^{-i\theta}\Psi^*_n \tag{21.42}$$

and apply the Maeda–Noether theorem discussed above. This gives the invariant of the motion

$$Q^n\equiv\int d^3\mathbf{x}\;\Psi^*_n\left(\frac{1}{2}+\frac{iT}{6\hbar}\,\overrightarrow{H}\right)\Psi_{n+1}+\Psi^*_{n+1}\left(\frac{1}{2}-\overleftarrow{H}\frac{iT}{6\hbar}\right)\Psi_n. \tag{21.43}$$

Q^n reduces to the standard total probability in the limit $T \to 0$, if we are dealing with ordinary solutions to the DT equations. With the equations of motion (21.9) we find

$$Q^n \underset{c}{=} Q^{n-1}. \tag{21.44}$$

From the charge (21.43) we construct the charge and current densities, which are given by

$$\begin{aligned} \rho_n &= \Psi_n^* \left(\frac{1}{2} + \frac{iT}{6\hbar} \overrightarrow{H} \right) \Psi_{n+1} + \Psi_{n+1}^* \left(\frac{1}{2} - \overleftarrow{H} \frac{iT}{6\hbar} \right) \Psi_n, \\ \mathbf{j}_n &= -\frac{i\hbar}{12m} \{ \Psi_{n+1}^* \overleftrightarrow{\nabla} \Psi_n + 4\Psi_n^* \overleftrightarrow{\nabla} \Psi_n + \Psi_n^* \overleftrightarrow{\nabla} \Psi_{n-1} \}, \end{aligned} \tag{21.45}$$

which satisfy the DT equation of continuity

$$\bar{\mathbb{D}}_n \rho_n + T\, \nabla \cdot \mathbf{j}_n \underset{c}{=} 0. \tag{21.46}$$

Then, provided that the current $\mathbf{j}_n$ vanishes suitably at spatial infinity, integration over all space plus an application of the summation theorem (4.13) gives charge conservation, (21.44).

The density ρ_n above is not positive definite except in the limit $T \to 0$ taken for ordinary solutions, and then the usual positive density $\Psi^* \Psi$ is recovered. This is analogous to the situation in CT mechanics regarding the charge density for the charged Klein–Gordon equation, before the non-relativistic limit $c \to \infty$ is taken. One of the reasons occasionally cited for Schrödinger's rejection of a relativistic wave equation in 1925–6 is because of this point: there was no evidence at that time for positive electron (i.e., positronic) charge densities. We imagine that if DT mechanics had been the accepted classical mechanical paradigm prior to wave mechanics then the Born probability interpretation of Schrödinger wave mechanics might not have been proposed.

If we wish to couple electromagnetic fields to the Schrödinger equation in a gauge-invariant way then we need to choose a modified virtual path. Such a path is used for the charged Klein–Gordon equation discussed in Chapter 24.

22
The discrete time Klein–Gordon equation

22.1 Introduction

When, in 1926, Schrödinger came to publish his non-relativistic wave equation (Schrödinger, 1926), namely the equation we discretized in the previous chapter, he had previously considered a special-relativistic wave equation, but discarded it on account of some of its properties, which he believed were unphysical. In particular, that relativistic equation has a conserved current density that cannot be interpreted as a classical probability current density because it can take on negative values, something that a true probability density would not do. If Schrödinger had been aware of antiparticles, which were discovered several years later, it is conceivable that he would have persisted with that relativistic wave equation. The equation he discarded is sometimes referred to as the *Schrödinger–Fock–Klein–Gordon* equation, but more commonly is known as just the *Klein–Gordon* (K–G) equation.

The significance of Schrödinger's rejection of the K–G equation lies in his decision to forgo Lorentz covariance in favour of an intuitive, albeit non-relativistic, interpretation of the wave equation that now bears his name. Schrödinger's non-relativistic wave equation was extraordinarily successful when applied to atomic physics, which greatly contributed to the rise of quantum physics.

Within two years, however, the situation was dramatically restored in favour of special relativity. In 1928 Dirac published his famous Lorentz covariant wave equation for the electron (Dirac, 1928). Not much later, the equation that Schrödinger had rejected, the K–G equation, was successfully reinterpreted as a relativistic quantum field theory with a completely satisfactory interpretation of positive-energy particles as antiparticles.

In this chapter we shall consider the temporal discretization of the K–G equation for a system of spinless particles of non-zero mass m and electric charge zero. This is almost the simplest interesting relativistic quantum field system possible (the case $m = 0$ must qualify as being the simplest), but it is not completely trivial. The quantized theory is fully Lorentz-covariant and describes states with arbitrary numbers of particles, each particle of which can carry quanta of energy and momentum. This system will serve us as a test bed for the temporal discretization of more complicated field theories.

We start off by defining our conventions and notation. In anticipation of temporal discretization, we will work in a preferred inertial frame $\mathcal{F}$ in four-dimensional Minkowski spacetime, with standard Cartesian coordinates $x^\mu \equiv (t, \mathbf{x})$. For convenience we shall use the natural unit system, in which $c = \hbar = 1$. The matrix $[\eta_{\mu\nu}]$ of metric tensor components will have non-zero components down the main diagonal only, with values $(1, -1, -1, -1)$.

The CT K–G equation is based on a real scalar field $\varphi(t, \mathbf{x})$ with the Lagrange density

$$\mathcal{L} \equiv \frac{1}{2}\, \partial_\mu \varphi\, \partial^\mu \varphi - \frac{1}{2}\mu^2 \varphi^2, \tag{22.1}$$

where μ is the rest mass of the corresponding particles. The equation of motion derived from (22.1) via the Weiss action principle (Weiss, 1936) is given by

$$(\Box + \mu^2)\varphi \underset{c}{=} 0. \tag{22.2}$$

As with all our equations of motion, we discretize at the Lagrange density level rather than attempting to discretize the equation of motion (22.2) directly. We shall employ a Type-1, regular, temporal discretization, taking $t_n \equiv nT$, where the chronon T is positive and independent of n and spatial position. The field is taken to be a node field, so we define

$$\varphi_n(\mathbf{x}) \equiv \varphi(t_n, \mathbf{x}). \tag{22.3}$$

For the free-particle system given by (22.1) the virtual paths are chosen to be

$$\varphi_{n\lambda}(\mathbf{x}) \equiv \lambda \varphi_{n+1}(\mathbf{x}) + \bar{\lambda}\varphi_n(\mathbf{x}), \quad 0 \leq \lambda \leq 1, \tag{22.4}$$

where $\bar{\lambda} = 1 - \lambda$. With this choice of virtual path the time derivative ∂_t is replaced by the operator $T^{-1}\, \mathbb{D}_n$, which gives

$$T^{-1}\, \mathbb{D}_n \varphi_n(\mathbf{x}) = \frac{\varphi_{n+1}(\mathbf{x}) - \varphi_n(\mathbf{x})}{T}. \tag{22.5}$$

The system function $F^n \equiv F[\varphi_n, \varphi_{n+1}]$ is the spatial integral

$$F^n \equiv \int d^3\mathbf{x}\, \mathcal{F}^n(\mathbf{x}) \tag{22.6}$$

of the system function density $\mathcal{F}^n(\mathbf{x}) \equiv T\langle \mathcal{L}_{n\lambda} \rangle$, where, as in previous chapters, the angular brackets denote integration over λ from zero to one and $\mathcal{L}_{n\lambda}$ is the Lagrange density (22.1) with the CT fields and field derivatives replaced by their virtual-path equivalents. Integrating over the virtual-path parameter λ gives the system function density

$$\mathcal{F}^n = \frac{(\varphi_{n+1} - \varphi_n)^2}{2T} - \frac{1}{2}T\langle \nabla\varphi_{n\lambda} \cdot \nabla\varphi_{n\lambda} \rangle - \frac{1}{2}\mu^2 T \langle {\varphi_{n\lambda}}^2 \rangle. \tag{22.7}$$

Applying the DT field equations (20.18) then gives the equation of motion

$$\mathbb{D}_n \bar{\mathbb{D}}_n \varphi_n + T^2 \overrightarrow{K} \mathbb{S}_n \varphi_n \underset{c}{=} 0, \tag{22.8}$$

where $\overrightarrow{K} \equiv -\nabla^2 + \mu^2$ and the operators $\mathbb{D}$, $\bar{\mathbb{D}}$ and $\mathbb{S}$ are defined in Chapter 4.

The equation of motion (22.8) can also be written in the form

$$T\ \mathbb{S}_n \overrightarrow{D} \varphi_n \underset{c}{=} \varphi_n, \tag{22.9}$$

where the operator

$$\overrightarrow{D} \equiv \frac{6 + T^2\ \overrightarrow{K}}{6T} \tag{22.10}$$

turns out to be useful for proving the conservation of various important quantities. We shall use this operator in the next section to show that total linear momentum is conserved in this system.

22.2 Linear momentum

The Lagrange density (22.1) is explicitly independent of spatial position, so, by application of Noether's theorem, there is a conserved three-vector, the linear momentum. Using (20.27), we find

$$\begin{aligned} \mathbf{P}^n &\equiv \int d^3\mathbf{x}\ \varphi_n\ \overleftarrow{\nabla} \left\{ \frac{\partial \mathcal{F}^n}{\partial \varphi_n} - \nabla \cdot \frac{\partial \mathcal{L}^n}{\partial \nabla \varphi_n} \right\} \\ &= \int d^3\mathbf{x}\ \varphi_n\ \overleftarrow{\nabla} \left\{ \frac{\varphi_n - \varphi_{n+1}}{T} - \frac{T}{6}\ \overrightarrow{K}(\varphi_{n+1} + 2\varphi_n) \right\}. \end{aligned} \tag{22.11}$$

The trick here is to split up the second factor into a term involving φ_n and another involving φ_{n+1}, i.e., we write

$$\mathbf{P}^n \equiv \int d^3\mathbf{x}\ \varphi_n \overleftarrow{\nabla} \{A + B\}, \tag{22.12}$$

where

$$A \equiv \left(\frac{1}{T} - \frac{T}{3}\ \overrightarrow{K} \right) \varphi_n, \quad B \equiv -\overrightarrow{D} \varphi_{n+1}. \tag{22.13}$$

Then the term involving A vanishes, if the field tends to zero sufficiently rapidly at spatial infinity. To show this, we integrate by parts and discard surface terms:

$$\begin{aligned} \int d^3\mathbf{x}\ \varphi_n\ \overleftarrow{\nabla} A &\equiv \int d^3\mathbf{x}\ \varphi_n\ \overleftarrow{\nabla} \left\{ \frac{1}{T} - \frac{T}{3}\ \overrightarrow{K} \right\} \varphi_n \\ &= -\int d^3\mathbf{x}\ \varphi_n\ \overrightarrow{\nabla} \left\{ \frac{1}{T} - \frac{T}{3}\ \overrightarrow{K} \right\} \varphi_n \\ &= -\int d^3\mathbf{x}\ \varphi_n \left\{ \frac{1}{T} - \frac{T}{3}\ \overleftarrow{K} \right\} \overrightarrow{\nabla} \varphi_n = -\int d^3\mathbf{x}\ \nabla \varphi_n\ A, \end{aligned} \tag{22.14}$$

so the integral involving the A term must be zero. Hence we conclude that

$$\mathbf{P}^n = -\int d^3\mathbf{x}\ \varphi_n\ \overleftarrow{\nabla}\,\overrightarrow{D}\varphi_{n+1}. \tag{22.15}$$

It is useful to check such expressions for consistency with the CT analogue theory. In the CT limit (CTL), where we take $T \to 0$, $n \to \infty$, with $nT \equiv t$ fixed, assuming we can perform a Taylor expansion in the chronon and integrating by parts as before, we have

$$\begin{aligned}\mathbf{P} \equiv \lim_{\text{CTL}} \mathbf{P}^n &= -\lim_{\text{CTL}} \int d^3\mathbf{x}\ \varphi\ \overleftarrow{\nabla} \left\{ \frac{6 + T^2\ \overrightarrow{K}}{6T} \right\} \{\varphi + T\dot{\varphi} + O(T^2)\} \\ &= -\lim_{\text{CTL}} \int d^3\mathbf{x}\ \varphi\ \overleftarrow{\nabla}\dot{\varphi},\end{aligned} \tag{22.16}$$

which agrees precisely with the linear momentum given by Bjorken and Drell (1965).

Conservation of linear momentum is proved as follows:

$$\begin{aligned}\mathbf{P}^n &\equiv -\int d^3\mathbf{x}\ \varphi_n\ \overleftarrow{\nabla}\overrightarrow{D}\varphi_{n+1}. \\ &= -\int d^3\mathbf{x}\ \varphi_n\ \overleftarrow{\nabla} \left\{ -4\ \overrightarrow{D}\varphi_n - \overrightarrow{D}\varphi_{n-1} + \frac{6}{T}\varphi_n \right\} \qquad \text{using (22.9)} \\ &= \int d^3\mathbf{x}\ \varphi_n\ \overleftarrow{\nabla}\overrightarrow{D}\varphi_{n-1} = -\int d^3\mathbf{x}\ \varphi_{n-1}\ \overleftarrow{\nabla}\overrightarrow{D}\varphi_n \\ &\underset{c}{=} \mathbf{P}^{n-1},\end{aligned} \tag{22.17}$$

as expected.

22.3 Orbital angular momentum

Likewise, the orbital angular momentum is conserved and given by

$$\mathbf{L}^n = -\int d^3\mathbf{x}(\mathbf{x} \times \nabla\varphi_n)\overrightarrow{D}\varphi_{n+1}. \tag{22.18}$$

These invariants of the motion exist because we have not destroyed Euclidean invariance; that is, there is still spatial translational and rotational invariance in our approach. This will occur in all our DT field models that are based on special relativistic Lagrangians. We note that a similar equation was found for the scalar field in the κ-Poincaré theory of Lukierski *et al.* (1992), with the important differences that their lattice parameter corresponding to our T is imaginary and that they are dealing with CT throughout.

Particle-like solutions

We turn now to particle-like solutions to the equation of motion. Consider the Fourier transform

$$\tilde{\varphi}_n(\mathbf{p}) = \int d^3\mathbf{x}\, e^{-i\mathbf{p}\cdot\mathbf{x}}\varphi_n(\mathbf{x}). \tag{22.19}$$

The equation of motion (22.9) becomes

$$T\beta\, \mathbb{S}_n \tilde{\varphi}_n \underset{c}{=} \tilde{\varphi}_n, \tag{22.20}$$

where $\beta \equiv (6 + T^2E^2)/(6T)$ and $E \equiv \sqrt{\mu^2 + \mathbf{p}\cdot\mathbf{p}}$. Now define the momentum functions

$$\begin{aligned} a_n(\mathbf{p}) &= i\beta e^{in\theta}\left[\tilde{\varphi}_{n+1} - e^{i\theta}\tilde{\varphi}_n\right], \\ a_n^*(\mathbf{p}) &= -i\beta e^{-in\theta}\left[\tilde{\varphi}_{n+1}^* - e^{-i\theta}\tilde{\varphi}_n^*\right], \end{aligned} \tag{22.21}$$

where $\cos\theta = \eta$ and $\sin\theta = \sqrt{1-\eta^2}$, with η given by the ratio

$$\eta = \frac{\alpha}{\beta}, \qquad \alpha = \frac{6 - 2T^2E^2}{6T}. \tag{22.22}$$

Particle states in DT field theory correspond to elliptic-type wave behaviour, which requires $|\eta| < 1$. Otherwise, physically unacceptable hyperbolic behaviour occurs, as discussed by Jaroszkiewicz and Norton (1997a). This is equivalent to the condition

$$TE < 2\sqrt{3}, \tag{22.23}$$

i.e.,

$$c\sqrt{\mathbf{p}\cdot\mathbf{p} + \mu^2c^2} < \frac{2\sqrt{3}\hbar}{T} \tag{22.24}$$

in SI units. The conclusion from this analysis is that energy and momentum are bounded from above for physical particle states in our approach. In DT field theory, the spectrum of acceptable particle states has a natural cutoff that can be made as large as necessary to avoid a violation of observed particle data by choosing the time interval T small enough.

An interesting point is that, for a given T, there can be no scalar particle species with rest mass greater than $2\sqrt{3}\hbar/(Tc^2)$.

By construction, the momentum fields are time-independent and satisfy

$$a_n(\mathbf{p}) \underset{c}{=} a_{n-1}(\mathbf{p}), \quad a_n^*(\mathbf{p}) \underset{c}{=} a_{n-1}^*(\mathbf{p}), \tag{22.25}$$

which leads us to construct an invariant of the motion

$$C^n \equiv \int \frac{d^3\mathbf{p}}{(2\pi)^3}\theta\left(\frac{12}{T^2} - \mu^2 - \mathbf{p}\cdot\mathbf{p}\right) a_n^*(\mathbf{p}) a_n(\mathbf{p}). \tag{22.26}$$

In the limit $T \to 0$ this becomes the standard CT field theory Hamiltonian.

22.4 The free-charged Klein–Gordon equation

For this system we use the CT Lagrange density

$$\mathcal{L} = \partial_\mu\varphi^*\, \partial^\mu\varphi - \mu^2\varphi^*\varphi. \tag{22.27}$$

The virtual paths are given as before by

$$\begin{aligned}\tilde{\varphi} &\equiv \lambda\varphi_{n+1} + \bar{\lambda}\varphi_n,\\ \tilde{\varphi}^* &\equiv \lambda\varphi^*_{n+1} + \bar{\lambda}\varphi^*_n, \qquad 0 \leq \lambda \leq 1,\end{aligned} \tag{22.28}$$

where we suppress the dependence on $\mathbf{x}$. The system function density is

$$\begin{aligned}\mathcal{F}^n = \frac{|\varphi_{n+1} - \varphi_n|^2}{T} &- \frac{T}{6}\{2|\nabla\varphi_{n+1}|^2 + 2\,|\nabla\varphi_n|^2 + \nabla\varphi^*_{n+1}\cdot\nabla\varphi_n\\ &\qquad + \nabla\varphi^*_n\cdot\nabla\varphi_{n+1}\}\\ -\frac{\mu^2 T}{6}&\{2|\varphi_{n+1}|^2 + 2\,|\varphi_n|^2 + \varphi^*_{n+1}\varphi_n + \varphi^*_n\varphi_{n+1}\},\end{aligned} \tag{22.29}$$

giving the equations of motion

$$\begin{aligned}\mathbb{T}_n\bar{\mathbb{T}}_n\varphi_n + \mathbb{S}_n\overrightarrow{K}\varphi_n &\underset{c}{=} 0,\\ \mathbb{T}_n\bar{\mathbb{T}}_n\varphi^*_n + \mathbb{S}_n\overrightarrow{K}\varphi^*_n &\underset{c}{=} 0,\end{aligned} \tag{22.30}$$

which can be written in the form

$$T\ \mathbb{S}_n\overrightarrow{D}\varphi_n \underset{c}{=} \varphi_n, \quad T\ \mathbb{S}_n\overrightarrow{D}\varphi^*_n \underset{c}{=} \varphi^*_n, \tag{22.31}$$

where $\overrightarrow{D}$ is given by (22.10).

The linear and angular momenta are easy to construct so we turn to the new feature, global gauge invariance. The system function density (22.29) is invariant with respect to a global gauge transformation of the fields, i.e.,

$$\begin{aligned}\varphi_n &\to \varphi'_n = e^{-i\chi}\varphi_n,\\ \varphi^*_n &\to \varphi^{*\prime}_n = e^{i\chi}\varphi^*_n,\end{aligned} \tag{22.32}$$

where χ is independent of time and space, so, using the DT Noether theorem discussed previously, we find the conserved charge

$$\begin{aligned}Q^n &\equiv i\int d^3\mathbf{x}\left\{\varphi^*_n\ \overrightarrow{D}\varphi_{n+1} - \varphi^*_{n+1}\ \overleftarrow{D}\ \varphi_n\right\}\\ &= i\int d^3\mathbf{x}\left\{\varphi^*_n\ \overleftarrow{D}\varphi_{n+1} - \varphi^*_{n+1}\ \overrightarrow{D}\varphi_n\right\}.\end{aligned} \tag{22.33}$$

This is real and global-gauge-invariant. Using the equations of motion, we find

$$Q^n \underset{c}{=} Q^{n-1}. \tag{22.34}$$

By inspection there are two possible candidates for a charge density, which are denoted by $\rho^{(-)}$ and $\rho^{(+)}$, given by

$$\begin{aligned}\rho^{(-)}_n &\equiv i\varphi^*_n\ \overrightarrow{D}\varphi_{n+1} - i\varphi^*_{n+1}\ \overleftarrow{D}\varphi_n,\\ \rho^{(+)}_n &\equiv i\varphi^*_n\ \overleftarrow{D}\varphi_{n+1} - i\varphi^*_{n+1}\ \overrightarrow{D}\varphi_n.\end{aligned} \tag{22.35}$$

These are related by a total divergence; that is,

$$\rho_n^{(+)} = \rho_n^{(-)} + \frac{iT}{6}\nabla\cdot\left[\varphi_n^* \overleftrightarrow{\nabla}\varphi_{n+1} + \varphi_{n+1}^* \overleftrightarrow{\nabla}\varphi_n\right], \tag{22.36}$$

so that they give the same total charge Q^n. Using the equations of motion, we find

$$\begin{aligned} \bar{\mathbb{T}}_n\rho_n^{(-)} + \nabla\cdot\mathbf{j}_n^{(-)} &\underset{c}{=} 0, \\ \bar{\mathbb{T}}_n\rho_n^{(+)} + \nabla\cdot\mathbf{j}_n^{(+)} &\underset{c}{=} 0, \end{aligned} \tag{22.37}$$

which are DT versions of the charge continuity equation, with corresponding charge currents given by

$$\begin{aligned} \mathbf{j}_n^{(-)} &= -\frac{i}{6}[\varphi_{n-1}^* \overleftrightarrow{\nabla}\varphi_n + \varphi_n^* \overleftrightarrow{\nabla}\varphi_{n-1} + 4\varphi_n^* \overleftrightarrow{\nabla}\varphi_n], \\ \mathbf{j}_n^{(+)} &= -\frac{i}{6}[\varphi_n^* \overleftrightarrow{\nabla}\varphi_{n+1} + \varphi_{n+1}^* \overleftrightarrow{\nabla}\varphi_n + 4\varphi_n^* \overleftrightarrow{\nabla}\varphi_n]. \end{aligned} \tag{22.38}$$

23

The discrete time Dirac equation

23.1 Introduction

In this chapter we apply temporal discretization to mechanical systems described by anticommuting variables rather than the commuting variables normally used in classical mechanics (CM). The anticommuting numbers representing such variables are called *Grassmann(ian) numbers* by mathematicians, after Hermann G. Grassmann (1809–1877), who did some pioneering work in linear algebra. On that account we shall refer to anticommuting numbers as *g-numbers*, in contrast to *c-numbers* (c for *commuting*[1]) when we refer to ordinary real or complex variables.

The significant difference between g-numbers and c-numbers is that, where as any two c-numbers x and y satisfy the commutative multiplication rule $xy = +yx$, any two g-numbers θ and ϕ satisfy the anticommutation rule $\theta\phi = -\phi\theta$. G-numbers and c-numbers can be multiplied together in any order, i.e., g-numbers and c-numbers commute. The product $z\theta$ of a c-number z and a g-number θ is a g-number.

The values of classical variables in ordinary CM are c-numbers, so we may refer to such variables as c-variables. It is possible to construct forms of CM where the classical variables are g-numbers, in which case we shall refer to such variables as g-variables.

G-variables should not be thought of as quantized versions of c-variables. They are just as classical as c-variables but much less familiar in terms of their applications to the real world. G-variables should not be confused with Dirac's *q-numbers* either. A q-number (q for *quantum*) was Dirac's name for a quantum operator, particularly any one of the four anticommuting symbols that occur in

[1] The c in *c-number* may be interpreted as *classical*, in which case real and complex numbers are examples of commuting c-numbers whilst Grassmann numbers are anticommuting c-numbers.

his famous equation for the spinning electron (Dirac, 1928). We shall refer to quantized g-variables as *fermions*, and to quantized c-variables as *bosons.*

G-variables need not be identified per se as matrices, although it is always possible to represent g-variables by matrices. This becomes particularly useful in the case of electron spin.

One of the curiosities of g-variables is that any g-variable always has a square equal to zero. This property, referred to as *nilpotency*, follows immediately from the anticommutation rule $\theta\phi = -\phi\theta$ on setting $\phi = \theta$.

G-variables can be multiplied by c-numbers without affecting their nilpotency or anticommutation properties. Nilpotency of a g-variable θ means that every function of it is always of the general form

$$f(\theta) = \alpha + \beta\theta, \tag{23.1}$$

where either α is a c-number and β is a g-number, in which case $f(\theta)$ is a c-number, or α is a g-number and β is a c-number, in which case $f(\theta)$ is a g-number. In general we will rule out mixed addition, i.e., we do not allow the sum of a c-number and a g-number in (23.1).

Because they can always be represented by matrices, g-variables have the associativity property of matrices: given any three g-variables α, β and γ, then $(\alpha\beta)\gamma = \alpha(\beta\gamma)$. Because the ordering of bracketing does not matter, we can drop the brackets and write $\alpha\beta\gamma$ to mean either $(\alpha\beta)\gamma$ or $\alpha(\beta\gamma)$.

The product $\alpha\beta$ of any two g-numbers α and β is always a c-number, a fact that allows us to construct quantum observables out of products of quantized g-variables. An example of such an observable is the electric charge current $j^\mu \equiv q\bar{\psi}\gamma^\mu\psi$ in quantum electrodynamics, where ψ and $\bar{\psi}$ are anticommuting quantized electron fields.

23.2 Grassmann variables in mechanics

The history of g-variable mechanics is almost the time reverse of what happened with c-variable mechanics, which was developed several centuries before c-variables were quantized (Dirac, 1925). Operators equivalent to quantized g-variables were postulated and applied to quantum mechanics (Jordan and Wigner, 1928; Candlin, 1956; Martin, 1959a, 1959b) several decades before g-variable CM was developed in the systematic form frequently referred to as *pseudomechanics* (Casalbuoni, 1976a).

The sequence of events is approximately as follows. In antiquity, c-variables in the form of real numbers found a natural application in classical geometry, being used to describe position and length. All measurable quantities are expressible in terms of real numbers, so it was natural to use them in Galilean and Newtonian mechanics to represent spatial coordinates and time. Then, as quantum mechanics started to be developed, *canonical quantization* was invoked, particularly by

Dirac (1925), as a recipe for developing a quantum analogue theory from a given classical mechanical model.

Canonical quantization follows a standard prescription: (i) postulate a Lagrangian; (ii) use it to develop Hamiltonian mechanics in classical phase space; and (iii) replace phase-space c-variables with quantum operators having *commutation* relations read off either from Poisson brackets (Dirac, 1925), or from Dirac brackets, in the case when second-class constraints occur (Dirac, 1964).

Quantum mechanics, as developed in its original form by Heisenberg (1925), Schrödinger (1926) and others, was based on the canonical quantization of c-variables. According to this form of mechanics, there would be three real numbers associated with each electron in any atomic state. Experiments, however, suggested that something was preventing the occurrence of many potential sets of numbers predicted by quantized c-number mechanics. To classify the observed states, Pauli proposed in 1925 the famous exclusion principle now named after him (Pauli, 1925, 1946). According to this principle, electrons are associated with an additional new type of variable, now known as *electron spin*, so that in fact each electron in a given atom should be associated with four numbers, not three. Pauli's principle was that no two electrons associated with the same state of a system under observation can have identical sets of four quantum numbers. This is a veto on electron degeneracy, i.e., the situation where two electrons have identical c-number properties.

It was not long before it was realized that Pauli's *ad hoc* veto on electron c-number degeneracy could be elegantly replaced in the theory by using g-variables for electronic degrees of freedom. Electron degeneracy is eliminated automatically in g-number mechanics by nilpotency.

This led to the far-reaching realization that quantum-mechanical degrees of freedom can be divided into two classes: *bosonic* and *fermionic*. Bosonic fields such as electromagnetic fields may be thought of as quantized c-numbers, whilst fermionic fields such as electron fields may be thought of as quantized g-numbers. The particles associated with quantized c-number fields are known as *bosons* on account of their multi-particle statistics, where as particles associated with quantized g-number fields are known as *fermions*, because their multi-particle states obey Fermi–Dirac (anticommuting) statistics (Fermi, 1926). Fermionic states have properties quite different from those of bosonic states. In particular, the fact that ordinary solid objects cannot penetrate each other is explained by the exclusion-principle properties of fermions.

Within a few years, quantum field theory had developed to the point where the exclusion principle had been fully incorporated into the formalism by the use of anticommuting quantum field operators.

Dirac used such objects, which he referred to as 'q-numbers' (quantum numbers), to explain electron spin (Dirac, 1928). By the time quantum field theory had developed, it had became routine to use anticommuting quantum fields to represent fermionic degrees of freedom.

Feynman initially developed the path-integral approach to standard quantum mechanics starting from quantized c-variables (Feynman, 1948; Feynman and Hibbs, 1965). At this point, g-variables entered the story. A number of theorists explored the use of g-variables in path integrals involving fermions (Candlin, 1956; Martin, 1959a), with some suggestion that such variables could occur in some form of CM (Martin, 1959b). Classical g-variable sources were introduced by Schwinger in source theory (Schwinger, 1969), a powerful approach in the study of the many-point Green function in quantum field theory, representing classical interventions on otherwise isolated quantum systems by external observers.

Eventually, the circle was completed when Casalbuoni published his approach to the CM of g-variables (Casalbuoni, 1976a), which he referred to as *pseudomechanics*. Casalbuoni also discussed the equivalent of canonical quantization of such a mechanics (Casalbuoni, 1976b).

Casalbuoni's pseudomechanics does not require any spatial dimensions. Indeed, in the next section we shall discuss it in $1+0$ spacetime, which means one time dimension and zero spatial dimensions.

23.3 The Grassmannian oscillator in continuous time

It is possible to construct classical g-mechanics, i.e., a mechanics that is based on g-variables, despite the fact that g-variables are nilpotent. What allows this possibility is that space and time coordinates serve as distinguishing labels that differentiate different instances of a g-variable occurring at either different times or different positions. For example, if we had a time-dependent g-variable $\theta(t)$, then $\theta(t)\theta(t) = 0$ by definition, but $\theta(t_1)\theta(t_2)$ need not be zero for $t_1 \neq t_2$.

In this section we shall study a g-mechanical system described by a single complex, time-dependent g-variable $\psi(t)$, satisfying the multiplication rule

$$\psi(t_1)\psi(t_2) = -\psi(t_2)\psi(t_1). \tag{23.2}$$

We shall in general choose to work with complex g-variables, which are defined by analogy with complex numbers. A real g-variable ψ_{R} is one that is invariant with respect to complex conjugation, i.e., $(\alpha\psi_{\mathrm{R}})^* = \alpha^*\psi_{\mathrm{R}}$, where α is a complex number and $*$ denotes complex conjugation. Given two real g-variables ψ_{R} and ψ_{I}, we construct the complex g-variable ψ as the sum $\psi \equiv \psi_{\mathrm{R}} + i\psi_{\mathrm{I}}$ and define its complex conjugate ψ^* to be $\psi^* \equiv \psi_{\mathrm{R}} - i\psi_{\mathrm{I}}$. The rule we adopt for the complex conjugation of products of complex g-variables θ and ψ is

$$(\theta\psi)^* = \psi^*\theta^*. \tag{23.3}$$

Complex g-variables are useful because products such as $\psi^*\psi$ do not necessarily vanish by nilpotency, which allows us to include mass terms and other interactions in Lagrangians with g-variable degrees of freedom. For instance, using the nilpotency of ψ_{R} and ψ_{I}, we find

$$\psi^*\psi \equiv (\psi_{\mathrm{R}} - i\psi_{\mathrm{I}})(\psi_{\mathrm{R}} + i\psi_{\mathrm{I}}) = 2i\psi_{\mathrm{R}}\psi_{\mathrm{I}}, \tag{23.4}$$

which need not vanish.

Another important construction in continuous time (CT) g-mechanics is the time derivative of a g-variable ψ, denoted by $\dot{\psi}$. We define this by analogy with the usual derivative of real- or complex-valued functions of a single real variable, i.e.,

$$\dot{\psi}(t) \equiv \lim_{\varepsilon \to 0} \frac{\psi(t+\varepsilon) - \psi(t)}{\varepsilon}, \tag{23.5}$$

assuming that this process makes mathematical sense. We cannot go here into the concepts of continuity and differentiability of g-numbers; we shall assume that such concepts make mathematical sense.

We define the following Lagrangian for a g-mechanical oscillator described by one complex g-variable ψ to be

$$L(\psi, \psi^*, \dot{\psi}, \dot{\psi}^*) = \frac{1}{2}i\psi^*\dot{\psi} - \frac{1}{2}i\dot{\psi}^*\psi - \omega\psi^*\psi, \tag{23.6}$$

where ω is some real, positive constant. By inspection, this Lagrangian is a real c-number, i.e., $L^* = L$. We choose to consider the g-variables ψ and ψ^* to be independent variables *before* we constrain them by the equations of motion, which we are to derive next. However, if we so wished, we could rewrite this Lagrangian in terms of the real fields ψ_{R} and ψ_{L}, where $\psi \equiv \psi_{\mathrm{R}} + i\psi_{\mathrm{L}}$ and $\psi^* \equiv \psi_{\mathrm{R}} - i\psi_{\mathrm{L}}$. However, it is much more convenient to work with ψ and ψ^*, and this choice is generally followed in fermionic field theory. In accordance with convention, we have arranged for ψ^* to be on the left in any term and ψ on the right in the above Lagrangian.

Given the Lagrangian (23.6), we construct the action integral in the usual way:

$$A_{\mathrm{fi}}[\psi, \psi^*] \equiv \int_{t_{\mathrm{i}}}^{t_{\mathrm{f}}} L(\psi, \psi^*, \dot{\psi}, \dot{\psi}^*)dt. \tag{23.7}$$

Next we perform an arbitrary variation of the g-variables ψ and ψ^* defined by

$$\begin{aligned} \psi(t) &\to \psi_\varepsilon(t) \equiv \psi(t) + \varepsilon u(t), \\ \psi^*(t) &\to \psi^*_\varepsilon(t) \equiv \psi^*(t) + \varepsilon u^*(t), \end{aligned} \tag{23.8}$$

where ε is a real parameter and $u(t)$ and $u^*(t)$ are independent, arbitrary g-variable functions of time. Then we find, to order ε,

$$\begin{aligned}\delta A_{\mathrm{fi}}[\psi,\psi^*] &\equiv A_{\mathrm{fi}}[\psi_\varepsilon,\psi^*_\varepsilon] - A_{\mathrm{fi}}[\psi,\psi^*] \\ &= \varepsilon\left[u^*\frac{\overrightarrow{\partial}}{\partial\dot{\psi}^*}L + L\frac{\overleftarrow{\partial}}{\partial\dot{\psi}}u\right]_{t_i}^{t_f} \\ &\quad + \int_{t_i}^{t_f} u^*\left\{\frac{\overrightarrow{\partial}}{\partial\psi^*}L - \frac{d}{dt}\left(\frac{\overrightarrow{\partial}}{\partial\dot{\psi}^*}L\right)\right\}dt \\ &\quad + \int_{t_i}^{t_f}\left\{L\frac{\overleftarrow{\partial}}{\partial\psi} - \frac{d}{dt}\left(L\frac{\overleftarrow{\partial}}{\partial\dot{\psi}}\right)\right\}u\,dt + O(\varepsilon^2). \end{aligned} \tag{23.9}$$

We note here the need to distinguish between g-variable differentiation from the left and that from the right.

Imposing the Weiss action principle (Weiss, 1936, 1938) on the variation (23.9) then leads to the equations of motion

$$\frac{d}{dt}\left\{\frac{\overrightarrow{\partial}}{\partial\dot{\psi}^*}L\right\} \underset{c}{=} \frac{\overrightarrow{\partial}}{\partial\psi^*}L, \quad \frac{d}{dt}\left\{L\frac{\overleftarrow{\partial}}{\partial\dot{\psi}}\right\} \underset{c}{=} L\frac{\overleftarrow{\partial}}{\partial\psi}. \tag{23.10}$$

Applied to the Lagrangian (23.6), we find

$$i\dot{\psi} \underset{c}{=} \omega\psi, \quad i\dot{\psi}^* \underset{c}{=} -\omega\psi^*. \tag{23.11}$$

Differentiating each side of these equations with respect to time and then rearranging terms leads to the second-order harmonic oscillator equations of motion

$$\frac{d^2}{dt^2}\psi \underset{c}{=} -\omega^2\psi, \quad \frac{d^2}{dt^2}\psi^* \underset{c}{=} -\omega^2\psi^*, \tag{23.12}$$

which justifies calling this system the Grassmannian oscillator.

The Weiss action principle can be used to find invariants of the motion. In this case, we note that the Lagrangian is invariant with respect to the time-independent phase change in the g-variables given by

$$\begin{aligned}\psi \to \psi' &\equiv e^{i\theta}\psi = \psi + i\theta\psi + O(\theta^2), \\ \psi^* \to \psi^{*\prime} &\equiv e^{-i\theta}\psi^* = \psi^* - i\theta\psi^* + O(\theta^2),\end{aligned} \tag{23.13}$$

where θ is an arbitrary real parameter. Hence modulo the equations of motion (23.11) we deduce that the quantity $C \equiv \psi^*\psi$ is conserved. The analogue of this invariant in the case of the Dirac equation is the total electric charge.

23.4 The Grassmannian oscillator in discrete time

We now turn to the temporal discretization of the above system. As with the temporal discretization c-variables discussed earlier, we define $\psi_n \equiv \psi(nT)$, where T is the chronon and n is an integer. Introducing the virtual paths

$$\psi_{n\lambda} \equiv \lambda\psi_{n+1} + \bar{\lambda}\psi_n, \quad \psi^*_{n\lambda} = \lambda\psi^*_{n+1} + \bar{\lambda}\psi^*_n, \tag{23.14}$$

replacing the CT variables in the Lagrangian (23.6) in the usual way and integrating with respect to λ from zero to one gives the system function

$$F^n = \frac{1}{2}i\left[\psi_n^*\psi_{n+1} - \psi_{n+1}^*\psi_n\right] - \frac{1}{6}\omega T\{2\psi_{n+1}^*\psi_{n+1} + \psi_n^*\psi_{n+1} + \psi_{n+1}^*\psi_n + 2\psi_n^*\psi_n\}. \tag{23.15}$$

The DT equations of motion are given by

$$\overrightarrow{\frac{\partial}{\partial\psi_n^*}}\{F^n + F^{n-1}\} \underset{c}{=} 0, \quad \{F^n + F^{n-1}\}\overleftarrow{\frac{\partial}{\partial\psi_n}} \underset{c}{=} 0, \tag{23.16}$$

which, applied to (23.15), lead to the equation of motion

$$i\frac{\psi_{n+1} - \psi_{n-1}}{2T} \underset{c}{=} \omega\frac{\psi_{n+1} + 4\psi_n + \psi_{n-1}}{6}, \tag{23.17}$$

and similarly for its complex conjugate.

These equations are similar to the DT Schrödinger equation discussed in Chapter 23, with similar, though not identical, properties. The essential reason for this is that both systems are described by differential equations that are of first order in CT.

At this stage, equation (23.17) does not look like the temporal discretization of the harmonic oscillator equation of motion (13.13), but we can readily show that DT oscillator behaviour occurs by the following method.

The first step is to rearrange (23.17) into the form

$$(3 + i\kappa)\psi_{n+1} \underset{c}{=} (3 - i\kappa)\psi_{n-1} - 4i\kappa\psi_n, \tag{23.18}$$

where the dimensionless quantity $\kappa \equiv \omega T$ will be referred to as the *reduced energy*. Since we are concerned only with ω and T both positive, we can safely assume $\kappa > 0$.

Next we define the skew angle λ by

$$\cos\lambda \equiv \frac{3}{\sqrt{9+\kappa^2}}, \quad \sin\lambda \equiv \frac{\kappa}{\sqrt{9+\kappa^2}}. \tag{23.19}$$

We encountered an analogous angle in Chapter 21 in our discussion of the DT Schrödinger equation. By inspection, we see that the skew angle can be taken to lie in the open interval $(0, \pi/2)$.

The skew angle turns out to be a critical feature of the DT Dirac equation, as we shall see. Its precise value will depend on three factors: (i) that time is discrete, which means that we have not taken the limit $T \to 0$; (ii) the fact that we are discretizing according to some prescription, such as the virtual path approach we favour; and (iii), in the case of field theory, the skew angle will depend on the spatial momentum of particle-like states. Whatever the details, however, we can choose the skew angle so that $0 < \lambda < \pi/2$.

The final step is to reparametrize the dynamical variables according to the rule

$$\psi_n \equiv e^{-in(\pi/2+\lambda)}\phi_n, \quad \psi_n^* \equiv e^{in(\pi/2+\lambda)}\phi_n^*, \tag{23.20}$$

where the ϕ_n and ϕ_n^* are g-variables. Then the equation of motion becomes

$$\phi_{n+1} + \phi_{n-1} \underset{c}{=} 2\eta\phi_n, \tag{23.21}$$

where $\eta \equiv 2\kappa/\sqrt{9+\kappa^2} = 2\sin\lambda$. From our analysis of the DT oscillator in Chapter 13, the elliptic regime, i.e., bounded oscillatory behaviour occurs only for $|\eta| < 1$. This leads to the conclusion that our Grassmannian oscillator has stable oscillations only if $\lambda < \pi/6$, which in turn means that $\kappa \equiv \omega T < \sqrt{3}$. This is exactly half of the upper limit found for the bosonic DT oscillator in Chapter 13.

23.5 The discrete time free Dirac equation

We turn now to the temporal discretization of the Dirac equation in $(1+3)$-dimensional spacetime. We shall use the natural unit system, in which $c = \hbar = 1$, working in an inertial frame with metric signature $(+1, -1, -1, -1)$. In this section we shall deal with the free-particle equation, reserving the gauge-invariant charged equations for the next section.

The original free Dirac equation in CT takes the form

$$i\,\partial_t\psi = \overrightarrow{H}_{\mathrm{D}}\psi, \tag{23.22}$$

where

$$\overrightarrow{H}_{\mathrm{D}} \equiv -i\boldsymbol{\alpha}\cdot\overrightarrow{\nabla} + \beta m \tag{23.23}$$

is the Dirac Hamiltonian and $\{\beta, \alpha^1, \alpha^2, \alpha^3\}$ are Dirac's q-numbers satisfying the anticommutation rules

$$\{\beta, \beta\} = 2, \quad \{\beta, \alpha^i\} = 0, \quad \{\alpha^i, \alpha^j\} = 2\delta_{ij}. \tag{23.24}$$

Throughout this chapter, $\{A, B\} \equiv AB + BA$, the anticommutator of A and B.

As with the Schrödinger equation discussed in Chapter 21, we can recover the Dirac equation from a classical Lagrange density. In this case, we define

$$\mathcal{L}_{\mathrm{D}} \equiv \frac{1}{2}i\bar{\psi}\gamma^\mu\,\overrightarrow{\partial}_\mu\psi - \frac{1}{2}i\bar{\psi}\,\overleftarrow{\partial}_\mu\gamma^\mu\psi - m\bar{\psi}\psi, \tag{23.25}$$

where we follow the standard conventions (Bjorken and Drell, 1964, 1965)

$$\gamma^0 \equiv \beta, \quad \gamma^i \equiv \beta\alpha^i, \quad \bar{\psi} \equiv \psi^\dagger\gamma^0, \tag{23.26}$$

which take into account that the 'wavefunction' ψ is represented by a column array with four complex components. In such a representation, the Dirac q-numbers are complex 4×4 matrices.

The Dirac equation (23.22) is recovered from the Euler–Lagrange equation

$$\frac{\partial}{\partial\bar{\psi}}\mathcal{L}_{\mathrm{D}} \underset{c}{=} \partial_\mu \frac{\partial}{\partial\left(\partial_\mu\bar{\psi}\right)}\mathcal{L}_{\mathrm{D}}, \tag{23.27}$$

with a similar derivation for the conjugate equation.

For the free-particle wavefunction, temporal discretization is implemented by the following virtual paths:

$$\psi_{n\lambda} \equiv \lambda\psi_{n+1} + \bar{\lambda}\psi_n, \quad \bar{\psi}_{n\lambda} \equiv \lambda\bar{\psi}_{n+1} + \bar{\lambda}\bar{\psi}_n, \tag{23.28}$$

which gives the system function

$$\begin{aligned}\mathcal{F}^n &= \frac{i}{2}\left\{\psi_n^+\psi_{n+1} - \psi_{n+1}^+\psi_n\right\}\\ &\quad - \frac{T}{6}\left\{2\psi_{n+1}^+ \overleftrightarrow{H}_{\mathrm{D}}\psi_{n+1} + \psi_{n+1}^+ \overleftrightarrow{H}_{\mathrm{D}}\psi_n + \psi_n^+ \overleftrightarrow{H}_{\mathrm{D}}\psi_{n+1} + 2\psi_n^+ \overleftrightarrow{H}_{\mathrm{D}}\psi_n\right\},\end{aligned} \tag{23.29}$$

where $\overleftrightarrow{H}_{\mathrm{D}} \equiv -\frac{1}{2}i\boldsymbol{\alpha}\cdot(\overrightarrow{\nabla} - \overleftarrow{\nabla}) + \beta m$. The formal equation of motion

$$\frac{\partial}{\partial\bar{\psi}_n}\{\mathcal{F}^n + \mathcal{F}^{n-1}\} \underset{c}{=} \nabla\cdot\frac{\partial}{\partial\nabla\bar{\psi}_n}\{\mathcal{F}^n + \mathcal{F}^{n-1}\} \tag{23.30}$$

gives the equation of motion

$$i\frac{\psi_{n+1} - \psi_{n-1}}{2T} \underset{c}{=} \overrightarrow{H}_{\mathrm{D}}\frac{\psi_{n+1} + 4\psi_n + \psi_{n-1}}{6}, \tag{23.31}$$

and similarly for the conjugate field $\psi_n^\dagger$. This equation is structurally identical to the DT Schrödinger wave equation (21.9).

To investigate solutions to equation (23.31), we use the same technique as the one we used on the DT Grassmannian oscillator equation (23.17).

First, we Fourier transform (23.31), defining

$$\tilde{\psi}_n(\mathbf{p}) \equiv \int d^3\mathbf{x}\, e^{-i\mathbf{p}\cdot\mathbf{x}}\psi_n(\mathbf{x}). \tag{23.32}$$

Then (23.31) becomes

$$\Omega_{\mathbf{p}}\psi_{n+1} \underset{c}{=} \Omega_{\mathbf{p}}^\dagger\psi_{n-1} - 4iK_{\mathbf{p}}\psi_n, \tag{23.33}$$

where $K_{\mathbf{p}} \equiv \boldsymbol{\alpha}\cdot\mathbf{p} + \beta m$ and $\Omega_{\mathbf{p}} \equiv 3 + iK_{\mathbf{p}}$. We note

$$K_{\mathbf{p}}K_{\mathbf{p}} = \mathbf{p}\cdot\mathbf{p} + m^2 = E_{\mathbf{p}}^2, \tag{23.34}$$

with $E_{\mathbf{p}}$ corresponding to the conventional definition of the particle's energy.

We now investigate possible stable, particle-like solutions. There are two properties such solutions should have: (i) stability and (ii) sharp momentum. To this end, we define the momentum eigenstate

$$\psi_{n,\mathbf{p}r}(\mathbf{x}) \equiv e^{-i\varpi_{\mathbf{p}}n + i\mathbf{p}\cdot\mathbf{x}}u(\mathbf{p}, r), \tag{23.35}$$

where $u(\mathbf{p}, r)$ is a momentum spinor (Bjorken and Drell, 1964) satisfying the equation

$$K_{\mathbf{p}} u(\mathbf{p}, r) = \kappa_{\mathbf{p}} u(\mathbf{p}, r), \tag{23.36}$$

where $\kappa_{\mathbf{p}} \equiv TE_{\mathbf{p}}$ and $\varpi_{\mathbf{p}}$ is the *vibrancy*. We introduced the notion of vibrancy in Chapter 21 dealing with the DT Schrödinger equation.

The momentum spinors are also eigenstates of the operators $\Omega_{\mathbf{p}}$ and $\Omega^{\dagger}_{\mathbf{p}}$. Specifically,

$$\Omega_{\mathbf{p}} u(\mathbf{p}, r) = \sqrt{9 + \kappa^2_{\mathbf{p}}} e^{i\lambda_{\mathbf{p}}} u(\mathbf{p}, r), \quad \Omega^{\dagger}_{\mathbf{p}} u(\mathbf{p}, r) = \sqrt{9 + \kappa^2_{\mathbf{p}}} e^{-i\lambda_{\mathbf{p}}} u(\mathbf{p}, r), \tag{23.37}$$

where $\lambda_{\mathbf{p}}$ is the spinor skew angle defined by

$$\cos \lambda_{\mathbf{p}} = \frac{3}{\sqrt{9 + \kappa^2_{\mathbf{p}}}}, \quad \sin \lambda_{\mathbf{p}} = \frac{\kappa_{\mathbf{p}}}{\sqrt{9 + \kappa^2_{\mathbf{p}}}}. \tag{23.38}$$

Inserting the momentum eigenstate (23.35) into the equation of motion (23.35) then gives an equation for the vibrancy, viz.,

$$\sin(\varpi_{\mathbf{p}} - \lambda_{\mathbf{p}}) = \frac{2\kappa_{\mathbf{p}}}{\sqrt{9 + \kappa^2_{\mathbf{p}}}}. \tag{23.39}$$

For a given $\kappa_{\mathbf{p}}$, there are two possible vibrancies, $\varpi_1(\mathbf{p})$ and $\varpi_2(\mathbf{p})$, given by relations (21.27) and (21.28).

From these relations and the properties of the momentum spinors, we can deduce the following.

1. For a given reduced energy $\kappa_{\mathbf{p}}$, there are positive- and negative-energy solutions corresponding to particles and antiparticles.
2. For a given reduced energy $\kappa_{\mathbf{p}}$, there are two vibrancies, $\varpi_1(\mathbf{p})$ and $\varpi_2(\mathbf{p})$. ω_1 corresponds to ordinary particle states, whilst ω_2 corresponds to the oscillon solutions.
3. There is a total of *four* solutions for each linear momentum $\mathbf{p}$: particle normal, particle oscillon, antiparticle normal and antiparticle oscillon.

23.6 Charge and charge density

We shall leave the construction of conserved quantities such as the linear momentum and other invariants built up from the ladder operators $a(\mathbf{p}), b(\mathbf{p})$, etc., as exercises. However, we will discuss here the charge and charge current densities.

First, we rewrite the DT equations of motion (23.31) in the more useful form

$$\begin{aligned} \left(-3i + T \overrightarrow{H_{\mathrm{D}}}\right) \psi_{n+1} &\underset{c}{=} -4T \overrightarrow{H_{\mathrm{D}}} \psi_n - \left(3i + T \overrightarrow{H_{\mathrm{D}}}\right) \psi_{n-1} \\ \psi^{\dagger}_{n+1} \left(3i + \overleftarrow{H_{\mathrm{D}}} T\right) &\underset{c}{=} -4\psi^{\dagger}_n \overleftarrow{H_{\mathrm{D}}} T - \psi^{\dagger}_{n-1} \left(-3i + \overleftarrow{H_{\mathrm{D}}} T\right), \end{aligned} \tag{23.40}$$

where $\overleftarrow{H_{\mathrm{D}}} \equiv i\boldsymbol{\alpha} \cdot \overleftarrow{\boldsymbol{\nabla}} + \beta m$.

Now the system function (23.29) is invariant with respect to the infinitesimal global gauge transformation

$$\begin{aligned} \psi_n &\to e^{i\theta}\psi_n, \quad \psi_n^\dagger \to \psi_n^\dagger e^{-i\theta}, \\ \delta\psi_n &= i\theta\psi_n, \quad \delta\psi^\dagger = -i\theta\psi_n^\dagger, \end{aligned} \tag{23.41}$$

so that there is a conserved charge given by

$$Q^n \equiv \int d^3\mathbf{x} \left\{ \psi_n^\dagger \left(\frac{1}{2} + \frac{iT}{6} \overrightarrow{H_{\mathrm{D}}} \right) \psi_{n+1} + \psi_{n+1}^\dagger \left(\frac{1}{2} - \overleftarrow{H_{\mathrm{D}}} \frac{T}{6} \right) \psi_n \right\}. \tag{23.42}$$

It is easy to use the equations of motion (23.40) to show that

$$Q^n \underset{c}{=} Q^{n-1}. \tag{23.43}$$

The charge and current densities are then given by

$$\begin{aligned} \rho_n &= \psi_n^\dagger \left(\frac{1}{2} + \frac{iT}{6} \overrightarrow{H_D} \right) \psi_{n+1} + \psi_{n+1}^\dagger \left(\frac{1}{2} - \overleftarrow{H_{\mathrm{D}}} \frac{iT}{6} \right) \psi_n \\ \mathbf{j}_n &= \frac{1}{6} \{ \psi_n^\dagger \boldsymbol{\alpha} \psi_{n-1} + 4\psi_n^\dagger \boldsymbol{\alpha} \psi_n + \psi_{n-1}^\dagger \boldsymbol{\alpha} \psi_n \}. \end{aligned} \tag{23.44}$$

These collapse to the standard densities in the limit $T \to 0$, and satisfy the DT equation of continuity

$$\bar{\mathbb{T}}_n \rho_n + \nabla \cdot \mathbf{j}_n \underset{c}{=} 0, \tag{23.45}$$

as required.

23.6.1 Wavepacket charge

To see how the various particle modes contribute to the charge of a wavepacket, consider a general free-particle wavepacket solution of the form

$$\begin{aligned} \psi_n = \sum_r \int \frac{d^3\mathbf{p}}{(2\pi)^3 2E_{\mathbf{p}}} \{ & [A_{\mathbf{p}r} e^{-i\varpi_1 n} + C_{\mathbf{p}r} e^{-i\varpi_2 n}] e^{i\mathbf{p}\cdot\mathbf{x}} u(\mathbf{p}r) \\ & + [B_{\mathbf{p}r} e^{i\varpi_1 n} + D_{\mathbf{p}r} e^{i\varpi_2 n}] e^{-i\mathbf{p}\cdot\mathbf{x}} v(\mathbf{p}r) \}. \end{aligned} \tag{23.46}$$

Then a lengthy calculation gives

$$\begin{aligned} Q_n = \frac{q}{6} \sum_r \int \frac{d^3\mathbf{p}}{(2\pi)^3 2E_{\mathbf{p}}} 2\sqrt{9 + \kappa_{\mathbf{p}}^2} \{ & [|A_{\mathbf{p}r}|^2 + |B_{\mathbf{p}r}|^2] \cos\theta_1 \\ & + [|C_{\mathbf{p}r}|^2 + |D_{\mathbf{p}r}|^2] \cos\theta_2 \}, \end{aligned} \tag{23.47}$$

where $\theta_1 \equiv \varpi_1 - \lambda$ and $\theta_2 \equiv \varpi_2 - \lambda$.

The interpretation is as follows. The A term is a normal particle amplitude and the B term is a normal antiparticle amplitude. The factor $\cos\theta_1$ is positive, so the normal solutions give a net positive contribution. This is not actually physical charge. Recall that the so-called 'charge density' $\rho \equiv \bar{\psi}\gamma^0\psi$ is non-negative in the CT wave-*function* theory, even for negative-energy solutions. It requires anticommutation rules in the quantized field theory to recover antiparticle charge contributions to the net charge.

The C and D oscillon terms in (23.47) contribute a net negative effect to Q_n, because $\cos\theta_2$ is negative. This tells us that the oscillons behave very much like particles with negative norm-squared, i.e., they lie outside the normal Hilbert space of particle–antiparticle states. This is to be expected, since the DT Dirac equation is of second order, rather than first order.

24

Discrete time Maxwell equations

24.1 Classical electrodynamical fields

In this chapter we develop a discrete time (DT) formulation of electrodynamics that is based on the well-known continuous time (CT) Maxwell equations. We shall work in natural units, where $c = \hbar = 1$ and the CT Minkowski-space metric tensor has diagonal components given by $(1, -1, -1, -1)$ relative to standard inertial coordinates. As before, temporal discretization is carried out in the preferred local inertial frame. We shall make extensive use of the DT notation

$$\begin{aligned} &\mathbb{T}_n \equiv T^{-1}\,\mathbb{D}_n, \quad \bar{\mathbb{T}}_n \equiv T^{-1}\,\bar{\mathbb{D}}_n, \quad \mathbb{A}_n \equiv \frac{1}{2}(\mathbb{U}_n + 1), \\ &\mathbb{S}_n \equiv (\mathbb{U}_n + 4 + \bar{\mathbb{U}}_n)/6, \quad \Box_n \equiv \mathbb{T}_n\bar{\mathbb{T}}_n - \mathbb{S}_n\nabla^2. \end{aligned} \tag{24.1}$$

In classical electrodynamics, the important physical fields are the electric field $\mathbf{E}$ and the magnetic field $\mathbf{B}$. These interact with the electric charge density ρ and the electric charge density current $\mathbf{j}$ according to the homogeneous Maxwell equations

$$\nabla\cdot\mathbf{B} = 0, \qquad \nabla\times\mathbf{E} + \partial_t\mathbf{B} = \mathbf{0} \tag{24.2}$$

and the inhomogeneous Maxwell equations

$$\nabla\cdot\mathbf{E} = \rho, \qquad \nabla\times\mathbf{B} - \partial_t\mathbf{E} = \mathbf{j}. \tag{24.3}$$

These equations are consistent with the continuity equation

$$\partial_t\rho + \nabla\cdot\mathbf{j} = 0 \tag{24.4}$$

for electric charge conservation if the electric and magnetic fields are sufficiently differentiable vector fields.

The first step in any temporal discretization is to establish whether the dynamical variables are node or link variables. It is not immediately obvious what $\mathbf{E}$ and $\mathbf{B}$ are from the above equations (24.3). To answer this question we turn to the CT Lagrange formulation of electrodynamics, deriving a system function for DT electrodynamics from the electromagnetic Lagrange density.

In the Lagrangian approach, the physical electric and magnetic fields are given in terms of the scalar and magnetic potential fields ϕ and $\mathbf{A}$ by

$$\mathbf{E} \equiv -\nabla\phi - \partial_t \mathbf{A}, \qquad \mathbf{B} \equiv \nabla \times \mathbf{A}. \tag{24.5}$$

Then the homogeneous Maxwell equations (24.2) are satisfied identically. The CT Lagrange density is given by

$$\mathcal{L}_{\text{EM}} \equiv -\frac{1}{4} F^{\mu\nu} F_{\mu\nu} - j^\mu A_\mu, \tag{24.6}$$

where $A^\mu \equiv (\phi, \mathbf{A})$ are the components of the electromagnetic four-vector potential, the $F^{\mu\nu} \equiv \partial^\mu A^\nu - \partial^\nu A^\mu$ are the components of the Maxwell–Faraday tensor and $j^\mu \equiv (\rho, \mathbf{j})$ are the components of the four-vector charge current. Then the CT Euler–Lagrange equations of motion

$$\partial_\mu \frac{\partial \mathcal{L}_{\text{EM}}}{\partial \partial_\mu A_\nu} \underset{c}{=} \frac{\partial \mathcal{L}_{\text{EM}}}{\partial A_\nu} \tag{24.7}$$

give the equations of motion

$$\Box A^\nu - \partial^\nu(\partial_\mu A^\mu) \underset{c}{=} j^\nu, \tag{24.8}$$

where $\Box \equiv \partial^\mu \partial_\mu = \partial_t^2 - \nabla^2$ is the CT d'Alembertian operator.

To settle the node-versus-link question, we first write out the Lagrange density in frame-dependent form, i.e.,

$$\begin{aligned}
\mathcal{L}_{\text{EM}} &= \frac{1}{2}\mathbf{E}\cdot\mathbf{E} - \frac{1}{2}\mathbf{B}\cdot\mathbf{B} - \phi\rho + \mathbf{A}\cdot\mathbf{j} \\
&= \frac{1}{2}\{\partial_t \mathbf{A} + \nabla\phi\}\cdot\{\partial_t \mathbf{A} + \nabla\phi\} \\
&\quad - \frac{1}{2}\{\partial_i A^j \, \partial_i A^j - \partial_i A^j \, \partial_j A^i\} - \phi\rho + \mathbf{A}\cdot\mathbf{j}.
\end{aligned} \tag{24.9}$$

By inspection, we see that there are no time derivatives of the scalar potential ϕ, which prompts us to identify it as a link variable. Looking at the coupling of the charge density ρ to ϕ, we then take the charge density ρ to be a link variable also. Our notation is that we associate the scalar field $\phi_n(\mathbf{x})$ and the electric charge density $\rho_n(\mathbf{x})$ with the link from time n to time $n+1$.

On the other hand, the vector potential $\mathbf{A}$ appears with a time derivative in (24.9), so we are led to identify it as a node variable. Consequently, the three-vector charge current density $\mathbf{j}$ is identified as a node variable.

In order to incorporate gauge invariance, we define the following virtual path fields:

$$\begin{aligned}
\phi_{n\lambda}(\mathbf{x}) &\equiv \phi_n(\mathbf{x}), \\
\mathbf{A}_{n\lambda}(\mathbf{x}) &\equiv \lambda \mathbf{A}_{n+1}(\mathbf{x}) + \bar{\lambda}\mathbf{A}_n(\mathbf{x}), \\
\partial_t \mathbf{A}_{n\lambda}(\mathbf{x}) &\equiv T^{-1}\, \partial_\lambda \mathbf{A}_{n\lambda}(\mathbf{x}) = \mathbb{T}_n \mathbf{A}_n,
\end{aligned} \tag{24.10}$$

where $\mathbb{T}_n \equiv T^{-1}\,\mathbb{D}_n$. We define

$$\mathcal{L}_{n\lambda} \equiv \frac{1}{2}\{\mathbb{T}_n \mathbf{A}_n + \nabla\phi_n\}\cdot\{\mathbb{T}_n \mathbf{A}_n + \nabla\phi_n\} - \frac{1}{2}\{\partial_i A^j_{n\lambda}\,\partial_i A^j_{n\lambda} - \partial_i A^j_{n\lambda}\,\partial_j A^i_{n\lambda}\} - \phi_n\rho_n + \mathbb{A}_n(\mathbf{A}_n\cdot\mathbf{j}_n) \tag{24.11}$$

and then the system function density $\mathcal{F}^n \equiv T\langle\mathcal{L}_{n\lambda}\rangle$ is given by

$$\mathcal{F}^n = \frac{T}{2}\mathbf{E}_n\cdot\mathbf{E}_n - \frac{T}{6}(\mathbf{B}_{n+1}\cdot\mathbf{B}_{n+1} + \mathbf{B}_{n+1}\cdot\mathbf{B}_n + \mathbf{B}_n\cdot\mathbf{B}_n) - T\phi_n\rho_n + T\mathbb{A}_n(\mathbf{A}_n\cdot\mathbf{j}_n), \tag{24.12}$$

where the DT electric and magnetic fields are given by

$$\mathbf{E}_n \equiv -\mathbb{T}_n\mathbf{A}_n - \nabla\phi_n, \quad \mathbf{B}_n \equiv \nabla\times\mathbf{A}_n. \tag{24.13}$$

These satisfy the DT analogues of the homogeneous Maxwell equations, i.e.,

$$\nabla\times\mathbf{E}_n + \mathbb{T}_n\mathbf{B}_n = \mathbf{0}, \quad \nabla\cdot\mathbf{B}_n = 0. \tag{24.14}$$

24.1.1 Electromagnetic duality

In CT charge-free electromagnetism, Maxwell's equations are invariant with respect to the interchange $\mathbf{E}\to\mathbf{B}$, $\mathbf{B}\to-\mathbf{E}$, a symmetry known as electromagnetic duality. The introduction of electric charges gives rise to the inhomogeneous equations (24.3), but this breaks electromagnetic duality, there being no magnetic current in Maxwell's equations.

The absence of magnetic current terms in Maxwell's equations has long been speculated to be of fundamental significance, prompting theoretical and experimental research into the possibility of finding isolated magnetic monopoles in the Universe. At this time, there is no evidence for such particles, but the question remains at the forefront of modern cosmology because magnetic monopoles are predicted to occur in grand unified field theories. It has been proposed that their absence can be accounted for by the theory of cosmological inflation.

According to our formulation of DT electrodynamics above, the electric field is a link variable, whilst the magnetic field is a node variable. This means that, on this level, electromagnetic duality does not hold even in the charge-free situation.

24.2 Gauge invariance

The use of electromagnetic potentials raises the issue of gauge invariance. The physically important equations of classical electrodynamics (24.3) are invariant with respect to the gauge transformation

$$\phi\to\phi' \equiv \phi + \partial_t\chi, \quad \mathbf{A}\to\mathbf{A}' \equiv \mathbf{A} - \nabla\chi, \tag{24.15}$$

where χ is an arbitrary differentiable function of class C^2 known as a gauge function.

The Lagrange density (24.6) is not gauge-invariant, however. The problem arises with the coupling term $j^\mu A_\mu$, which transforms under a gauge transformation according to the rule

$$j^\mu A_\mu \to j'^\mu A'_\mu = j^\mu (A_\mu + \partial_\mu \chi), \tag{24.16}$$

where it is assumed that the charge density four-vector is gauge-invariant. However, we note that

$$j^\mu \,\partial_\mu \chi = \partial_\mu (j^\mu \chi) - \chi \,\partial_\mu j^\mu = \partial_\mu (j^\mu \chi), \tag{24.17}$$

provided that the equation of continuity $\partial^\mu j_\mu = 0$ holds, which will generally be the case. Hence the gauge-transformed Lagrange density differs from the original (24.6) by a total four-divergence, $\partial_\mu (j^\mu \chi)$. Since the equations of motion are derived from an action integral over a four-volume of spacetime, the integral over spacetime of this four-divergence can be rewritten as a surface integral over the boundary of this spacetime volume by a generalization of Gauss' theorem. This tells us that the equations of motion will be gauge-independent.

The same issue arises in DT. The DT analogue of the gauge transformation (24.15) is chosen to be

$$\phi_n \to \phi'_n \equiv \phi_n + \mathbb{T}_n \chi_n, \quad \mathbf{A}_n \to \mathbf{A}'_n \equiv \mathbf{A}_n - \nabla \chi_n, \tag{24.18}$$

which assumes that χ_n is a node variable. The DT electric and magnetic fields (24.13) are invariant with respect to this transformation, which will be referred to as DT gauge invariance.

On the other hand, the coupling term $\mathcal{C}_n \equiv -\phi_n \rho_n + \mathbb{A}_n(\mathbf{A}_n \cdot \mathbf{j}_n)$ in the system function density (24.12) is not gauge-invariant, transforming according to the rule

$$\mathcal{C}_n \to \mathcal{C}'_n = \mathcal{C}_n - \{\mathbb{T}_n \chi_n\}\rho_n - \mathbb{A}_n(\nabla \chi_n \cdot \mathbf{j}_n). \tag{24.19}$$

However, as with the action integral in the CT case, the important quantity which determines the equations of motion is the action sum

$$A_{NM} \equiv \sum_{n=M}^{N-1} \int d^3\mathbf{x}\, \mathcal{F}^n, \tag{24.20}$$

so our task at this point is to find out how it transforms under the gauge transformation (24.18). We find

$$A_{NM} \to A'_{NM} = A_{NM} + \mathrm{ST} + T \sum_{n=M+1}^{N-1} \int d^3\mathbf{x}\; \chi_n \{\bar{\mathbb{T}}_n \rho_n + \nabla \cdot \mathbf{j}_n\}, \tag{24.21}$$

where ST represents surface terms in the summation and integration. These surface terms do not contribute to the equations of motion, which are therefore gauge-invariant provided that the DT continuity equations

$$\bar{\mathbb{T}}_n \rho_n + \nabla \cdot \mathbf{j}_n \underset{c}{=} 0 \tag{24.22}$$

hold over the temporal interval under consideration.

24.3 The inhomogeneous equations

Because it is a link variable, the DT equation of motion for the scalar potential ϕ_n is given by (20.35), i.e.,

$$\frac{\partial}{\partial \phi_n}\mathcal{F}^n \underset{c}{=} \nabla \frac{\partial}{\partial \nabla \phi_n}\mathcal{F}^n. \tag{24.23}$$

From (24.12) we find

$$\frac{\partial}{\partial \phi_n}\mathcal{F}^n = -T\rho_n, \quad \frac{\partial}{\partial \nabla \phi_n}\mathcal{F}^n = T\{\mathbb{T}_n \mathbf{A}_n + \nabla \phi_n\}, \tag{24.24}$$

so we arrive at the equation of motion

$$\nabla \cdot \mathbf{E}_n \underset{c}{=} \rho_n. \tag{24.25}$$

Since it is a node variable, the equation of motion for the vector potential $\mathbf{A}_n$ is derived from

$$\frac{\partial}{\partial \mathbf{A}_n}\{\mathcal{F}^n + \mathcal{F}^{n-1}\} \underset{c}{=} \partial_i \frac{\partial}{\partial \partial_i \mathbf{A}_n}\{\mathcal{F}^n + \mathcal{F}^{n-1}\}, \tag{24.26}$$

giving

$$\nabla \times \mathbb{S}_n \mathbf{B}_n - \bar{\mathbb{T}}_n \mathbf{E}_n \underset{c}{=} \mathbf{j}_n. \tag{24.27}$$

Equations (24.25) and (24.27) are consistent with the equation of continuity (24.22).

As for the electromagnetic potentials themselves, they satisfy the equations of motion

$$\begin{aligned} -\nabla^2 \phi_n - \mathbb{T}_n \nabla \cdot \mathbf{A}_n &\underset{c}{=} \rho_n, \\ \Box_n \mathbf{A}_n + \nabla \Lambda_n &\underset{c}{=} \mathbf{j}_n. \end{aligned} \tag{24.28}$$

These are consistent with the equation of continuity (24.22). Here

$$\Lambda_n \equiv \bar{\mathbb{T}}_n \phi_n + \mathbb{S}_n \nabla \cdot \mathbf{A}_n \tag{24.29}$$

is the DT Lorentz function.

24.3.1 Lorentz gauges

The equations of electrodynamics are often simplified by working in a specific gauge. We define a DT Lorentz gauge by the condition $\Lambda_n = 0$. Specifically, suppose we start with electromagnetic potentials $\phi_n^{(0)}$ and $\mathbf{A}_n^{(0)}$ such that $\Lambda_n^{(0)} \equiv \bar{\mathbb{T}}_n \phi_n^{(0)} + \mathbb{S}_n \nabla \cdot \mathbf{A}_n^{(0)} \neq 0$. Now suppose we make a gauge transformation from $(\phi_n^{(0)}, \mathbf{A}_n^{(0)})$ to $(\phi_n, \mathbf{A}_n)$, i.e., we choose a gauge function $\tilde{\chi}_n$ such that

$$\phi_n^{(0)} \to \phi_n \equiv \phi_n^{(0)} + \mathbb{T}_n \tilde{\chi}_n, \quad \mathbf{A}_n^{(0)} \to \mathbf{A}_n \equiv \mathbf{A}_n^{(0)} - \nabla \tilde{\chi}_n. \tag{24.30}$$

Then

$$\Lambda_n^{(0)} = \Lambda_n + \Box_n \tilde{\chi}_n, \tag{24.31}$$

where $\Box_n$ is given in (24.1). If now we arrange $\tilde{\chi}_n$ to be any solution to the equation

$$\Box_n \tilde{\chi}_n = \Lambda_n^{(0)}, \tag{24.32}$$

then we deduce that $\Lambda_n = 0$.

In a DT Lorentz gauge the potentials satisfy the equations

$$\Box_n \mathbf{A}_n \underset{c}{=} \mathbf{j}_n, \quad \Box_n \phi_n \underset{c}{=} \mathbb{S}_n \rho_n, \tag{24.33}$$

which are consistent with charge conservation.

24.4 The charge-free equations

Setting the charge density and current to zero in (24.25) and (24.27) gives the charge-free equations

$$\nabla \cdot \mathbf{E}_n \underset{c}{=} 0, \quad \bar{\mathbb{T}}_n \mathbf{E}_n \underset{c}{=} \mathbb{S}_n \nabla \times \mathbf{B}_n. \tag{24.34}$$

Then, using the homogeneous DT equations (24.14), we find that the physical electromagnetic fields $\mathbf{E}_n$ and $\mathbf{B}_n$ satisfy the DT massless Klein–Gordon equations

$$\Box_n \mathbf{E}_n \underset{c}{=} \mathbf{0}, \quad \Box_n \mathbf{B}_n \underset{c}{=} \mathbf{0}, \tag{24.35}$$

where $\Box_n$ is the DT d'Alembertian given above.

Turning to the electromagnetic potentials, we find their equations of motion to be

$$\Box_n \mathbf{A}_n + \nabla \Lambda_n \underset{c}{=} \mathbf{0}, \quad \Box_n \phi_n - \mathbb{T}_n \Lambda_n \underset{c}{=} 0, \tag{24.36}$$

so in a Lorentz gauge they satisfy the DT massless Klein–Gordon equations

$$\Box_n \mathbf{A}_n \underset{c}{=} \mathbf{0}, \quad \Box_n \phi_n \underset{c}{=} 0. \tag{24.37}$$

The total linear momentum for free electromagnetic fields is found to be

$$\mathbf{P}^n = \int d^3\mathbf{x} \left\{ \mathbf{E}_n \times \mathbf{B}_n + \frac{T}{6} B_n^i \overleftrightarrow{\nabla} B_{n+1}^i \right\}. \tag{24.38}$$

Then, using the equations of motion, we find

$$\mathbf{P}^n \underset{c}{=} \mathbf{P}^{n-1} \tag{24.39}$$

as expected. In the limit $T \to 0$ the above expression reduces to the Poynting vector.

24.5 Gauge transformations and virtual paths

We are now in a position to discuss the coupling of the Maxwell potentials to the various matter fields such as the Schrödinger wavefunction, the charged Klein–Gordon field and the Dirac electron field. In this scenario the virtual paths for charged fields and their complex conjugates have to be modified so as to ensure that system functions are DT gauge-invariant.

Experience with DT gauge invariance gives an important insight into temporal discretization, which is that the obvious strategy of replacing fluxions with simple node differences everywhere, such as $\dot{x} \to (x_{n+1} - x_n)/T$, is too naive and does not work. Gauge invariance has to be maintained in DT mechanics if charge is to be conserved, and this means that virtual paths have to be carefully constructed. Indeed, the reason why we adopted the virtual-path strategy in the first place was because we could think of no simpler method of constructing DT gauge-invariant system functions.

The same issue arises in any theory involving non-abelian gauge fields. In that scenario there are additional complications arising from the fact that the gauge fields couple to themselves. That is a problem we do not address in this book. Note, however, that our discussion of the Skyrme model in Chapter 25 involves a non-abelian gauge symmetry. For quantum electrodynamics, however, the construction of U(1) gauge-invariant system functions is relatively straightforward.

We introduce additional notation to aid us in this construction. We define the gauge link factor $W_n(\mathbf{x})$ and the gauge node function $X_n(\mathbf{x})$ by

$$W_n(\mathbf{x}) \equiv \exp\{iq\phi_n(\mathbf{x})T\}, \quad X_n(\mathbf{x}) \equiv \exp\{-iq\chi_n(x)\}, \tag{24.40}$$

where q is the electric charge of the matter field being discussed.

In the presence of a generic charged field Φ_n, the DT gauge transformations (24.18) are extended to the set

$$\begin{aligned} \phi_n \to \phi_n' \equiv \phi_n + \mathbb{T}_n\chi_n, \quad \mathbf{A}_n \to \mathbf{A}_n' \equiv \mathbf{A}_n - \nabla\chi_n, \\ \Phi_n \to \Phi_n' \equiv X_n\Phi_n, \quad \Phi_n^* \to \Phi_n^{*\prime} \equiv X_n^*\Phi_n^*. \end{aligned} \tag{24.41}$$

The real problem is the temporal discretization of the CT gauge-covariant derivatives, $D_\mu\Phi \equiv (\partial_\mu + iqA_\mu)\Phi$. We need to discuss the temporal and spatial derivatives separately.

The first step is to construct virtual paths for all fields and operators. We use the following definitions:

scalar potential,

$$\phi \to \phi_{n\lambda} \equiv \phi_n;$$

vector potential,

$$\mathbf{A} \to \mathbf{A}_{n\lambda} \equiv \lambda\mathbf{A}_{n+1} + \bar{\lambda}\mathbf{A}_n;$$

gauge function,

$$\chi \to \chi_{n\lambda} \equiv \lambda\chi_{n+1} + \bar{\lambda}\chi_n;$$

generic matter field,

$$\Phi \to \Phi_{n\lambda} \equiv \lambda W_n^{\bar{\lambda}}\Phi_{n+1} + \bar{\lambda}W_n^{\lambda}\Phi_n;$$

gauge node function,

$$\exp\{-iq\chi\} \to X_{n\lambda} \equiv \exp\{-iq\chi_{n\lambda}\};$$

temporal derivative operators,

$$\overrightarrow{D_0} \equiv \overrightarrow{\partial_0} + iq\phi \to \overrightarrow{D}_{n\lambda} \equiv T^{-1}\,\overrightarrow{\partial}_\lambda + iq\phi_{n\lambda},$$
$$\overleftarrow{D_0} \equiv \overleftarrow{\partial_0} - iq\phi \to \overleftarrow{D}_{n\lambda} \equiv T^{-1}\,\overleftarrow{\partial}_\lambda - iq\phi_{n\lambda};$$

spatial derivative operators,

$$\overrightarrow{\mathbf{D}} \equiv \overrightarrow{\nabla} - iq\mathbf{A} \to \overrightarrow{\mathbf{D}}_{n\lambda} \equiv \overrightarrow{\nabla} - iq\mathbf{A}_{n\lambda},$$
$$\overleftarrow{\mathbf{D}} \equiv \overleftarrow{\nabla} + iq\mathbf{A} \to \overleftarrow{\mathbf{D}}_{n\lambda} \equiv \overleftarrow{\nabla} + iq\mathbf{A}_{n\lambda}.$$

Then, under the general DT gauge transformation (24.41), we have the transformations

$$\begin{aligned}
\overrightarrow{D}_{n\lambda}\Phi_{n\lambda} &\to \overrightarrow{D'}_{n\lambda}\Phi'_{n\lambda} = X_{n\lambda}\,\overrightarrow{D}_{n\lambda}\Phi_{n\lambda},\\
\Phi^*_{n\lambda}\overleftarrow{D}_{n\lambda} &\to \Phi^{*\prime}_{n\lambda}\,\overleftarrow{D'}_{n\lambda} = \Phi^*_{n\lambda}\,\overleftarrow{D}_{n\lambda}X^*_{n\lambda},\\
\overrightarrow{\mathbf{D}}_{n\lambda}\Phi_{n\lambda} &\to \overrightarrow{\mathbf{D}'}_{n\lambda}\Phi'_{n\lambda} = X_{n\lambda}\,\overrightarrow{\mathbf{D}}_{n\lambda}\Phi_{n\lambda},\\
\Phi^*_{n\lambda}\,\overleftarrow{\mathbf{D}}_{n\lambda} &\to \Phi^{*\prime}_{n\lambda}\,\overleftarrow{\mathbf{D}'}_{n\lambda} = \Phi^*_{n\lambda}\,\overleftarrow{\mathbf{D}}_{n\lambda}X^*_{n\lambda}.
\end{aligned} \tag{24.42}$$

24.6 Coupling to matter fields

The general strategy for coupling matter fields to the electromagnetic fields is as follows.

1. Write down the CT free Lagrange density:
$$\mathcal{L}_0 \equiv \mathcal{L}(\Phi, \Phi^*, \overrightarrow{\partial_0}\Phi, \Phi^*\,\overleftarrow{\partial_0}, \overrightarrow{\nabla}\Phi, \Phi^*\,\overleftarrow{\nabla}). \tag{24.43}$$
2. Replace all fields by virtual-path equivalents according to the list of definitions given in the previous section:
$$\mathcal{L}_0 \to \mathcal{L}_{n\lambda} \equiv \mathcal{L}(\Phi_{n\lambda}, \Phi^*_{n\lambda}, \overrightarrow{D}_{n\lambda}\Phi_{n\lambda}, \Phi^*_{n\lambda}\,\overleftarrow{D}_{n\lambda}, \overrightarrow{\mathbf{D}}_{n\lambda}\Phi_{n\lambda}, \Phi^*_{n\lambda}\,\overleftarrow{\mathbf{D}}_{n\lambda}). \tag{24.44}$$
3. Add the term $T\langle\mathcal{L}_{n\lambda}\rangle$ to the system function density $\mathcal{F}^n_{EM}$ for the free electromagnetic fields:
$$\begin{aligned}
\mathcal{F}^n_{\mathrm{EM}} \equiv T\langle \tfrac{1}{2}\{\mathbb{T}_n\mathbf{A}_n + \nabla\phi_n\}\cdot\{\mathbb{T}_n\mathbf{A}_n + \nabla\phi_n\}\\
- \tfrac{1}{2}\{\partial_i A^j_{n\lambda}\,\partial_i A^j_{n\lambda} - \partial_i A^j_{n\lambda}\,\partial_j A^i_{n\lambda}\}\rangle.
\end{aligned} \tag{24.45}$$

We are now in position to write down gauge-invariant system functions for the charged Schrödinger fields, the charged Klein–Gordon field and the charged Dirac field.

24.6.1 The charged DT Schrödinger equation

The free CT Schrödinger equation was derived from the Lagrange density

$$\mathcal{L}_{\mathrm{SCH}} \equiv \frac{1}{2} i\hbar \left\{ \Psi^* \overrightarrow{\partial_t} \Psi - \Psi^* \overleftarrow{\partial_t} \Psi \right\} - \frac{\hbar^2}{2m} \Psi^* \overleftarrow{\nabla} \cdot \overrightarrow{\nabla} \Psi. \tag{24.46}$$

When an external electromagnetic field is introduced, the above density takes the gauge-invariant form

$$\mathcal{L}_{\mathrm{SCH}} \equiv \frac{1}{2} i\hbar \left\{ \Psi^* \overrightarrow{D_0} \Psi - \Psi^* \overleftarrow{D_0} \Psi \right\} - \frac{\hbar^2}{2m} \Psi^* \overleftarrow{\mathbf{D}} \cdot \overrightarrow{\mathbf{D}} \Psi. \tag{24.47}$$

It is this new equation that is discretized, using the definitions given in Section 24.5. The DT gauge-invariant system function density is therefore

$$\begin{aligned} \mathcal{F}^n_{\mathrm{SCH}} = T \Big\langle & \frac{1}{2} i\hbar \left\{ \Psi^*_{n\lambda} \overrightarrow{D}_{n\lambda} \Psi_{n\lambda} - \Psi^*_{n\lambda} \overleftarrow{D}_{n\lambda} \Psi_{n\lambda} \right\} \\ & - \frac{\hbar^2}{2m} \Psi^*_{n\lambda} \overleftarrow{\mathbf{D}}_{n\lambda} \cdot \overrightarrow{\mathbf{D}}_{n\lambda} \Psi_{n\lambda} \Big\rangle . \end{aligned} \tag{24.48}$$

24.6.2 The charged DT Klein–Gordon equation

The free charged Klein–Gordon Lagrangian

$$\mathcal{L}_{\mathrm{KG}} \equiv \varphi^* \overleftarrow{\partial_\mu} \overrightarrow{\partial^\mu} \varphi - m^2 \varphi^* \varphi \tag{24.49}$$

when coupled to the electromagnetic field is transformed into the gauge-invariant CT Lagrange density

$$\mathcal{L}_{\mathrm{KG}} \equiv \varphi^* \overleftarrow{D_\mu} \overrightarrow{D^\mu} \varphi - m^2 \varphi^* \varphi = \varphi \overleftarrow{D_0} \overrightarrow{D_0} \varphi - \varphi^* \overleftarrow{\mathbf{D}} \cdot \overrightarrow{\mathbf{D}} \varphi - m^2 \varphi^* \varphi. \tag{24.50}$$

Hence, according to the above prescription, the gauge-invariant system function density is given by

$$\mathcal{F}^n_{\mathrm{KG}} \equiv T \langle \varphi^*_{n\lambda} \overleftarrow{D}_{n\lambda} \overrightarrow{D}_{n\lambda} \varphi_{n\lambda} - \varphi^*_{n\lambda} \overleftarrow{\mathbf{D}}_{n\lambda} \cdot \overrightarrow{\mathbf{D}}_{n\lambda} \varphi_{n\lambda} - m^2 \varphi^*_{n\lambda} \varphi_{n\lambda} \rangle, \tag{24.51}$$

which is equivalent to a very long expression, given by Jaroszkiewicz and Norton (1997b), once the integration over λ has been carried out. This system function is gauge-invariant.

24.6.3 The charged DT Dirac equation

The CT free Dirac Lagrange density (23.25) is replaced by

$$\mathcal{L}_{\mathrm{D}} \equiv \frac{1}{2} i \bar{\psi} \gamma^\mu \overrightarrow{D_\mu} \psi - \frac{1}{2} i \bar{\psi} \overleftarrow{D_\mu} \gamma^\mu \psi - m \bar{\psi} \psi, \tag{24.52}$$

and therefore the discretized system function is

$$\mathcal{F}_{\mathrm{D}}^{n} \equiv T\langle \tfrac{1}{2}i\bar{\psi}_{n\lambda}\gamma^{0}\,\overrightarrow{D}_{n\lambda}\psi_{n\lambda} - \tfrac{1}{2}i\bar{\psi}_{n\lambda}\,\overleftarrow{D}_{n\lambda}\gamma^{0}\psi_{n\lambda}$$
$$- \tfrac{1}{2}i\bar{\psi}_{n\lambda}\boldsymbol{\gamma}\cdot\overrightarrow{\mathbf{D}}_{n\lambda}\psi_{n\lambda} + \tfrac{1}{2}i\bar{\psi}_{n\lambda}\,\overleftarrow{\mathbf{D}}_{n\lambda}\cdot\boldsymbol{\gamma}\psi_{n\lambda} - m\bar{\psi}_{n\lambda}\psi_{n\lambda}\rangle. \quad (24.53)$$

24.6.4 Comments

It is clear from these system functions that DT QED will be more complex than CT QED. A considerable notational economy can be made by not evaluating virtual-path integrals until absolutely necessary.

25
The discrete time Skyrme model

In this chapter, the virtual-path approach to temporal discretization is applied to the Skyrme model (Jaroszkiewicz and Nikolaev, 2001), to demonstrate an application to a field theory with non-abelian gauge group symmetry.

25.1 The Skyrme model

The Skyrme model (Skyrme, 1961) stands somewhere between scalar field theories and non-abelian gauge theories of vector bosons, having soliton solutions with a number of properties suggestive of baryon physics. Its basic dynamical degrees of freedom are spacetime fields $U(x)$ that take values in SU(2). In order to deal with the constraints this theory has, it is convenient to parametrize these fields in terms of an unconstrained isotopic triplet of real scalar fields $\boldsymbol{\pi} \equiv (\pi^1, \pi^2, \pi^3)$:

$$U \equiv \exp\{i\boldsymbol{\tau} \cdot \boldsymbol{\pi}\} = \cos(|\boldsymbol{\pi}|) + i\frac{\boldsymbol{\tau} \cdot \boldsymbol{\pi}}{|\boldsymbol{\pi}|} \sin(|\boldsymbol{\pi}|), \tag{25.1}$$

where $\boldsymbol{\tau} \equiv (\tau^1, \tau^2, \tau^3)$ are the Pauli matrices. With the definitions

$$U_\mu \equiv \partial_\mu U, \quad L_\mu \equiv U^\dagger \, \partial_\mu U = -\partial_\mu U^\dagger \, U \tag{25.2}$$

the Lagrange density may be written in the form

$$\mathcal{L}_{\mathrm{S}} \equiv -\frac{F_\Pi^2}{16} \operatorname{Tr}(L_\mu L^\mu) - \frac{1}{32e^2} \operatorname{Tr}[L_\mu, L_\nu][L^\mu, L^\nu], \tag{25.3}$$

where F_Π is the pion coupling constant and the second term is known as the *Skyrme term*. In terms of the U fields this is equivalent to

$$\mathcal{L}_{\mathrm{S}} = \frac{F_\Pi^2}{16} \operatorname{Tr}\big(U_\mu^\dagger U^\mu\big) - \frac{1}{16e^2} \operatorname{Tr}\big\{U_\mu^\dagger U_\nu U^{\dagger\mu} U^\nu - U_\mu^\dagger U_\nu U^{\dagger\nu} U^\mu\big\}. \tag{25.4}$$

In addition to the standard Poincaré symmetries, an important symmetry of this Lagrangian is invariance under separate left and right SU(2) transformations;

$$U \to U' \equiv AUB^\dagger, \tag{25.5}$$

where A and B are spacetime-independent elements of SU(2). This generates the so-called axial and vector charges.

In Appendix C the quaternion approach to the parametrization of the U variables is given. With this approach, a given U may be written in the form

$$U = q_\mu \varphi^\mu = \varphi^0 + i\tau^i \varphi^i, \tag{25.6}$$

where $q_0 \equiv I_2$, the 2×2 identity matrix, and $q_i \equiv i\tau^i$, $i = 1, 2, 3$, have the properties of the quaternions i, j and k. The four real fields φ^μ are read off from (25.1) to be

$$\varphi^0 \equiv \cos(|\boldsymbol{\pi}|), \qquad \varphi^i \equiv \sin(|\boldsymbol{\pi}|)n^i, \qquad \mathbf{n} \cdot \mathbf{n} = 1. \tag{25.7}$$

Since there are only three independent parameters describing the elements of SU(2), the four components φ^μ are constrained to the surface of S^3, the unit sphere in four dimensions, i.e.,

$$\varphi^\mu \varphi^\mu = 1. \tag{25.8}$$

With this reparametrization the Lagrange density becomes

$$\begin{aligned} \mathcal{L}_S = & \frac{\alpha^2}{2}\, \partial_\mu \varphi^\alpha\, \partial^\mu \varphi^\alpha - \frac{\beta^2}{4}\, \partial_\mu \varphi^\alpha\, \partial_v \varphi^\beta \{\partial^\mu \varphi^\beta\, \partial^v \varphi^\alpha - \partial^\mu \varphi^\alpha\, \partial^\nu \varphi^\beta\} \\ & + \frac{1}{2}\mu(\varphi^\alpha \varphi^\alpha - 1), \end{aligned} \tag{25.9}$$

where $\alpha^2 \equiv F_\Pi^2/4$, $\beta^2 \equiv e^{-2}$ and the Lagrange multiplier μ enforces the S^3 constraint (25.8). Then the conjugate momenta are given by

$$\pi^\alpha \equiv \frac{\partial \mathcal{L}}{\partial \dot{\varphi}^\alpha} = M_{\alpha\beta} \dot{\varphi}^\beta, \tag{25.10}$$

where

$$M_{\alpha\beta} = (\alpha^2 - \beta^2\, \partial_i \varphi^\mu\, \partial_i \varphi^\mu)\delta_{\alpha\beta} + \beta^2\, \partial_i \varphi^\alpha\, \partial_i \varphi^\beta. \tag{25.11}$$

The constraints turn out to be second class in the terminology of Dirac (1964) and are given by

$$\chi_1 \equiv \varphi^\alpha \varphi^\alpha - 1 \approx 0, \quad \chi_2 \equiv \varphi^\alpha \pi^\alpha \approx 0. \tag{25.12}$$

The non-zero Dirac brackets are evaluated to be

$$\begin{aligned} \{\pi^\alpha_{\mathbf{x}}, \varphi^\beta_{\mathbf{y}}\}_{DB} &= \left(\varphi^\alpha_{\mathbf{x}} \varphi^\beta_{\mathbf{x}} - \delta_{\alpha\beta}\right) \delta^3(\mathbf{x} - \mathbf{y}), \\ \{\pi^\alpha_{\mathbf{x}}, \pi^\beta_{\mathbf{y}}\}_{DB} &= \left(\pi^\alpha_{\mathbf{x}} \varphi^\beta_{\mathbf{x}} - \pi^\beta_{\mathbf{x}} \varphi^\alpha_{\mathbf{x}}\right) \delta^3(\mathbf{x} - \mathbf{y}). \end{aligned} \tag{25.13}$$

It is these which should be used in the quantization of the fields.

25.2 The SU(2) particle

25.2.1 Continuous time

In this section the most basic variant of the Skyrme model is considered, which is to drop the Skyrme term and the spatial degrees of freedom. The Lagrange density then reduces to

$$L = \frac{1}{4}\alpha^2 \operatorname{Tr}(\dot{U}^\dagger \dot{U}), \tag{25.14}$$

where $U \equiv U(t)$ is a time-dependent element of SU(2) and $\alpha = F_{\pi/2}$. The number of independent real dynamical variables is three and there are two alternative formulations.

The π fields

The U fields may be parametrized using three unconstrained real fields: $U(t) = \exp\{i\boldsymbol{\tau} \cdot \boldsymbol{\pi}(t)\}$, where $\boldsymbol{\pi}(t) \equiv F(t)\mathbf{n}(t)$ is an element of $\mathcal{R}^3$ and $\mathbf{n}(t)$ is a unit three-vector. Then

$$U(t) = \cos F + i \sin F \,\boldsymbol{\tau} \cdot \mathbf{n}. \tag{25.15}$$

The mapping from the $\mathcal{R}^3$ space of the parameters $\boldsymbol{\pi}$ to SU(2) is many to one, with the vectors $(F + 2k\pi)\mathbf{n}$, k an integer, mapping into the same point of SU(2). The Lagrangian (25.14) then reduces to

$$L = \frac{1}{2}\alpha^2\left\{\dot{F}^2 + \dot{\mathbf{n}} \cdot \dot{\mathbf{n}} \sin^2 F\right\} + \frac{1}{2}\mu(\mathbf{n} \cdot \mathbf{n} - 1), \tag{25.16}$$

where a Lagrange multiplier is included to enforce the normalization condition on the unit vector $\mathbf{n}$. The equations of motion are

$$\begin{gathered} \ddot{F} - \sin F \cos F \,\dot{\mathbf{n}} \cdot \dot{n} \underset{\text{c}}{=} 0, \\ \alpha^2 \sin^2 F \,\ddot{\mathbf{n}} + 2\alpha^2 \sin F \cos F \,\dot{F}\dot{\mathbf{n}} \underset{\text{c}}{=} \mu\mathbf{n}, \quad \mathbf{n} \cdot \mathbf{n} \underset{\text{c}}{=} 1. \end{gathered} \tag{25.17}$$

In phase space the system has two second-class constraints in the language of Dirac (1964). Now define p and $\mathbf{p}$ to be the momenta conjugate to F and $\mathbf{n}$, respectively. Then these constraints take the form

$$\chi_1 \equiv \mathbf{n} \cdot \mathbf{n} - 1 \approx 1, \quad \chi_2 \equiv \mathbf{n} \cdot \mathbf{p} \approx 0. \tag{25.18}$$

Following Dirac (1964), the Dirac brackets can be constructed in the standard way, giving the non-zero brackets

$$\{p, F\}_{\text{D}} = -1, \quad \left\{p^i, n^j\right\}_{\text{D}} = -\delta_{ij} + n^i n^j, \quad \left\{p^i, p^j\right\}_{\text{D}} = p^i n^j - p^j n^i. \tag{25.19}$$

The total Hamiltonian is given by

$$H_{\text{T}} = \frac{p^2}{2\alpha^2} + \frac{\mathbf{p} \cdot \mathbf{p}}{2\alpha^2 \sin^2 F}, \tag{25.20}$$

which is an invariant of the motion.

Two additional invariants of the motion can be found using Noether's theorem by observing that the transformation $U \to U' \equiv AUB^\dagger$ is a symmetry of the Lagrangian, where A and B are space- and time-independent elements of SU(2). On writing $A \simeq 1 + i\boldsymbol{\tau}\cdot\mathbf{a}$, $B \simeq 1 + i\boldsymbol{\tau}\cdot\mathbf{b}$, where $\mathbf{a}$ and $\mathbf{b}$ are infinitesimal,

$$\delta F = (\mathbf{a}-\mathbf{b})\cdot\mathbf{n}, \quad \delta\mathbf{n} = \cot F\{\mathbf{a}-\mathbf{b}-\mathbf{n}\cdot(\mathbf{a}-\mathbf{b})\mathbf{n}\} + \mathbf{n}\times(\mathbf{a}+\mathbf{b}) \tag{25.21}$$

to lowest order in the infinitesimal parameters. Now an application of Noether's theorem gives the conserved left and right charges

$$\mathbf{L} \equiv \mathbf{n}\times\mathbf{p} - p\mathbf{n} - \cot F\,\mathbf{p}, \quad \mathbf{R} \equiv \mathbf{n}\times\mathbf{p} + p\mathbf{n} + \cot F\,\mathbf{p} \tag{25.22}$$

in phase space. In configuration space they take the form

$$\begin{aligned}\mathbf{L} = &\ \alpha^2\Big\{\sin^2\!F\,\mathbf{n}\times\dot{\mathbf{n}} - \dot{F}\mathbf{n} - \cos F\sin F\,\dot{\mathbf{n}}\Big\} \equiv \mathbf{A} - \mathbf{V},\\ \mathbf{R} = &\ \alpha^2\Big\{\sin^2\!F\,\mathbf{n}\times\dot{\mathbf{n}} + \dot{F}\mathbf{n} + \cos F\sin F\,\dot{\mathbf{n}}\Big\} \equiv \mathbf{A} + \mathbf{V},\end{aligned} \tag{25.23}$$

where

$$\mathbf{A} \equiv \alpha^2\sin^2 F\,\mathbf{n}\times\dot{\mathbf{n}}, \quad V \equiv \alpha^2\Big(\dot{F}\mathbf{n} + \cos F\sin F\,\dot{\mathbf{n}}\Big) \tag{25.24}$$

are conserved separately. These are known conventionally as the *vector* and *axial* charges, respectively.

The φ fields

The Lagrangian takes the form

$$L \equiv \frac{1}{2}\alpha^2\dot{\varphi}^\alpha\dot{\varphi}^\alpha + \frac{1}{2}\mu(\varphi^\alpha\varphi^\alpha - 1) \tag{25.25}$$

and gives equations of motion

$$\alpha^2\ddot{\varphi}^\alpha \underset{c}{=} \mu\varphi^\alpha, \quad \varphi^\alpha\varphi^\alpha \underset{c}{=} 1, \tag{25.26}$$

i.e.,

$$\ddot{\varphi}^\alpha \underset{c}{=} \left(\varphi^\beta\ddot{\varphi}^\beta\right)\varphi^\alpha. \tag{25.27}$$

The Lagrangian is invariant with respect to the global SU(2) transformation $U' = AUB^\dagger$, where A and B are time-independent elements of SU(2). Now suppose that

$$A \simeq 1 + i\boldsymbol{\tau}\cdot\mathbf{a}, \quad B \simeq 1 + i\boldsymbol{\tau}\cdot\mathbf{b}, \tag{25.28}$$

where $\mathbf{a}$ and $\mathbf{b}$ are infinitesimal. Then

$$\delta\varphi^0 = (\mathbf{b}-\mathbf{a})\cdot\boldsymbol{\varphi}, \quad \delta\boldsymbol{\varphi} = (\mathbf{a}-\mathbf{b})\varphi^0 + \boldsymbol{\varphi}\times(\mathbf{a}+\mathbf{b}).$$

Hence the conserved charge is given by

$$\delta\rho = (\mathbf{a}-\mathbf{b})\cdot\mathbf{V} + (\mathbf{a}+\mathbf{b})\cdot\mathbf{A}, \tag{25.29}$$

where

$$\mathbf{V} \equiv \alpha^2(\varphi^0\dot{\boldsymbol{\varphi}} - \dot{\varphi}^0\boldsymbol{\varphi}), \quad \mathbf{A} \equiv \alpha^2\dot{\boldsymbol{\varphi}} \times \boldsymbol{\varphi}, \tag{25.30}$$

with $\dot{\mathbf{V}} \underset{c}{=} \dot{\mathbf{A}} \underset{c}{=} \mathbf{0}$.

25.2.2 Temporal discretization

Turning to the temporal discretization of the SU(2) particle system, the problem reduces to choosing a suitable virtual path between temporal notes. Appropriate paths are of the form

$$U_{n\lambda} \equiv \lambda U_{n+1} + (1-\lambda)U_n, \tag{25.31}$$

where the parameter $\lambda \in [0,1]$ interpolates temporal nodes and

$$U_n \equiv q_\alpha \varphi_n^\alpha, \qquad \varphi_n^\alpha \varphi_n^\alpha = 1. \tag{25.32}$$

Given the Lagrangian

$$L \equiv \frac{1}{4}\alpha^2 \, \mathrm{Tr}(\dot{U}^\dagger \dot{U}), \tag{25.33}$$

the system function is given by

$$F^n = \frac{1}{2}T\alpha^2 \, \mathbb{T}_n\varphi_n^\beta . \mathbb{T}_n\varphi_n^\beta + \frac{1}{4}\mu_n T(\varphi_n^\alpha\varphi_n^\alpha - 1) + \frac{1}{4}\mu_{n+1}T\big(\varphi_{n+1}^\alpha\varphi_{n+1}^\alpha - 1\big). \tag{25.34}$$

The DT equations of motion are

$$\frac{\alpha^2}{T}\big(\varphi_{n+1}^\alpha - 2\varphi_n^\alpha + \varphi_{n-1}^\alpha\big) \underset{c}{=} \mu_n T \varphi_n^\alpha, \quad \varphi_n^\alpha\varphi_n^\alpha \underset{c}{=} 1. \tag{25.35}$$

This is equivalent to

$$\varphi_{n+1}^\alpha + \varphi_{n-1}^\alpha \underset{c}{=} \varphi_n^\beta\left(\varphi_{n+1}^\beta + \varphi_{n-1}^\beta\right)\varphi_n^\alpha. \tag{25.36}$$

There is invariance under the transformation

$$U_n \to U_n' \equiv AU_nB^\dagger, \tag{25.37}$$

where A and B are SU(2) transformations. For infinitesimal transformations $A \simeq 1 + i\boldsymbol{\tau}\cdot\mathbf{a}$ and $B \simeq 1 + i\boldsymbol{\tau}\cdot\mathbf{b}$, we find the conserved charges

$$\begin{aligned}\mathbf{A}^n &\equiv \alpha^2 \, \mathbb{T}_n\boldsymbol{\varphi}_n \times \boldsymbol{\varphi}_n = \frac{\alpha^2}{T}\boldsymbol{\varphi}_{n+1} \times \boldsymbol{\varphi}_n, \\ \mathbf{V}^n &\equiv \alpha^2[\varphi_n^0 \, \mathbb{T}_n\boldsymbol{\varphi}_n - \mathbb{T}_n\varphi_n^0 . \boldsymbol{\varphi}_n] = \frac{\alpha^2}{T}\big(\varphi_n^0\boldsymbol{\varphi}_{n+1} - \varphi_{n+1}^0\boldsymbol{\varphi}_n\big).\end{aligned} \tag{25.38}$$

If these charges are non-zero, then there is a natural time-independent frame in isospace given by the directions $(\mathbf{V}^n, \mathbf{A}^n, \mathbf{A}^n \times \mathbf{V}^n)$.

The following argument simplifies the equations of motion and establishes the existence of an infinite hierarchy of quadratic invariants. First note that, regardless of the equations of motion, the quantity $U_{n+1}^{\dagger}U_n$ is an element of SU(2), and so may be written in the form

$$U_{n+1}^{\dagger}U_n = q_\lambda \Phi_n^\lambda \tag{25.39}$$

using the quaternionic notation discussed in Appendix C, where

$$\Phi_n^\lambda \equiv c^{\alpha\beta}{}_\lambda \varphi_{n+1}^\alpha \varphi_n^\beta \tag{25.40}$$

and

$$\Phi_n^\lambda \Phi_n^\lambda = 1. \tag{25.41}$$

In detail, the components of Φ^λ turn out to be

$$\Phi_n^0 = \varphi_n^\alpha \varphi_{n+1}^\alpha, \qquad \mathbf{\Phi}_n = \mathbf{v}^n - \mathbf{a}^n, \tag{25.42}$$

where

$$\mathbf{v}^n \equiv \frac{T}{\alpha^2}\mathbf{V}^n = \varphi_n^0 \boldsymbol{\varphi}_{n+1} - \varphi_{n+1}^0 \boldsymbol{\varphi}_n, \quad \mathbf{a}^n \equiv \frac{T}{\alpha^2}\mathbf{A}^n = \boldsymbol{\varphi}_{n+1} \times \boldsymbol{\varphi}_n. \tag{25.43}$$

From this it follows that $(\Phi_n^0)^2$ is an invariant of the motion, namely

$$(\Phi_n^0)^2 \underset{c}{=} (\Phi_{n-1}^0)^2, \tag{25.44}$$

so we may write

$$\varphi_n^\alpha \varphi_{n+1}^\alpha \underset{c}{=} C\varepsilon_n, \tag{25.45}$$

for some real constant C and where $\varepsilon_n = \pm 1$. It turns out that

$$C^2 + \mathbf{v}_n^2 + \mathbf{a}_n^2 = 1, \tag{25.46}$$

which means that $-1 \leq C \leq 1$. Hence the equation of motion can be written in the form

$$\varphi_{n+1}^\alpha + \varphi_{n-1}^\alpha \underset{c}{=} C(\varepsilon_n + \varepsilon_{n-1})\varphi_n^\alpha, \tag{25.47}$$

where the ε_n are of magnitude $+1$ but otherwise arbitrary. This arbitrariness can be traced to the use of the Lagrange multipliers μ_n in the system function (25.34) and is not a feature that exists in the CT limit $T \to 0$.

In the special case that $\varepsilon_n = +1$ for all n, the equation is recognized to be equivalent to the DT harmonic oscillator discussed in Chapter 13. Moreover, the bounds on the constant C mean that the motion is never hyperbolic. In this case it is found that

$$\varphi_n^\alpha \varphi_n^\alpha = 1, \qquad \varphi_{n+1}^\alpha \varphi_n^\alpha = C, \qquad \varphi_{n+2}^\alpha \varphi_n^\alpha = 2C^2 - 1 \tag{25.48}$$

and generalizing this result gives the infinite set of invariants

$$\varphi_{n+m}^\alpha \varphi_n^\alpha = T_m(C), \tag{25.49}$$

where T_m is a Chebyshev polynomial of Type 1 (Arfken, 1985).

In terms of the $F, \mathbf{n}$ description, the parametrization is given by

$$U_n = c_n + i s_n \mathbf{n}_n, \tag{25.50}$$

where $c_n \equiv \cos F_n$, $s_n \equiv \sin F_n$ and $\mathbf{n}_n \cdot \mathbf{n}_n = 1$, and then

$$\mathbf{a}^n = s_n s_{n+1} \mathbf{n}_n \times \mathbf{n}_{n+1}, \quad \mathbf{v}^n = s_n c_{n+1} \mathbf{n}_n - s_{n+1} c_n \mathbf{n}_{n+1}. \tag{25.51}$$

25.3 The σ model

25.3.1 Continuous time

The model is now extended to include spatial dependence, but not the quartic terms in the original Lagrangian. We shall call this the σ model. The CT Lagrange density is now

$$\mathcal{L} = \frac{1}{4}\alpha^2 \operatorname{Tr}\left(\partial_\mu U^\dagger \, \partial^\mu U\right), \tag{25.52}$$

which is equivalent to

$$\mathcal{L} = \frac{1}{2}\alpha^2 \, \partial_\mu \varphi^\alpha \, \partial^\mu \varphi^\alpha + \frac{1}{2}\mu(\varphi^\alpha \varphi^\alpha - 1). \tag{25.53}$$

The equations of motion are

$$\alpha^2 \, \Box \varphi^\alpha \underset{c}{=} \mu \varphi^\alpha, \qquad \varphi^\alpha \varphi^\alpha \underset{c}{=} 1, \tag{25.54}$$

which reduce to

$$\Box \varphi^\alpha \underset{c}{=} \left(\varphi^\beta \, \Box \varphi^\beta\right) \varphi^\alpha. \tag{25.55}$$

The conserved energy-momentum tensor density is

$$\mathcal{T}^{\mu\nu} = \alpha^2 \left\{ \partial^\mu \varphi^\alpha \, \partial^\nu \varphi^\alpha - \frac{1}{2}\eta^{\mu\nu} \, \partial_\beta \varphi^\alpha \, \partial^\beta \varphi^\alpha \right\}. \tag{25.56}$$

The invariance of the Lagrange density with respect to the same transformation as before allows the vector and axial currents to be determined. Under the infinitesimal transformation

$$U \to U' \equiv (1 + i\boldsymbol{\tau} \cdot \mathbf{a}) U (1 - i\boldsymbol{\tau} \cdot \mathbf{b}) \tag{25.57}$$

the fields change according to the rule

$$\delta\varphi^0 = (\mathbf{b} - \mathbf{a}) \cdot \boldsymbol{\varphi}, \quad \delta\boldsymbol{\varphi} = (\mathbf{a} - \mathbf{b})\varphi^0 + \boldsymbol{\varphi} \times (\mathbf{a} + \mathbf{b}), \tag{25.58}$$

giving the conserved currents

$$\mathbf{V}^\mu = \varphi^0 \, \partial^\mu \boldsymbol{\varphi} - \partial^\mu \varphi^0 \, \boldsymbol{\varphi}, \quad A^\mu = \partial^\mu \boldsymbol{\varphi} \times \boldsymbol{\varphi}. \tag{25.59}$$

25.3.2 Discrete time

The system function density for the σ model is taken to be

$$\mathcal{F}^n = \frac{1}{2}T\alpha^2\{\mathbb{T}_n\varphi_n^\alpha . \mathbb{T}_n\varphi_n^\alpha - \tfrac{1}{2}\ \nabla\varphi_n^\alpha \cdot \nabla\varphi_n^\alpha - \nabla\varphi_{n+1}^\alpha \cdot \nabla\varphi_{n+1}^\alpha\} \\ + \frac{1}{4}T\mu_n(\varphi_n^\alpha\varphi_n^\alpha - 1) + \frac{1}{4}T\mu_{n+1}(\varphi_{n+1}^\alpha\varphi_{n+1}^\alpha - 1), \qquad (25.60)$$

which gives the equation of motion

$$\frac{\alpha^2}{T}\{\varphi_{n+1}^\alpha - 2\varphi_n^\alpha + \varphi_{n-1}^\alpha\} - T\alpha^2\ \nabla^2\varphi_n^\alpha \underset{c}{=} T\mu_n\varphi_n^\alpha, \qquad \varphi_n^\alpha\varphi_n^\alpha \underset{c}{=} 1. \qquad (25.61)$$

There is invariance of the system function density under the infinitesimal transformation (25.57), and this gives two conserved currents:

$$\mathbb{T}_n\mathbf{V}_n^0 + \partial_i\mathbf{V}_n^i \underset{c}{=} \mathbf{0}, \quad \mathbb{T}_n\mathbf{A}_n^0 + \partial_i\mathbf{A}_n^i \underset{c}{=} \mathbf{0}, \qquad (25.62)$$

where

$$\begin{aligned} \mathbf{V}_n^0 &\equiv \frac{\alpha^2}{T}\{\varphi_n^0\boldsymbol{\varphi}_{n+1} - \boldsymbol{\varphi}_n\varphi_{n+1}^0\} + \frac{1}{2}T\alpha^2\{\boldsymbol{\varphi}_n\ \nabla^2\varphi_n^0 - \varphi_n^0\ \nabla^2\boldsymbol{\varphi}_n\}, \\ \mathbf{V}_n^i &\equiv \frac{1}{2}\alpha^2\{\boldsymbol{\varphi}_n\ \partial_i\varphi_n^0 - \varphi_n^0\ \partial_i\boldsymbol{\varphi}_n + \boldsymbol{\varphi}_{n+1}\ \partial_i\varphi_{n+1}^0 - \varphi_{n+1}^0\ \partial_i\boldsymbol{\varphi}_{n+1}\} \\ \mathbf{A}_n^0 &\equiv \frac{\alpha^2}{T}\boldsymbol{\varphi}_{n+1} \times \boldsymbol{\varphi}_n - \frac{1}{2}T\alpha^2\ \nabla^2\boldsymbol{\varphi}_n \times \boldsymbol{\varphi}_n, \\ \mathbf{A}_n^i &\equiv -\frac{1}{2}\alpha^2\ \partial_i\boldsymbol{\varphi}_n \times \boldsymbol{\varphi}_n - \frac{1}{2}\alpha^2\ \partial_i\boldsymbol{\varphi}_{n+1} \times \boldsymbol{\varphi}_{n+1}. \end{aligned} \qquad (25.63)$$

25.4 Further considerations

The full Skyrme Lagrange density equation (25.9) may be written in the form

$$\mathcal{L}_S = \frac{1}{2}\dot{\varphi}^\alpha M^{\alpha\beta}(\partial_i\varphi^\mu)\dot{\varphi}^\beta - W(\partial_i\varphi^\mu) + \frac{1}{2}\mu(\varphi^\alpha\varphi^\alpha - 1), \qquad (25.64)$$

where $M^{\alpha\beta}(\partial_i\varphi^\mu)$ is given by equation (25.11) and the potential function $W(\partial_i\varphi^\mu)$ contains quadratic and quartic terms in the derivatives of the fields. There are several possible discretizations of the Lagrange density (25.64), such as

$$\mathcal{F}^n = \frac{1}{2}T\ \mathbb{T}_n\varphi_n^\alpha M^{\alpha\beta}\left(\frac{1}{2}\ \partial_i\varphi_n^\mu + \frac{1}{2}\ \partial_i\varphi_{n+1}^\mu\right)\mathbb{T}_n\varphi_n^\beta - TW\left(\frac{1}{2}\ \partial_i\varphi_n^\mu + \frac{1}{2}\ \partial_i\varphi_{n+1}^\mu\right) \\ + \frac{1}{4}T\mu(\varphi_n^\alpha\varphi_n^\alpha - 1) + \frac{1}{4}T\mu(\varphi_{n+1}^\alpha\varphi_{n+1}^\alpha - 1) \qquad (25.65)$$

and

$$\begin{aligned} \mathcal{F}^n = &\frac{1}{4}T\ \mathbb{T}_n\varphi_n^\alpha\left\{M^{\alpha\beta}(\partial_i\varphi_n^\mu) + M^{\alpha\beta}(\partial_i\varphi_{n+1}^\mu)\right\}\mathbb{T}_n\varphi_n^\beta \\ &- \frac{1}{2}T\{W(\partial_i\varphi_n^\mu) + W(\partial_i\varphi_n^\mu)\} \\ &+ \frac{1}{4}T\mu(\varphi_n^\alpha\varphi_n^\alpha - 1) + \frac{1}{4}T\mu(\varphi_{n+1}^\alpha\varphi_{n+1}^\alpha - 1). \end{aligned} \qquad (25.66)$$

Whichever form is chosen, the symmetries discussed in previous sections will hold, because the variations are global. The conserved DT vector and axial charges can be readily worked out, and this is left as an exercise. There is now a guarantee that these are dynamical invariants, even though it might not be possible to put the equation of motion into an explicit form.

To investigate a conserved energy, the Type-3 discretization of Lee may be considered (Lee, 1983). The system function density is changed according to the rule

$$\begin{aligned}\mathcal{F}^n &\equiv \mathcal{F}\big(\varphi_n^\alpha, \varphi_{n+1}^\alpha, \nabla\varphi_n^\alpha, \nabla\varphi_{n+1}^\alpha, T\big)\\ &\to \mathcal{F}^n(T_n) \equiv \mathcal{F}\big(\varphi_n^\alpha, \varphi_{n+1}^\alpha, \nabla\varphi_n^\alpha, \nabla\varphi_{n+1}^\alpha, T_n\big),\end{aligned} \tag{25.67}$$

where now T_n is the dynamical temporal interval between nodes n and $n+1$. The equations of motion for the extended variables will be SU(2) gauge-invariant and there will be an analogue of a conserved energy.

Part V

Discrete time quantum field theory

26

Discrete time quantum field theory

26.1 Introduction

In the following, we work in natural units, where $c = \hbar = 1$. In this section, we discuss general principles for calculating n-point functions in discrete time (DT) quantum field theory. Then, in the sections following, we discuss the DT field quantization for the Klein–Gordon field, the Dirac field and the electromagnetic fields.

We make extensive use of the quantized DT oscillator discussed in previous chapters, because it is the basis for the construction of quantized free-particle states in continuous time (CT) quantum field theory.

The source-functional techniques of Schwinger, which were developed originally for CT field theory (Schwinger, 1969), have proved remarkably adaptable to DT field theory. The system function for a system under observation with a scalar field φ degree of freedom coupled to an external source j is chosen to be

$$F^n[j] = F^n + \tfrac{1}{2}T \int d^3\mathbf{x}\{j_{n+1}\varphi_{n+1} + j_n\varphi_n\}, \tag{26.1}$$

where $F^n \equiv \int d^3\mathbf{x}\, \mathcal{F}^n$ is the system function in the absence of the source. There are other ways of introducing sources into the system, but the above method has been found to be the most practical. Since these sources are eventually set to zero, the details of the coupling appear not to matter, provided that all fields are dealt with consistently according to the principles of DT mechanics.

The DT equation of motion is the functional derivative

$$\frac{\delta}{\delta\varphi_n(\mathbf{x})}\{F^n + F^{n-1}\} + Tj_n(\mathbf{x}) \underset{c}{=} 0, \tag{26.2}$$

which reduces to

$$\frac{\partial}{\partial\varphi_n}\{\mathcal{F}^n + \mathcal{F}^{n-1}\} - \nabla \cdot \frac{\partial}{\partial\nabla\varphi_n}\{\mathcal{F}^n + \mathcal{F}^{n-1}\} + Tj_n \underset{c}{=} 0, \tag{26.3}$$

where $\mathcal{F}^n$ is the system function density in the absence of sources.

The action sum $A_{NM}[j]$ for evolution between times MT and NT in the presence of sources is

$$\begin{aligned} A_{NM}[j] = A_{NM} &+ \frac{1}{2}T\int d^3\mathbf{x}\{j_M\varphi_M + j_N\varphi_N\} \\ &+ T\sum_{n=M+1}^{N-1}\int d^3\mathbf{x}\; j_n\varphi_n, \qquad M < N. \end{aligned} \tag{26.4}$$

The DT Schwinger action principle (Jaroszkiewicz and Norton, 1997a)

$$\delta\langle\phi, N|\psi, M\rangle^j = i\langle\phi, N|\delta\hat{A}_{NM}[j]|\psi, M\rangle^j, \qquad M < N, \tag{26.5}$$

leads to the functional derivatives

$$\begin{aligned} -\frac{i}{T}\frac{\delta}{\delta j_M(\mathbf{x})}\langle\phi, N|\psi, M\rangle^j &= \frac{1}{2}\langle\phi, N|\hat{\varphi}_M(\mathbf{x})|\psi, M\rangle^j, \\ -\frac{i}{T}\frac{\delta}{\delta j_n(\mathbf{x})}\langle\phi, N|\psi, M\rangle^j &= \langle\phi, N|\hat{\varphi}_n(\mathbf{x})|\psi, M\rangle^j, \qquad M < n < N, \\ -\frac{i}{T}\frac{\delta}{\delta j_N(\mathbf{x})}\langle\phi, N|\psi, M\rangle^j &= \frac{1}{2}\langle\phi, N|\hat{\varphi}_N(\mathbf{x})|\psi, M\rangle^j. \end{aligned} \tag{26.6}$$

Schwinger's function approach provides an elegant and powerful way of generating all of the relevant n-point functions (Green functions) by the further application of functional differentiation. We find, for example,

$$\frac{-i}{T}\frac{\delta}{\delta j_n(\mathbf{x})}\frac{-i}{T}\frac{\delta}{\delta j_m(\mathbf{y})}\langle\phi, N|\psi, M\rangle^j = \langle\phi, N|\tilde{T}\hat{\varphi}_n(\mathbf{x})\hat{\varphi}_m(\mathbf{y})|\psi, M\rangle^j \tag{26.7}$$

($M < m, n < N$), where $\tilde{T}$ denotes the DT ordering operator. Specifically,

$$\tilde{T}\hat{\varphi}_n(\mathbf{x})\hat{\varphi}_m(\mathbf{y}) = \begin{cases} \hat{\varphi}_n(\mathbf{x})\hat{\varphi}_m(\mathbf{y}), & n > m, \\ \frac{1}{2}(\hat{\varphi}_n(\mathbf{x})\hat{\varphi}_n(\mathbf{y}) + \hat{\varphi}_n(\mathbf{y})\hat{\varphi}_n(\mathbf{x})), & n = m, \\ \hat{\varphi}_m(\mathbf{y})\hat{\varphi}_n(\mathbf{x}), & m > n, \end{cases} \tag{26.8}$$

and similarly for higher-order n-point functions.

In common with CT field theory, our interest will be in scattering amplitudes taken over infinite interaction time, parametrized by the scattering limit $N = -M \to \infty$. We shall follow the general principles of S-matrix theory (Bjorken and Drell, 1965; Eden *et al.*, 1966), working with matrix elements involving the '*in*' and '*out*' vacua. We shall restrict our calculations to such matters. This means that we will discuss the r-point functions defined by

$$\begin{aligned} G^j_{n_1 n_2 \ldots n_r}(\mathbf{x}_1, \ldots, \mathbf{x}_r) &= \frac{\langle 0^{\text{out}}|\tilde{T}\hat{\varphi}_{n_1}(\mathbf{x}_1)\ldots\hat{\varphi}_{n_r}(\mathbf{x}_r)|0^{\text{in}}\rangle^j}{\langle 0^{\text{out}}|0^{\text{in}}\rangle^j} \\ &= \frac{1}{Z[j]}\frac{-i\delta}{T\delta j_{n_1}(\mathbf{x}_1)}\cdots\frac{-i\delta}{T\delta j_{n_r}(\mathbf{x}_r)}Z[j], \end{aligned} \tag{26.9}$$

where $Z[j] = \langle 0^{\text{out}}|0^{\text{in}}\rangle^j$ is the ground-state (vacuum) functional in the presence of the sources and $\tilde{T}$ denotes DT ordering.

An important question here concerns the existence of the ground state. In common with most CT field theories, we have no general proof that a ground state exists for interacting DT field theories. Moreover, in DT mechanics there is no Hamiltonian as such, so the question of the meaning of a 'ground state' becomes problematical.

Fortunately, for free DT fields, there will be a compatible operator corresponding to some appropriate DT Logan invariant that will be the nearest analogue to the Hamiltonian in CT theory. Moreover, the appropriate compatible operator for free neutral scalar fields is positive definite and this allows a meaning for the DT in and out vacua to be given.

26.2 The discrete time free quantized scalar field

26.2.1 The propagator

Given the CT Lagrange density $\mathcal{L}_0 = \frac{1}{2}\partial_\mu\varphi\,\partial^\mu\varphi - \frac{1}{2}\mu^2\varphi^2$, we use the virtual-path approach discussed above to find the system function density

$$\begin{aligned}\mathcal{F}_0^n &= \frac{(\varphi_{n+1}-\varphi_n)^2}{2T} - \frac{T}{6}\left(|\nabla\varphi_{n+1}|^2 + \nabla\varphi_{n+1}\cdot\nabla\varphi_n + |\nabla\varphi_n|^2\right) \\ &\quad - \frac{\mu^2 T}{6}\left(\varphi_{n+1}^2 + \varphi_{n+1}\varphi_n + \varphi_n^2\right). \end{aligned} \tag{26.10}$$

In the presence of the sources we take

$$F_0^n[j] = F_0^n + \frac{1}{2}T\int d^3\mathbf{x}\{j_{n+1}\varphi_{n+1} + j_n\varphi_n\} \tag{26.11}$$

and then the DT equation of motion is

$$\mathbb{T}_n\bar{\mathbb{T}}_n\varphi_n + \mathbb{S}_n(\mu^2 - \nabla^2)\varphi_n \underset{c}{=} j_n. \tag{26.12}$$

We define the spatial Fourier transforms

$$\begin{aligned}\tilde{\varphi}_n(\mathbf{p}) &\equiv \int d^3\mathbf{x}\, e^{-i\mathbf{p}\cdot\mathbf{x}}\varphi_n(\mathbf{x}), \\ \tilde{j}_n(\mathbf{p}) &\equiv \int d^3\mathbf{x}\, e^{-i\mathbf{p}\cdot\mathbf{x}} j_n(\mathbf{x}),\end{aligned} \tag{26.13}$$

and then the equation of motion becomes

$$\{\mathbb{T}_n\bar{\mathbb{T}}_n + \mathbb{S}_n E_{\mathbf{p}}^2\}\tilde{\varphi}_n(\mathbf{p}) \underset{c}{=} \tilde{j}_n(\mathbf{p}), \tag{26.14}$$

where $E_{\mathbf{p}} \equiv \sqrt{\mathbf{p}\cdot\mathbf{p}+\mu^2}$. The solution to (26.14) with Feynman scattering boundary conditions is

$$\tilde{\varphi}_n(\mathbf{p}) = \tilde{\varphi}_n^{(0)}(\mathbf{p}) - T\sum_{m=-\infty}^{\infty}\tilde{\Delta}_F^{n-m}(\mathbf{p})\tilde{j}_m(\mathbf{p}), \tag{26.15}$$

where $\tilde{\varphi}_n^{(0)}(\mathbf{p})$ is a solution to the homogeneous equation

$$\{\mathbb{T}_n\bar{\mathbb{T}}_n + \mathbb{S}_n E_{\mathbf{p}}^2\}\tilde{\varphi}_n^{(0)}(\mathbf{p}) = 0 \tag{26.16}$$

and $\tilde{\Delta}_{\mathrm{F}}^n(\mathbf{p})$ is the DT Feynman propagator in momentum space satisfying the equation

$$\{\mathbb{T}_n\bar{\mathbb{T}}_n + \mathbb{S}_n E_{\mathbf{p}}^2\}\tilde{\Delta}_{\mathrm{F}}^n(\mathbf{p}) = -\frac{\delta_n}{T}. \tag{26.17}$$

This equation for the propagator may be written in the form

$$\{\mathbb{U}_n - 2\eta_{\mathbf{p}} + \mathbb{U}_n^{-1}\}\tilde{\Delta}_{\mathrm{F}}^n(\mathbf{p}) = -\Gamma_{\mathbf{p}}\delta_n, \tag{26.18}$$

where

$$\Gamma_{\mathbf{p}} = \frac{6T}{6+\kappa_{\mathbf{p}}^2}, \qquad \eta_{\mathbf{p}} = \frac{6-2\kappa_{\mathbf{p}}^2}{6+\kappa_{\mathbf{p}}^2}, \qquad \kappa_{\mathbf{p}} \equiv TE_{\mathbf{p}}. \tag{26.19}$$

Using our knowledge gained with the DT harmonic oscillator propagator discussed previously, we may write down the solution for the propagator in the form

$$\begin{aligned}\tilde{\Delta}_{\mathrm{F}}^n(\mathbf{p}) &= \frac{\Gamma_{\mathbf{p}}}{2i\sin\theta_{\mathbf{p}}}e^{-i|n|\theta_{\mathbf{p}}}\\ &= \frac{\Gamma_{\mathbf{p}}}{2i\sin\theta_{\mathbf{p}}}\left\{e^{-in\theta_{\mathbf{p}}}\Theta_n + \delta_n + e^{in\theta_{\mathbf{p}}}\Theta_{-n}\right\},\end{aligned} \tag{26.20}$$

where $\eta_E \equiv \cos\theta_{\mathbf{p}}$. This expression holds for the elliptic and hyperbolic regimes with suitable analytic continuation. In the CT limit $T \to 0$, $nT \to t$, we recover the usual Feynman propagator in a spatially Fourier transformed form, viz.,

$$\begin{aligned}\lim_{T\to 0,\, n\to\infty,\, nT=t} \tilde{\Delta}_{\mathrm{F}}^n(\mathbf{p}) &= -\frac{i}{2E}\{e^{-itE_{\mathbf{p}}}\theta(t) + e^{itE_{\mathbf{p}}}\theta(-t)\}\\ &= \int \frac{d\omega}{2\pi}\frac{e^{-i\omega t}}{\omega^2 - \mathbf{p}\cdot\mathbf{p} - \mu^2 + i\epsilon}\\ &= \tilde{\Delta}_{\mathrm{F}}(\mathbf{p}, t) = \int d^3\mathbf{x}\, e^{-i\mathbf{p}\cdot\mathbf{x}}\Delta_{\mathrm{F}}(\mathbf{x}, t).\end{aligned} \tag{26.21}$$

Turning to the quantization process, the functional derivative satisfies the rule

$$\frac{\delta}{\delta j_n(\mathbf{x})}j_m(\mathbf{y}) = \delta_{n-m}\delta^3(\mathbf{x}-\mathbf{y}). \tag{26.22}$$

With the definition

$$\frac{\delta}{\delta\tilde{j}_n(\mathbf{p})} \equiv \frac{1}{(2\pi)^3}\int d^3\mathbf{x}\, e^{i\mathbf{p}\cdot\mathbf{x}}\frac{\delta}{\delta j_n(\mathbf{x})} \tag{26.23}$$

we find

$$\frac{\delta}{\delta\tilde{j}_n(\mathbf{p})}\tilde{j}_m(\mathbf{q}) = \delta_{n-m}\delta^3(\mathbf{p}-\mathbf{q}) \tag{26.24}$$

and then

$$\langle 0^{\mathrm{out}}|\varphi_n(\mathbf{x})|0^{\mathrm{in}}\rangle^j = -\frac{i}{T}\frac{\delta}{\delta j_n(\mathbf{x})}Z_0[j] \tag{26.25}$$

gives

$$\langle 0^{\text{out}}|\tilde{\varphi}_n(\mathbf{p})|0^{\text{in}}\rangle^j = -\frac{i\,(2\pi)^3}{T}\frac{\delta}{\delta\tilde{j}_n(-\mathbf{p})}Z_0[j]. \tag{26.26}$$

Hence we find

$$Z_0[j] = Z_0[0]\exp\left\{-\frac{1}{2}iT^2\sum_{n,m=-\infty}^{\infty}\int d^3\mathbf{x}\,d^3\mathbf{y}\,j_n(\mathbf{x})\Delta_{\mathrm{F}}^{n-m}(\mathbf{x}-\mathbf{y})j_m(\mathbf{y})\right\}, \tag{26.27}$$

where

$$\Delta_{\mathrm{F}}^n(\mathbf{x}) = \int\frac{d^3\mathbf{p}}{(2\pi)^3}e^{i\mathbf{p}\cdot\mathbf{x}}\tilde{\Delta}_{\mathrm{F}}^n(\mathbf{p}). \tag{26.28}$$

This propagator satisfies the equation

$$\{\Box_n + \mathbb{S}_n\mu^2\}\Delta_{\mathrm{F}}^n(\mathbf{x}) = -\frac{\delta_n}{T}\delta^3(\mathbf{x}), \tag{26.29}$$

where $\Box_n \equiv \mathbb{T}_n\bar{\mathbb{T}}_n - \mathbb{S}_n\nabla^2$.

26.2.2 The free scalar field commutators

In this section we use (26.27) to obtain the vacuum expectation value of the free scalar field commutators. An application of (26.9) gives

$$\langle 0|\hat{\varphi}_{n+1}(\mathbf{x})\hat{\varphi}_n(\mathbf{y})|0\rangle = i\Delta_{\mathrm{F}}^1(\mathbf{x}-\mathbf{y}), \tag{26.30}$$

from which we deduce

$$\begin{aligned}\langle 0|[\hat{\varphi}_{n+1}(\mathbf{x}),\hat{\varphi}_n(\mathbf{y})]|0\rangle &= -6iT\int\frac{d^3\mathbf{p}}{(2\pi)^3}\frac{e^{i\mathbf{p}\cdot(\mathbf{x}-\mathbf{y})}}{6+(\mathbf{p}\cdot\mathbf{p}+\mu^2)T^2}\\ &= -\frac{6i}{4\pi T|\mathbf{x}-\mathbf{y}|}e^{-\sqrt{\mu^2+6/T^2}|\mathbf{x}-\mathbf{y}|}.\end{aligned} \tag{26.31}$$

Both elliptic and hyperbolic regions of momentum space contribute to this result, which has the form of a Yukawa potential function.

We turn now to the direct approach to quantization. If we define the momentum $\pi_n(\mathbf{x})$ conjugate to $\varphi_n(\mathbf{x})$ via the rule

$$\pi_n(\mathbf{x}) \equiv -\frac{\delta}{\delta\varphi_n(\mathbf{x})}F_n \tag{26.32}$$

then (26.10) gives

$$\pi_n \equiv \mathbb{T}_n\varphi_n + \frac{T}{6}(\mu^2-\nabla^2)(\varphi_{n+1}+2\varphi_n) \tag{26.33}$$

for the free field. Now, naive canonical quantization for point-particle systems is equivalent in field theory terms to

$$[\hat{\pi}_n(\mathbf{x}),\hat{\varphi}_n(\mathbf{y})] = -i\delta^3(\mathbf{x}-\mathbf{y}), \tag{26.34}$$

from which we deduce

$$[\hat{\varphi}_{n+1}(\mathbf{x}), \hat{\varphi}_n(\mathbf{y})] = -6iT \int \frac{d^3\mathbf{p}}{(2\pi)^3} \frac{e^{i\mathbf{p}\cdot(\mathbf{x}-\mathbf{y})}}{6 + (\mu^2 + \mathbf{p}\cdot\mathbf{p})T^2}$$
$$= -\frac{6i}{4\pi T\,|\mathbf{x}-\mathbf{y}|} e^{-\sqrt{\mu^2+6/T^2}|\mathbf{x}-\mathbf{y}|}, \qquad (26.35)$$

assuming that $[\hat{\varphi}_n(\mathbf{x}), \hat{\varphi}_n(\mathbf{y})] = 0$. This is consistent with the approach to quantization via the Schwinger action principle from which we obtained (26.31).

The reason why this works is that the system function for a free field is an example of what we call a normal system. For interacting field theories this will no longer be the case and then the commutators analogous to the above will probably no longer be c-functions. We recall that, in CT field theories, interacting field commutators are not canonical in general either, so the analogies between DT and CT hold well here also.

For the free-field particle creation and annihilation operators we define the variables

$$a_n(\mathbf{p}) = i\Gamma_{\mathbf{p}}^{-1} e^{in\theta_{\mathbf{p}}} \left[\tilde{\varphi}_{n+1}(\mathbf{p}) - e^{i\theta_{\mathbf{p}}}\tilde{\varphi}_n(\mathbf{p})\right], \qquad (26.36)$$

where

$$\Gamma_{\mathbf{p}} = \frac{6T}{6+\kappa_{\mathbf{p}}^2}, \quad \tilde{\varphi}_n(\mathbf{p}) = \int d^3\mathbf{x}\, e^{-i\mathbf{p}\cdot\mathbf{x}} \varphi_n(\mathbf{x}) \qquad (26.37)$$

and the momentum $\mathbf{p}$ is restricted to the elliptic region. Then we find

$$\left[\hat{a}_n(\mathbf{p}), \hat{a}_n^\dagger(\mathbf{q})\right] = 2E_{\mathbf{p}}\sqrt{1 - T^2E_{\mathbf{p}}^2/12}(2\pi)^3 \delta^3(\mathbf{p}-\mathbf{q}) \qquad (26.38)$$

when we quantize and use (26.35). This tends to the correct CT limit as $T \to 0$.

If we interpret the factor $2E_{\mathbf{p}}\sqrt{1 - T^2E_{\mathbf{p}}^2/12}$ in the above as a particle flux density then this will be indistinguishable from the conventional density $2E_{\mathbf{p}}$ in CT field theory for normal momenta, but falls to zero as the parabolic barrier $TE_{\mathbf{p}_{\max}} = \sqrt{12}$ is approached from below. This suggests that there is in principle a physical limit to the possibility of creating particle states of extremely high momentum in the laboratory or of observing such particles in cosmic rays. This should have an effect on all discussions involving momentum space, such as particle decay lifetime and cross-section calculations, and, in the long term, on unified field theories.

26.3 The discrete time free quantized Dirac field

We turn now to the second quantization of the free Dirac equation in one time and three spatial dimensions (Norton and Jaroszkiewicz, 1998b). We expect some interesting structure to emerge in the DT version, because there are now four modes when the oscillon solutions are counted, rather than the two modes, namely the normal particle and normal antiparticle, found in the CT theory.

The source-free equations are discussed in Section 23.5, so we move straight on to the system function when an external fermionic source is included. In the presence of external fermionic sources η and $\bar{\eta}$, the system function density $\mathcal{F}^n$ becomes

$$\mathcal{F}^n_\eta(\mathbf{x}) = \mathcal{F}^n(\mathbf{x}) + \frac{1}{2}T\Big\{\bar{\eta}_n(\mathbf{x})\psi_n(\mathbf{x}) + \bar{\eta}_{n+1}(\mathbf{x})\psi_{n+1}(\mathbf{x}) + \bar{\psi}_n(\mathbf{x})\eta_n(\mathbf{x}) + \bar{\psi}_{n+1}(\mathbf{x})\eta_{n+1}(\mathbf{x})\Big\}, \tag{26.39}$$

which gives the Cadzow equation for ψ_n:

$$i\frac{\psi_{n+1} - \psi_{n-1}}{2T} \underset{c}{=} \frac{1}{6}\overrightarrow{H_D}\{\psi_{n+1} + 4\psi_n + \psi_{n-1}\} - \beta\eta_n, \tag{26.40}$$

where $\overrightarrow{H_D} = -i\boldsymbol{\alpha}\cdot\overrightarrow{\nabla} + \beta m$. A similar equation is readily derived for $\psi^\dagger$.

On taking Fourier transforms and with the definitions

$$\hat{\Omega}(\mathbf{p}) \equiv \frac{\kappa + 3i\hat{H}(\mathbf{p})}{\sqrt{9+\kappa^2}}, \qquad \kappa \equiv TE, \;\; \hat{H}(\mathbf{p}) \equiv \frac{H_D(\mathbf{p})}{E}, \qquad E \equiv \sqrt{\mathbf{p}\cdot\mathbf{p} + m^2}, \tag{26.41}$$

equation (26.40) becomes

$$\left(\hat{\Omega}^\dagger(\mathbf{p})\mathbb{U}_n - 2\eta + \hat{\Omega}(\mathbf{p})\mathbb{U}_n^{-1}\right)\tilde{\psi}_n(\mathbf{p}) \underset{c}{=} T\Gamma\hat{H}(\mathbf{p})\beta\tilde{\eta}_n(\mathbf{p}), \tag{26.42}$$

where

$$\Gamma \equiv \frac{6}{\sqrt{9+\kappa^2}}, \qquad \eta = -\frac{2\kappa}{\sqrt{9+\kappa^2}} \equiv \cos\theta. \tag{26.43}$$

A formal solution is

$$\tilde{\psi}_n(\mathbf{p}) = \tilde{\psi}_n^{(0)}(\mathbf{p}) - T\sum_{m=-\infty}^{\infty}\tilde{S}_F^{n-m}(\mathbf{p})\tilde{\eta}_m(\mathbf{p}), \tag{26.44}$$

where $\tilde{\psi}_n^{(0)}(\mathbf{p})$ is a solution to the free DT Dirac equation and the propagators

$$\tilde{S}_F^n(\mathbf{p}) \equiv \tilde{\Delta}_F^n(\mathbf{p})\hat{\Omega}^n(\mathbf{p})\hat{H}(\mathbf{p})\gamma^0, \qquad \tilde{\Delta}_F^n(\mathbf{p}) \equiv \frac{e^{-i|n|\theta}\Gamma}{2i\sin\theta} \tag{26.45}$$

satisfy the equations

$$\left(\hat{\Omega}^{-1}(\mathbf{p})\mathbb{U}_n - 2\eta + \hat{\Omega}(\mathbf{p})\mathbb{U}_n^{-1}\right)\tilde{S}_F^n(\mathbf{p}) = -\Gamma\hat{H}(\mathbf{p})\beta\delta_n, \tag{26.46}$$

$$(\mathbb{U}_n - 2\eta + \mathbb{U}_n^{-1})\tilde{\Delta}_F^n(\mathbf{p}) = -\Gamma\delta_n. \tag{26.47}$$

To investigate the nature of the propagator, define the Fourier-series transform

$$\tilde{S}_F(\mathbf{p},\Theta) \equiv T\sum_{n=-\infty}^{\infty} e^{in\Theta}\tilde{S}_F^n(\mathbf{p}), \tag{26.48}$$

where the parameter Θ is taken to be real. Then we find

$$\left[\kappa(\cos\Theta + 2)\hat{H} - 3\sin\Theta\right]\tilde{S}_F(\mathbf{p},\Theta) = -3T\beta. \tag{26.49}$$

We may now solve for $\tilde{S}_F(\mathbf{p},\Theta)$ if we give κ^2 a small imaginary term, according to the standard Feynman $m \rightarrow m - i\epsilon$ prescription. Hence we find

$$\tilde{S}_F(\mathbf{p},\Theta) = -\frac{3T\left[\kappa(\cos\Theta+2)\hat{H}+3\sin\Theta\right]\beta}{\left[\kappa^2(\cos\Theta+2)^2-9\sin^2\Theta-i\epsilon\right]}. \tag{26.50}$$

In this form the propagator looks quite different from the standard CT propagator, but a suitable reparametrization can change this. We introduce the parameter p_0 (which should **not** be confused with $E_\mathbf{p} \equiv \sqrt{\mathbf{p}\cdot\mathbf{p}+m^2}$) related to the parameter Θ by

$$\cos\Theta = \frac{6-2p_0^2T^2}{6+p_0^2T^2}, \quad \text{sign } \Theta = \text{sign } p_0. \tag{26.51}$$

Then we find

$$\tilde{S}_F(\mathbf{p},\Theta) = \frac{\gamma^0 p_0 + \gamma^i p_i + m}{p_0^2 - E_\mathbf{p}^2 + i\epsilon} + O(T^2) = \frac{\not{p}+m}{p^2-m^2+i\epsilon} + O(T^2). \tag{26.52}$$

This propagator is the same as the CT Feynman propagator for the Dirac field to lowest order in T. This is important for two reasons. First, the parameter Θ flowing through Feynman-diagram networks has the same representation and interpretation for fermions as it has for bosons, and can be justifiably regarded as the DT analogue of energy, up to a factor of T. Second, this shows us that Lorentz covariance for the Dirac equation emerges from DT mechanics at the same level of approximation as it does in bosonic theory.

26.3.1 Field anticommutators

After taking Fourier-series transforms, the action sum in the presence of external sources becomes

$$\begin{aligned}\hat{A}^{NM}[\eta] &\equiv \sum_{n=M}^{N-1}\int d\mathbf{x}\,\mathcal{F}^n[\eta]\\ &= \hat{A}^{NM} + T\sum_{n=M+1}^{N-1}\int\frac{d\mathbf{p}}{(2\pi)^3}\left[\tilde{\eta}_n^\dagger(\mathbf{p})\gamma^0\tilde{\psi}_n(\mathbf{p}) + \tilde{\psi}_n^\dagger(\mathbf{p})\gamma^0\tilde{\eta}_n(\mathbf{p})\right]\\ &\quad + \frac{1}{2}T\int\frac{d\mathbf{p}}{(2\pi)^3}\{\tilde{\eta}_N^\dagger(\mathbf{p})\gamma^0\tilde{\psi}_N(\mathbf{p}) + \tilde{\psi}_N^\dagger(\mathbf{p})\gamma^0\tilde{\eta}_N(\mathbf{p})\\ &\qquad + \tilde{\eta}_M^\dagger(\mathbf{p})\gamma^0\tilde{\psi}_M(\mathbf{p}) + \tilde{\psi}_M^\dagger(\mathbf{p})\gamma^0\tilde{\eta}_M(\mathbf{p})\}.\end{aligned} \tag{26.53}$$

For the Schwinger action principle, we first define the functional derivatives

$$\frac{\delta}{\delta\eta_n(\mathbf{x})}\eta_m(\mathbf{y}) = \delta_{n-m}\delta^3(\mathbf{x}-\mathbf{y}), \quad \frac{\delta}{\delta\eta_n^\dagger(\mathbf{x})}\eta_m^\dagger(\mathbf{y}) = \delta_{n-m}\delta^3(\mathbf{x}-\mathbf{y}) \tag{26.54}$$

and

$$\frac{\delta}{\delta\tilde{\eta}_n^\dagger(\mathbf{p})} \equiv \int d\mathbf{x}\, e^{-i\mathbf{p}\cdot\mathbf{x}}\frac{\delta}{\delta\eta_n^\dagger(\mathbf{x})}, \quad \frac{\delta}{\delta\tilde{\eta}_n(\mathbf{p})} \equiv \int d\mathbf{x}\, e^{i\mathbf{p}\cdot\mathbf{x}}\frac{\delta}{\delta\eta_n(\mathbf{x})} \tag{26.55}$$

so that

$$\frac{\delta}{\delta\tilde{\eta}_n(\mathbf{p})}\tilde{\eta}_m(\mathbf{q}) = (2\pi)^3\delta^3(\mathbf{p}-\mathbf{q})\delta_{n-m},$$
$$\frac{\delta}{\delta\tilde{\eta}_n^\dagger(\mathbf{p})}\tilde{\eta}_m^\dagger(\mathbf{q}) = (2\pi)^3\delta^3(\mathbf{p}-\mathbf{q})\delta_{n-m}. \tag{26.56}$$

Then the Schwinger action principle gives

$$-\frac{i}{T}\frac{\delta}{\delta\tilde{\eta}_n^\dagger(\mathbf{p})}\langle\alpha,N|\beta,M\rangle_\eta = \langle\alpha,N|\gamma^0\tilde{\psi}_n(\mathbf{p})|\beta,M,\rangle_\eta, \quad N>n>M,$$
$$\frac{i}{T}\frac{\delta}{\delta\tilde{\eta}_n(\mathbf{p})}\langle\alpha,N|\beta,M\rangle_\eta = \langle\alpha,N|\tilde{\psi}_n^\dagger(\mathbf{p})\gamma^0|\beta,M\rangle_\eta, \quad N>n>M, \tag{26.57}$$

and so on. With this and Cadzow's equations of motion, we find the vacuum functional

$$Z[\eta] = Z[0]\exp\left\{-iT^2\sum_{n,m=-\infty}^{\infty}\int\frac{d\mathbf{p}}{(2\pi)^3}\tilde{\eta}_n^\dagger(\mathbf{p})\gamma^0\tilde{S}_{\mathrm{F}}^{n-m}(\mathbf{p})\tilde{\eta}_m(\mathbf{p})\right\}, \tag{26.58}$$

where

$$\int\frac{d\mathbf{p}}{(2\pi)^3}e^{i\mathbf{p}\cdot\mathbf{x}}\tilde{S}_{\mathrm{F}}^n(\mathbf{p}) = S_{\mathrm{F}}^n(\mathbf{x}). \tag{26.59}$$

We may also write

$$Z[\eta] = Z[0]\exp\left\{-iT^2\sum_{n,m=-\infty}^{\infty}\int d\mathbf{x}\,d\mathbf{y}\;\eta_n^\dagger(\mathbf{x})\gamma^0S_{\mathrm{F}}^{n-m}(\mathbf{x}-\mathbf{y})\eta_m(\mathbf{x})\right\}, \tag{26.60}$$

where the propagators $S_{\mathrm{F}}^n(\mathbf{x})$ satisfy the equations

$$\left\{i\frac{\mathbb{U}_n^{-1}-\mathbb{U}_n}{2T}+\overrightarrow{H}\frac{\mathbb{U}_n+4+\mathbb{U}_n^{-1}}{6}\right\}S_{\mathrm{F}}^n(\mathbf{x}) = -\frac{\delta_n}{T}\gamma^0\delta^3(\mathbf{x}). \tag{26.61}$$

Using the rule

$$\langle 0|\tilde{T}\tilde{\psi}_{na}^\dagger(\mathbf{p})\tilde{\psi}_{mb}(\mathbf{q})|0\rangle = -i\left[\tilde{S}_{\mathrm{F}}^{m-n}(\mathbf{p})\gamma^0\right]_{ba}(2\pi)^3\delta^3(\mathbf{p}-\mathbf{q}) \tag{26.62}$$

we find the following vacuum expectation values:

$$\langle 0|\tilde{\psi}_{n+1a}^\dagger(\mathbf{p})\tilde{\psi}_{nb}(\mathbf{q})|0\rangle = \frac{18ie^{-i\theta}}{(9+\kappa^2)\sin\theta}\tilde{\Lambda}^{(+)}(\mathbf{p})_{ba}(2\pi)^3\delta^3(\mathbf{p}-\mathbf{q}),$$
$$\langle 0|\tilde{\psi}_{nb}^\dagger(\mathbf{p})\tilde{\psi}_{n+1a}(\mathbf{q})|0\rangle = -\frac{18ie^{i\theta}}{(9+\kappa^2)\sin\theta}\tilde{\Lambda}^{(-)}(\mathbf{p})_{ab}(2\pi)^3\delta^3(\mathbf{p}-\mathbf{q}),$$
$$\langle 0|\tilde{\psi}_{n+1b}(\mathbf{p})\tilde{\psi}_{na}^\dagger(\mathbf{q})|0\rangle = \frac{18ie^{-i\theta}}{(9+\kappa^2)\sin\theta}\tilde{\Lambda}^{(-)}(\mathbf{p})_{ba}(2\pi)^3\delta^3(\mathbf{p}-\mathbf{q}),$$
$$\langle 0|\tilde{\psi}_{na}(\mathbf{p})\tilde{\psi}_{n+1b}^\dagger(\mathbf{q})|0\rangle = -\frac{18ie^{i\theta}}{(9+\kappa^2)\sin\theta}\tilde{\Lambda}^{(+)}(\mathbf{p})_{ab}(2\pi)^3\delta^3(\mathbf{p}-\mathbf{q}), \tag{26.63}$$

where $\tilde{\Lambda}^{(\pm)}(\mathbf{p})$ are spin projection operators. We notice that these are singular at the parabolic barrier. However, by taking anticommutators we arrive at the fundamental quantization relations

$$\begin{aligned}\left\{\tilde{\psi}^{+}_{n+1a}(\mathbf{p}), \tilde{\psi}_{nb}(\mathbf{q})\right\} &= \frac{36}{9+\kappa^2}\tilde{\Lambda}^{(+)}(\mathbf{p})_{ba}(2\pi)^3\delta^3(\mathbf{p}-\mathbf{q}),\\ \left\{\tilde{\psi}^{+}_{nb}(\mathbf{p}), \tilde{\psi}_{n+1a}(\mathbf{q})\right\} &= \frac{36}{9+\kappa^2}\tilde{\Lambda}^{(-)}(\mathbf{p})_{ab}(2\pi)^3\delta^3(\mathbf{p}-\mathbf{q}),\\ \left\{\tilde{\psi}^{+}_{nb}(\mathbf{p}), \tilde{\psi}_{na}(\mathbf{q})\right\} &= 0,\end{aligned} \tag{26.64}$$

which are, remarkably, free of any singularities at the parabolic barrier. If we had taken commutators instead, we would find that the singularities still occurred at the parabolic barrier.

26.3.2 Ladder operators

Provided we are in the elliptic regime, the solution to the source-free Dirac equation

$$\left(\hat{\Omega}^{\dagger}(\mathbf{p})\tilde{\psi}_{n+1}(\mathbf{p}) - 2\eta_{\mathbf{p}}\tilde{\psi}_n(\mathbf{p}) + \hat{\Omega}(\mathbf{p})\tilde{\psi}_{n-1}(\mathbf{p})\right) \underset{c}{=} 0 \tag{26.65}$$

is given by

$$\begin{aligned}\tilde{\psi}_n(\mathbf{p}) = \frac{1}{2E_{\mathbf{p}}}\sum_{r=1}^{2}&\left\{\left[\hat{a}(\mathbf{p}r)\,e^{-in\bar{\omega}_1} + \hat{c}(-\mathbf{p}r)e^{-in\bar{\omega}_2}\right]u(\mathbf{p}r)\right.\\ &\left.+\left[\hat{d}^{\dagger}(\mathbf{p}r)e^{in\bar{\omega}_2} + \hat{b}^{\dagger}(-\mathbf{p}r)e^{in\bar{\omega}_1}\right]v(-\mathbf{p}r)\right\}.\end{aligned} \tag{26.66}$$

Then we find

for the particles

$$\begin{aligned}\hat{a}(\mathbf{p}r) &= \frac{i}{2\sin\theta}u^{\dagger}(\mathbf{p}r)\left\{\tilde{\psi}_{n+1}(\mathbf{p})e^{-i\xi} - \tilde{\psi}_n(\mathbf{p})e^{i\theta}\right\}e^{in\bar{\omega}_1},\\ \hat{a}^{\dagger}(\mathbf{p}r) &= -\frac{i}{2\sin\theta}\left\{\tilde{\psi}^{\dagger}_{n+1}(\mathbf{p})e^{i\xi} - \tilde{\psi}^{\dagger}_n(\mathbf{p})e^{-i\theta}\right\}u(\mathbf{p}r)e^{-in\bar{\omega}_1},\end{aligned} \tag{26.67}$$

$$\begin{aligned}\hat{b}(\mathbf{p}r) &= \frac{i}{2\sin\theta}\left\{\tilde{\psi}^{\dagger}_{n+1}(-\mathbf{p})e^{-i\xi} - \tilde{\psi}^{\dagger}_n(-\mathbf{p})e^{i\theta}\right\}v(\mathbf{p}r)e^{in\bar{\omega}_1},\\ \hat{b}^{\dagger}(\mathbf{p}r) &= -\frac{i}{2\sin\theta}v^{\dagger}(\mathbf{p}r)\left\{\tilde{\psi}_{n+1}(-\mathbf{p})e^{i\xi} - \tilde{\psi}_n(-\mathbf{p})e^{-i\theta}\right\}e^{-in\bar{\omega}_2},\end{aligned} \tag{26.68}$$

and for the oscillons

$$\begin{aligned}\hat{c}(\mathbf{p}r) &= \frac{i}{2\sin\theta}\left\{\tilde{\psi}^{\dagger}_{n+1}(-\mathbf{p})e^{i\xi} - \tilde{\psi}^{\dagger}_n(-\mathbf{p})e^{i\theta}\right\}u(-\mathbf{p}r)e^{-in\bar{\omega}_2},\\ \hat{c}^{\dagger}(\mathbf{p}r) &= -\frac{i}{2\sin\theta}u^{\dagger}(-\mathbf{p}r)\left\{\tilde{\psi}_{n+1}(-\mathbf{p})e^{-i\xi} - \tilde{\psi}_n(-\mathbf{p})e^{-i\theta}\right\}e^{in\bar{\omega}_2},\end{aligned} \tag{26.69}$$

$$\begin{aligned}\hat{d}(\mathbf{p}r) &= \frac{i}{2\sin\theta}v^{+}(-\mathbf{p}r)\left\{\tilde{\psi}_{n+1}(\mathbf{p})e^{i\xi} - \tilde{\psi}_n(\mathbf{p})e^{i\theta}\right\}e^{-in\bar{\omega}_2},\\ \hat{d}^{\dagger}(\mathbf{p}r) &= -\frac{i}{2\sin\theta}\left\{\tilde{\psi}^{\dagger}_{n+1}(\mathbf{p})e^{-i\xi} - \tilde{\psi}^{\dagger}_n(\mathbf{p})e^{-i\theta}\right\}v(-\mathbf{p}r)e^{in\bar{\omega}_2},\end{aligned} \tag{26.70}$$

where the vibrancies $\bar{\omega}_1$ and $\bar{\omega}_2$ are given by $\bar{\omega}_1 = \theta - \xi$ and $\bar{\omega}_2 = -\theta - \xi$.

Here we have used conventional Dirac momentum spinors defined by

$$u(\mathbf{p}r) \equiv \frac{p+m}{\sqrt{m+E}} u_r = \frac{1}{\sqrt{m+E}} \begin{bmatrix} (m+E)\chi_r \\ \mathbf{p}\cdot\boldsymbol{\sigma}\chi_r \end{bmatrix},$$
$$v(\mathbf{p}r) \equiv \frac{m-p}{\sqrt{m+E}} v_r = \frac{1}{\sqrt{m+E}} \begin{bmatrix} \mathbf{p}\cdot\boldsymbol{\sigma}\eta_r \\ (m+E)\eta_r \end{bmatrix}, \tag{26.71}$$

with

$$\hat{H}(\mathbf{p})u(\mathbf{p}r) = u(\mathbf{p}r), \quad \hat{H}(-\mathbf{p})v(\mathbf{p}r) = -v(\mathbf{p}r). \tag{26.72}$$

Now using the field anticommutators, we arrive at the following non-zero creation and annihilation relations: for the particles we have

$$\left\{\hat{a}(\mathbf{p}r), \hat{a}^\dagger(\mathbf{q}s)\right\} = \left\{\hat{b}(\mathbf{p}r), \hat{b}^\dagger(\mathbf{q}s)\right\} = \frac{6E}{\sqrt{9-3\kappa^2}}\delta_{rs}(2\pi)^3\delta^3(\mathbf{p}-\mathbf{q}), \tag{26.73}$$

whereas for the oscillons we find

$$\left\{\hat{c}(\mathbf{p}r), \hat{c}^\dagger(\mathbf{q}s)\right\} = \left\{\hat{d}(\mathbf{p}r), \hat{d}^\dagger(\mathbf{q}s)\right\} = -\frac{6E}{\sqrt{9-3\kappa^2}}\delta_{rs}(2\pi)^3\delta^3(\mathbf{p}-\mathbf{q}). \tag{26.74}$$

All other anticommutators are zero. It is clear that we should be in the elliptic regime in order for any of these anticommutators to make physical sense. Moreover, we see that, even though their linear momenta may be in the elliptic regime $TE_{\mathbf{p}} < \sqrt{3}$, oscillon and anti-oscillon particle states have a negative inner product and are therefore unphysical relative to the normal states. This confirms the results of previous sections.

26.4 The discrete time free quantized Maxwell fields

Before we embark on the quantization of the electromagnetic fields, we review free-field scalar propagators in DT.

Integration over space and discrete time is denoted by the sum/integral symbol

$$\sum\!\!\!\!\!\!\int_x \equiv T\sum_{n=-\infty}^{\infty}\int d^3\mathbf{x}. \tag{26.75}$$

Given a variable $f_x \equiv f_n(\mathbf{x})$ indexed by a discrete temporal index n and by a continuous spatial index $\mathbf{x}$, the DT Fourier transform $\tilde{f}_p$ of f_x is defined by

$$\tilde{f}_p \equiv \tilde{f}(z,\mathbf{p}) \equiv \sum\!\!\!\!\!\!\int_x e_{px} f_x, \tag{26.76}$$

where

$$e_{px} \equiv z^n e^{-i\mathbf{p}\cdot\mathbf{x}} \tag{26.77}$$

and z is complex and non-zero. It is assumed that such a transform exists and defines a function that is analytic in some annular region in the complex-z plane centred on the origin and including the unit circle. This imposes restrictions on the sequence defined by f_n. Although $\mathbf{p}$ is the direct equivalent of the spatial components of the momentum-space four-vector of CT mechanics, the analogue of the timelike component p^0 of the CT momentum-space four-vector p^μ is not z. Physical applications normally require z to lie on the unit circle, and then it is the principal argument of z which is related to p^0. The symbol p is used to denote the coordinates collectively, i.e., $p \equiv (z, \mathbf{p})$.

If the DT Fourier transform exists and has a Laurent expansion in some annular region centred on the origin in the complex-z plane and containing the unit circle, then the inverse transform can be constructed. This is given by

$$f_x = \oint_p \tilde{e}_{px} \tilde{f}_p, \tag{26.78}$$

where

$$\oint_p \equiv \frac{1}{2\pi i T} \oint \frac{dz}{z} \int \frac{d^3\mathbf{p}}{(2\pi)^3} \tag{26.79}$$

with the z integral being over the unit circle in the complex plane taken in the anticlockwise sense and

$$\tilde{e}_{px} \equiv z^{-n} e^{i\mathbf{p}\cdot\mathbf{x}}. \tag{26.80}$$

It would be possible to define the dynamics in the $(z, \mathbf{p})$ transform space, restricting all dynamical variables to be functions in the complex-z plane for which the inverse transform (26.78) exists. No new dynamical content should emerge from this approach, but it may be mathematically more secure. However, it will normally be assumed that coming first from the space-time direction is valid. This is the conventional way to define field theories.

The analogues of the identity operators (delta functions) in this formalism are defined as follows. If δ_n is the DT Kronecker delta satisfying

$$\delta_n = \begin{cases} 1, & n = 0, \\ 0, & n \neq 0, \end{cases} \tag{26.81}$$

where n is an integer, then the four-dimensional DT Dirac delta δ_x is defined by

$$\delta_{x-y} \equiv \frac{\delta_{n-m}}{T} \delta^3(\mathbf{x} - \mathbf{y}), \qquad T > 0, \tag{26.82}$$

where $x \equiv (n, \mathbf{x})$ and $y \equiv (m, \mathbf{y})$. Then

$$\sum\!\!\!\!\!\!\int_x f_x \delta_{x-y} = f_y. \tag{26.83}$$

An integral representation of δ_{x-y} is

$$\delta_{x-y} = \oint_p e_{px} \tilde{e}_{py}. \tag{26.84}$$

The corresponding operator $\tilde{\delta}_{p-q}$ in DT Fourier-transform space satisfies the relation

$$\oint_p \tilde{f}_p \tilde{\delta}_{p-q} = \tilde{f}_q, \quad q \equiv (u, \mathbf{q}), \tag{26.85}$$

where u is complex. A DT sum/integral representation of $\tilde{\delta}_{p-q}$ is given by

$$\tilde{\delta}_{p-q} = \oint_x \tilde{e}_{px} e_{qx}. \tag{26.86}$$

Given a polynomial function P of $\mathbb{U}_n$, the rearrangement

$$\oint_x f_x \overrightarrow{P(\mathbb{U}_n)} g_x = \oint_x f_x \overleftarrow{P(\mathbb{U}_n^{-1})} g_x \tag{26.87}$$

will be assumed to hold for any physically acceptable functions f and g. If P is a time-symmetric operator then

$$\oint_x f_x \overrightarrow{P(\mathbb{U}_n)} g_x = \oint_x f_x \overleftarrow{P(\mathbb{U}_n)} g_x \tag{26.88}$$

The DT Fourier transform gives the following useful result:

$$\oint_x e_{px} P(\mathbb{U}_n) f_x = P(z^{-1}) \tilde{f}_p, \qquad z \neq 0. \tag{26.89}$$

The above may be used to find the DT Feynman propagator $\Delta_{\mathrm{F}x}$, which satisfies the equation (Jaroszkiewicz and Norton, 1997b)

$$(\Box_x + m^2 \, \mathbb{S}_n)\Delta_{\mathrm{F}x} = -\delta_x. \tag{26.90}$$

Taking the DT Fourier transform of this equation gives

$$(p^2 - m^2 Sz)\tilde{\Delta}_{\mathrm{F}p} = 1, \tag{26.91}$$

where

$$p^2 \equiv -D^2 z - Sz\mathbf{p}\cdot\mathbf{p}, \tag{26.92}$$

with

$$D^2 z \equiv \frac{z - 2 + z^{-1}}{T^2}, \qquad Sz \equiv \frac{z + 4 + z^{-1}}{6}. \tag{26.93}$$

A solution of interest in particle theory is

$$\Delta_{\mathrm{F}x} = \oint_p \tilde{e}_{px} \frac{1}{p^2 - m^2 Sz + i\epsilon}, \tag{26.94}$$

which uses the DT analogue of the Feynman $+i\epsilon$ prescription. The singularity structure in the complex-z plane of the integrand in the above is particularly interesting. First, it can be shown that the equation

$$p^2 - m^2 Sz + i\epsilon = 0 \tag{26.95}$$

has no solution on the unit circle in the complex-z plane for any value of the linear momentum $\mathbf{p}$. To prove this, write $z = \exp(i\theta)$. Then (26.95) becomes

$$\frac{2(\cos\theta - 1)}{T^2} - \frac{\cos\theta + 2}{3}(\mathbf{p}\cdot\mathbf{p} + m^2) + i\epsilon = 0, \tag{26.96}$$

which has no solution for real θ if $\epsilon > 0$. This result is important because it means that integration is over a fully closed contour (the unit circle) in the complex-z plane, requiring no principal-value discussion.

Now (26.95) is a quadratic in z with roots z_1 and z_2 satisfying the relation

$$z_1 z_2 = 1. \tag{26.97}$$

From this it can be deduced that the denominator in (26.94) contributes one simple pole inside the unit circle and one outside. By looking carefully at the locations of these poles as the spatial momentum $\mathbf{p}$ varies, two distinct patterns of behaviour are seen (Jaroszkiewicz and Norton, 1997a). For momentum in the elliptic regime, corresponding to momenta bounded by $T|\mathbf{p}| < \sqrt{12}$, the simple pole inside the unit circle is just inside the contour of integration and gives the equivalent of a trigonometric solution when the calculus of residues is used to evaluate the integral. For momentum in the hyperbolic regime, on the other hand, given by $T|\mathbf{p}| > \sqrt{12}$, the simple pole interior to the unit circle starts to move towards $z = 0$, giving a damped exponential solution after complex integration.

If a slightly different prescription had been taken, viz.

$$\Delta_{\mathrm{F}x} \equiv \oint_p \tilde{e}_{px} \frac{1}{p^2 - (m^2 - i\epsilon)Sz}, \tag{26.98}$$

the same conclusions would be drawn.

26.4.1 Quantization of the electromagnetic field

Quantization is more convenient in the Coulomb gauge, where the condition

$$\nabla \cdot \mathbf{A}_n = 0 \tag{26.99}$$

is imposed. Then the electromagnetic equations of motion (24.28) become

$$\nabla^2 \phi_n \underset{c}{=} -\rho_n, \tag{26.100}$$

$$\square_n \mathbf{A}_n + \nabla \bar{\mathbb{T}}_n \phi_n \underset{c}{=} \mathbf{j}_n. \tag{26.101}$$

From these equations it is clear that the scalar potential ϕ_n cannot be regarded as a dynamical field in the same way as the components of the vector potential are. Equation (26.100) is of zeroth order in time whereas (26.101) is a full second-order dynamical equation. This is the DT analogue of the situation in CT electromagnetism, where a direct application of Dirac's constraint analysis shows that the momentum conjugate to the scalar potential vanishes. This has important consequences when quantization via the Schwinger action principle is considered, as will be discussed next.

26.4.2 The DT Schwinger action principle

The DT Schwinger action principle is used in the Heisenberg picture and matrix elements are taken between states associated with different times. These correspond to the stages of preparation and observation, that is, the initial and final states. If $|\Psi, M\rangle$ is the state prepared at time MT and $|\Phi, N\rangle$ is a state about which questions are being asked at time $NT > MT$, then the infinitesimal change $\delta\langle\Phi, N|\Psi, M\rangle$ in the transition amplitude $\langle\Phi, N|\Psi, M\rangle$ due to infinitesimal changes in the external sources is, by virtue of the DT Schwinger action principle (Jaroszkiewicz and Norton, 1997a), defined by

$$\delta\langle\Phi, N|\Psi, M\rangle = i\langle\Phi, N|\delta A^{NM}|\Psi, M\rangle, \quad N > M, \tag{26.102}$$

where δA^{NM} is the infinitesimal change in the action sum operator

$$A^{NM} \equiv \sum_{n=M}^{N-1} \int d^3\mathbf{x}\, \mathcal{F}^n. \tag{26.103}$$

In the case of the electromagnetic field, the free-field system function density is used with external sources ρ_n and $\mathbf{j}_n$ coupled in the manner of Schwinger (1963). Since these sources are arbitrary, care must be taken to ensure that the charges which couple to the electromagnetic fields satisfy the equation of continuity. Following Schwinger and anticipating the use of the Coulomb gauge, the system function density in the presence of the sources is taken to be

$$\mathcal{F}^n[j] = \mathcal{F}^n - T\phi_n\rho_n^{\mathrm{c}} + \tfrac{1}{2}T\mathbf{A}_{n+1}\cdot\mathbf{j}_{n+1}^{\mathrm{c}} + \tfrac{1}{2}T\mathbf{A}_n\cdot\mathbf{j}_n^{\mathrm{c}}, \tag{26.104}$$

where

$$\begin{aligned} \rho_n^{\mathrm{c}}(\mathbf{x}) &\equiv \rho_n(\mathbf{x}), \\ \mathbf{j}_n^{\mathrm{c}}(\mathbf{x}) &\equiv \mathbf{j}_n(\mathbf{x}) + \nabla_{\mathbf{x}} \sum\!\!\!\!\!\!\int_y G_{\mathrm{C}x-y}[D_m^-\rho_y + \nabla_{\mathbf{y}}\cdot\mathbf{j}_y] \end{aligned} \tag{26.105}$$

are the conserved charge densities constructed from the independent external densities ρ_n and $\mathbf{j}_n$, and G_{C} is the Coulomb Green function which satisfies the equation

$$\nabla_{\mathbf{x}}^2 G_{\mathrm{C}x} = -\delta_x. \tag{26.106}$$

This has the particular solution

$$G_{\mathrm{C}x} = \frac{\delta_n}{4\pi|\mathbf{x}|T}. \tag{26.107}$$

The conserved charge densities satisfy the DT equation of continuity

$$\bar{\mathbb{T}}_n\rho_n^{\mathrm{c}}(\mathbf{x}) + \nabla_{\mathbf{x}}\cdot\mathbf{j}_n^{\mathrm{c}}(\mathbf{x}) = 0, \tag{26.108}$$

regardless of the values of the independent densities. The above coupling ensures that ρ_n and $\mathbf{j}_n$ can be arbitrarily varied whilst still ensuring that the electromagnetic fields are coupled to conserved charges.

Functional differentiation is defined via

$$\frac{\delta}{\delta \rho_x}\rho_y \equiv \frac{\delta}{\delta \rho_n(\mathbf{x})}\rho_m(\mathbf{y}) = \delta_{x-y} \equiv \frac{\delta_{n-m}}{T}\delta^3(\mathbf{x}-\mathbf{y}) \tag{26.109}$$

and similarly for the currents. Then, with the external sources coupled as in (26.104), functional differentiation gives

$$\frac{i\delta}{\delta \rho_n(\mathbf{x})}\langle \Phi, n+1|\Psi, n\rangle_j = \langle \Phi, n+1|\phi_n(\mathbf{x})|\Psi, n\rangle_j, \tag{26.110}$$

$$-\frac{i\delta}{\delta j_n^i(\mathbf{x})}\langle \Phi, n|\Psi, n-1\rangle_j = \frac{1}{2}\langle \Phi, n|A_n^i(\mathbf{x})|\Psi, n-1\rangle_j, \tag{26.111}$$

$$-\frac{i\delta}{\delta j_n^i(\mathbf{x})}\langle \Phi, n+1|\Psi, n\rangle_j = \frac{1}{2}\langle \Phi, n+1|A_n^i(\mathbf{x})|\Psi, n\rangle_j, \tag{26.112}$$

using the Coulomb gauge (or transversality) condition

$$\langle \Phi|\nabla\cdot\mathbf{A}_n(\mathbf{x})|\Psi\rangle = 0 \tag{26.113}$$

for all states Ψ and Φ.

If the in and out vacua are assumed to exist, and if $Z[j] \equiv \langle 0^{\mathrm{out}}|0^{\mathrm{in}}\rangle_j$ denotes the vacuum functional in the presence of the external sources, then direct application of the DT Schwinger action principle gives the functional derivatives

$$\frac{i\delta}{\delta \rho_n(\mathbf{x})}Z[j] = \langle 0^{\mathrm{out}}|\phi_n(\mathbf{x})|0^{\mathrm{in}}\rangle_j, \tag{26.114}$$

$$-\frac{i\delta}{\delta j_n^i(\mathbf{x})}Z[j] = \langle 0^{\mathrm{out}}|A_n^i(\mathbf{x})|0^{\mathrm{in}}\rangle_j. \tag{26.115}$$

In the Coulomb gauge the quantum analogues of Cadzow's equations are

$$\begin{aligned} \nabla^2\phi_n(\mathbf{x}) &\underset{\mathrm{c}}{=} -\rho_n^{\mathrm{c}}(\mathbf{x}), \\ \Box_n\langle 0^{\mathrm{out}}|\mathbf{A}_n(\mathbf{x})|0^{\mathrm{in}}\rangle_j + \nabla_{\mathbf{x}}\bar{\mathbb{T}}_n\phi_n(\mathbf{x})Z[j] &\underset{\mathrm{c}}{=} \mathbf{j}_n^{\mathrm{c}}(\mathbf{x})Z[j], \end{aligned} \tag{26.116}$$

taking the scalar potential in the Coulomb gauge to be a c-number. From these equations it is found that

$$\phi_x = \oint_y G_{\mathrm{C}x-y}\rho_y \tag{26.117}$$

and

$$\Box_x\langle 0^{\mathrm{out}}|A_x^i|0^{\mathrm{in}}\rangle_j = j_x^{\mathrm{c}\,i}Z\left[j\right] + \partial_i^x\partial_j^x\oint_y G_{\mathrm{C}\,x-y}j_y^{\mathrm{c}\,j}Z[j]. \tag{26.118}$$

Here the notation has been switched, with the symbols x and y denoting $(n,\mathbf{x})$ and $(m,\mathbf{y})$, respectively.

The second functional derivative of the last equation gives the vacuum expectation value (VEV) of the DT time-ordered product (Jaroszkiewicz and Norton, 1997a; Norton and Jaroszkiewicz, 1998a)

$$\langle 0|\tilde{T}A_x^i A_y^j|0\rangle = i\Delta_{\mathrm{F}x-y}\delta_{ij} + i\oint\!\!\!\!\!\!\sum_z \Delta_{\mathrm{F}x-z}\,\partial_i^z\partial_j^z G_{\mathrm{C}z-y} \tag{26.119}$$

in the absence of the sources. This is equivalent to

$$\langle 0|\tilde{T}A_n^i(\mathbf{x})A_m^j(\mathbf{y})|0\rangle = i\int \frac{d^3\mathbf{p}}{(2\pi)^3}\left[\delta_{ij} - \frac{p^i p^j}{\mathbf{p}^2}\right]\tilde{\Delta}_{\mathrm{F}}^{n-m}(\mathbf{p})e^{-i\mathbf{p}\cdot(\mathbf{x}-\mathbf{y})}, \tag{26.120}$$

where $\tilde{\Delta}_{\mathrm{F}}^n(\mathbf{p})$ is the Fourier transform

$$\tilde{\Delta}_{\mathrm{F}}^n(\mathbf{p}) = \int d^3\mathbf{x}\;\Delta_{\mathrm{F}}^n(\mathbf{x})e^{-i\mathbf{p}\cdot\mathbf{x}} \tag{26.121}$$

of the temporally indexed Green function, which satisfies the DT massless field equation

$$\Box_n\Delta_{\mathrm{F}}^n(\mathbf{x}) = -\frac{\delta_n}{T}\delta^3(\mathbf{x}). \tag{26.122}$$

In the absence of external sources the radiation gauge may be chosen for convenience. This is defined by

$$\phi_n(\mathbf{x}) = 0, \quad \langle\Phi|\nabla\cdot\mathbf{A}_n(\mathbf{x})|\Psi\rangle = 0, \quad \forall\Psi,\Phi. \tag{26.123}$$

Using the above time-ordered products (26.120), the following VEVs of the unequal-time commutators in this gauge are

$$\langle 0|\left[\tilde{A}_{n+1}^i(\mathbf{p}),\tilde{A}_n^{j\dagger}(\mathbf{q})\right]|0\rangle = -i\Gamma_{\mathbf{p}}\left[\delta_{ij} - \frac{p^i p^j}{\mathbf{p}^2}\right](2\pi)^3\delta^3(\mathbf{p}-\mathbf{q}), \tag{26.124}$$

$$\langle 0|\left[\tilde{A}_{n+1}^{i\dagger}(\mathbf{p}),\tilde{A}_n^{j}(\mathbf{q})\right]|0\rangle = -i\Gamma_{\mathbf{p}}\left[\delta_{ij} - \frac{p^i p^j}{\mathbf{p}^2}\right](2\pi)^3\delta^3(\mathbf{p}-\mathbf{q}), \tag{26.125}$$

where

$$\Gamma_{\mathbf{p}} = \frac{6T}{6+T^2\mathbf{p}^2} \tag{26.126}$$

and

$$\tilde{A}_n^i(\mathbf{p}) \equiv \int d^3\mathbf{x}\;e^{-i\mathbf{p}\cdot\mathbf{x}}A_n^i(\mathbf{x}), \tag{26.127}$$

$$\tilde{A}_n^{i\dagger}(\mathbf{p}) \equiv \int d^3\mathbf{x}\;e^{i\mathbf{p}\cdot\mathbf{x}}A_n^i(\mathbf{x}). \tag{26.128}$$

This gives the result

$$\langle 0|\left[A_{n+1}^i(\mathbf{x}),A_n^j(\mathbf{y})\right]|0\rangle = -i\int \frac{d^3\mathbf{p}}{(2\pi)^3}e^{i\mathbf{p}\cdot(\mathbf{x}-\mathbf{y})}\left[\delta_{ij} - \frac{p^i p^j}{\mathbf{p}^2}\right]\frac{6T}{6+T^2\mathbf{p}^2}, \tag{26.129}$$

which is precisely the same as for the scalar field discussed by Norton and Jaroszkiewicz (1998a) apart from the modified Kronecker delta, which is necessary in order to preserve the transversality condition (26.113).

As a final step, it may be supposed that the commutators of the fields are c-numbers, in the language of Dirac, and then this gives the operator commutator

$$\left[A^i_{n+1}(\mathbf{x}), A^j_n(\mathbf{y})\right] = -i\int \frac{d^3\mathbf{p}}{(2\pi)^3} e^{i\mathbf{p}\cdot(\mathbf{x}-\mathbf{y})}\left[\delta_{ij} - \frac{p^i p^j}{\mathbf{p}^2}\right]\frac{6T}{6+T^2\mathbf{p}^2}. \tag{26.130}$$

This amounts to a DT quantization prescription for the electromagnetic field in the Coulomb gauge. Equation (26.130) and the operator equation of motion

$$\Box_n \mathbf{A}_n(\mathbf{x}) = \mathbf{0}, \tag{26.131}$$

may be used to deduce the equal-time commutators

$$\left[\tilde{A}^i_n(\mathbf{p}), \tilde{A}^{j\dagger}_n(\mathbf{q})\right] = 0, \tag{26.132}$$

which is equivalent to

$$\left[A^i_n(\mathbf{x}), A^j_n(\mathbf{y})\right] = 0. \tag{26.133}$$

Photon creation and annihilation operators are defined by

$$\begin{aligned} a_n(\mathbf{p},\lambda) &\equiv \frac{i}{\Gamma_{\mathbf{p}}}\int d^3\mathbf{x}\, e^{in\theta_{\mathbf{p}} - i\mathbf{p}\cdot\mathbf{x}}\boldsymbol{\epsilon}(\mathbf{p},\lambda)\cdot\left[\mathbf{A}_{n+1}(\mathbf{x}) - e^{i\theta_{\mathbf{p}}}\mathbf{A}_n(\mathbf{x})\right], \\ a^\dagger_n(\mathbf{p},\lambda) &\equiv -\frac{i}{\Gamma_{\mathbf{p}}}\int d^3\mathbf{x}\, e^{-in\theta_{\mathbf{p}} + i\mathbf{p}\cdot\mathbf{x}}\boldsymbol{\epsilon}(\mathbf{p},\lambda)\cdot\left[\mathbf{A}_{n+1}(\mathbf{x}) - e^{-i\theta_{\mathbf{p}}}\mathbf{A}_n(\mathbf{x})\right], \end{aligned} \tag{26.134}$$

where

$$\cos\theta_{\mathbf{p}} = \frac{6 - 2T^2\mathbf{p}^2}{6 + T^2\mathbf{p}^2} \tag{26.135}$$

and

$$\boldsymbol{\epsilon}(\mathbf{p},\lambda)\cdot\boldsymbol{\epsilon}(\mathbf{p},\lambda') = \delta_{\lambda\lambda'}, \qquad \boldsymbol{\epsilon}(\mathbf{p},\lambda)\cdot\mathbf{p} = 0. \tag{26.136}$$

In these definitions the momentum $\mathbf{p}$ is taken to be in the so-called elliptic regime (Norton and Jaroszkiewicz, 1998a). Given our discussion in previous chapters of the DT harmonic oscillator, we know that momentum in the hyperbolic region $|\mathbf{p}|T > \sqrt{12}$ leads to wavefunctions that either decay to zero or diverge for large time (Jaroszkiewicz and Norton, 1997a). The implications are that, as with the scalar and Dirac particles discussed above, there is a natural cutoff in this mechanics in the photon spectrum. This may be significant in discussions involving, for example, the black-body spectrum.

A direct application of the commutation rules (26.130) and (26.133) gives

$$\begin{aligned} \left[a_n(\mathbf{p},\lambda), a^\dagger_n(\mathbf{q},\lambda')\right] &= 2|\mathbf{p}|\sqrt{1 - \frac{T^2\mathbf{p}^2}{12}}\delta_{\lambda\lambda'}(2\pi)^3\delta^3(\mathbf{p}-\mathbf{q}), \\ \left[a_n(\mathbf{p},\lambda), a_n(\mathbf{q},\lambda')\right] &= 0, \end{aligned} \tag{26.137}$$

which shows explicitly that there is a spectrum of polarized photon states, but only up to the parabolic barrier $|\mathbf{p}|T < \sqrt{12}$, as discussed above. An expansion in powers of T gives

$$\left[a_n(\mathbf{p},\lambda), a_n^\dagger(\mathbf{q},\lambda')\right] = 2|\mathbf{p}|\delta_{\lambda\lambda'}(2\pi)^3\delta^3(\mathbf{p}-\mathbf{q}) + O(T^2), \tag{26.138}$$

which supports the result that Lorentz symmetry emerges very rapidly from this mechanics if, as may be imagined, T is of the order of the Planck time.

27

Interacting discrete time scalar fields

In this chapter we explore the discrete time (DT) quantum field theory (QFT) of a system under observation described via a self-interacting neutral scalar field, as an illustration of the sort of techniques and results we encounter in such a theory.

We have commented several times in other chapters on the importance of recognizing the architecture of an experiment or theory describing that experiment. The same is true here. Even in continuous time (CT) QFT, it was found necessary to modify the architecture used in scattering calculations. It was found to be no longer sufficient to keep to a single Hilbert-space architecture as in Figure 5.1(a), so the architecture had to be extended to incorporate the observers in some way. The usual abbreviation for the enhancement of this architecture is LSZ, after Lehmann, Symanzik and Zimmermann, who added an 'in' free-particle Hilbert space to model the processes of state preparation and an 'out' free-particle Hilbert space to model the process of state detection (Lehmann *et al.*, 1955).

An important factor in the LSZ formalism is the recognition that the ultimate objectives in QFT are scattering amplitudes, rather than solutions to operator equations of motion. The former are objects from which physical quantities can be calculated, whilst the latter are ill-defined at best and would have to be processed further anyway by the taking of matrix elements in order to be of any use. It is noteworthy that Feynman's path-integral approach (Abers and Lee, 1973) and Schwinger's action-principle and source-function approach (Schwinger, 1969) are all amplitude-based and refer only indirectly to raw quantum field operators.

The LSZ formalism starts off by assuming that the 'in' and 'out' free-particle Hilbert spaces represent observable physical particle states. Such states have sharp energy, definite rest mass, momentum, charge and so on. The traditional states used correspond to plane waves and so are not localized in space. However, it is generally believed that the plane-wave approximations used for the in and out states can always be superposed to form realistic localized packets. What helps here is the belief that the particle interactions are of short range compared with the dimensions of the state-preparation and -detection apparatus normally used in scattering experiments. Whilst this is valid for purely hadronic scattering, it is a questionable assumption when it comes to quantum electrodynamics (QED). However, perturbation theory seems to work in practice even for QED.

The price to pay for the LSZ architecture is substantial: the transition from the physically accessible 'in' and 'out' Hilbert spaces to the void,[1] or the intermediate Hilbert space representing the region between in and out apparatus, invariably requires mathematically ill-defined (divergent) renormalization constants to appear, even for those few interactions which are renormalizable. Most Lagrangians are non-renormalizable.

In the model theory we discuss here, questions of renormalization will not be discussed. However, what will be seen is that there are cutoffs in the theory, which appear to modify the couplings at vertices.

The first step in the analysis is to set up the so-called reduction formalism. This is the formal mechanism which takes us from 'in' or 'out' free-particle states into n-point functions evaluated over the void.

27.1 Reduction formulae

In DT scalar particle scattering theory, we will be interested in incoming and outgoing physical particle states with individual particle energies satisfying the elliptic condition $TE_{\mathbf{p}} < \sqrt{12}$. Here the 'energy' $E_{\mathbf{p}}$ is given by $E \equiv +\sqrt{\mathbf{p}\cdot\mathbf{p}+\mu^2}$, where $\mathbf{p}$ is the linear momentum of the particle concerned.

We shall use the annihilation and creation operators for the scalar field discussed in the previous chapter:

$$\begin{aligned} \hat{a}_n(\mathbf{p}) &= i\Gamma_{\mathbf{p}}^{-1}\int d^3\mathbf{x}\, e^{in\theta_{\mathbf{p}}-i\mathbf{p}\cdot\mathbf{x}}\left[\hat{\varphi}_{n+1}(\mathbf{x}) - e^{i\theta_{\mathbf{p}}}\hat{\varphi}_n(\mathbf{x})\right], \\ \hat{a}_n^{\dagger}(\mathbf{p}) &= -i\Gamma_{\mathbf{p}}^{-1}\int d^3\mathbf{x}\, e^{-in\theta_{\mathbf{p}}+i\mathbf{p}\cdot\mathbf{x}}\left[\hat{\varphi}_{n+1}(\mathbf{x}) - e^{-i\theta_{\mathbf{p}}}\hat{\varphi}_n(\mathbf{x})\right], \end{aligned} \tag{27.1}$$

where

$$\Gamma_{\mathbf{p}} = \frac{6T}{6+T^2E_{\mathbf{p}}^2}, \qquad \cos\theta_{\mathbf{p}} \equiv \eta_{\mathbf{p}} = \frac{6-2T^2E_{\mathbf{p}}^2}{6+T^2E_{\mathbf{p}}^2}. \tag{27.2}$$

A direct application of the standard reduction formalism (Bjorken and Drell, 1965; Gasiorowicz, 1967) gives the reduced matrix elements

$$\begin{aligned} &\langle\alpha^{\text{out}}|\left(\tilde{T}\hat{\zeta}\right)\hat{a}_{\text{in}}^{\dagger}(\mathbf{p})\,|\beta^{\text{in}}\rangle_{\text{R}} \\ &\quad = i\sum_{n=-\infty}^{\infty}\int d^3\mathbf{x}\; e^{-in\theta_{\mathbf{p}}+i\mathbf{p}\cdot\mathbf{x}}\;\overrightarrow{K_{n,\mathbf{p}}}\langle\alpha^{\text{out}}|\tilde{T}\hat{\zeta}\hat{\varphi}_n(\mathbf{x})|\beta^{\text{in}}\rangle \end{aligned} \tag{27.3}$$

[1] We prefer the term *void* rather than vacuum, for two reasons: (i) the space between state-preparation and state-detection apparatus is devoid of any information-gathering apparatus; and (ii), when particle excitations occur in the vacuum, then presumably it is no longer a vacuum.

and

$$\langle\alpha^{\text{out}}|\hat{a}_{\text{out}}(\mathbf{p})\tilde{T}\hat{\zeta}|\beta^{\text{in}}\rangle_{\text{R}} = i\sum_{n=-\infty}^{\infty}\int d^3\mathbf{x}\; e^{in\theta_{\mathbf{p}}-i\mathbf{p}\cdot\mathbf{x}}\;\overrightarrow{K_{n,\mathbf{p}}}\langle\alpha^{\text{out}}|\tilde{T}\hat{\zeta}\hat{\varphi}_n(\mathbf{x})|\beta^{\text{in}}\rangle, \tag{27.4}$$

where $\hat{\zeta}$ denotes any collection of field operators, $|\beta^{\text{in}}\rangle$ and $|\alpha^{\text{out}}\rangle$ are arbitrary 'in' and 'out' states, and

$$\overrightarrow{K_{n,\mathbf{p}}} \equiv \Gamma_{\mathbf{p}}^{-1}(\mathbb{U}_n - 2\eta_{\mathbf{p}} + \mathbb{U}_n^{-1}). \tag{27.5}$$

Using these results we can readily write down the scattering matrix for a process consisting of r incoming physical momentum particles with momenta $\mathbf{p}_1, \mathbf{p}_2, \ldots, \mathbf{p}_r$ and s outgoing particles with momenta $\mathbf{q}_1, \mathbf{q}_2, \ldots, \mathbf{q}_s$.

27.2 Interacting fields: scalar field theory

We turn now to interacting scalar field theories that are based on CT Lagrange densities of the form

$$\mathcal{L} = \mathcal{L}_0 - V(\varphi). \tag{27.6}$$

In order to illustrate what happens in DT quantum field theory, we shall discuss the details of a scalar field with a φ^3 interaction term, deriving the analogue of the Feynman rules.

In the presence of sources the above Lagrange density leads to the system function

$$F^n[j] = F^n_{(0)} - T\int_0^1 d\lambda\int d^3\mathbf{x}\; V(\varphi_{n\lambda}) + \frac{1}{2}T\int d^3\mathbf{x}\{j_n\varphi_n + j_{n+1}\varphi_{n+1}\}, \tag{27.7}$$

where we use the virtual paths

$$\varphi_{n\lambda}(\mathbf{x}) \equiv \mathbb{U}_n^{\lambda}\varphi(\mathbf{x}) = \lambda\varphi_{n+1}(\mathbf{x}) + \bar{\lambda}\varphi_n(\mathbf{x}), \quad 0 \le \lambda \le 1, \quad \bar{\lambda} \equiv 1-\lambda, \tag{27.8}$$

as discussed previously for neutral scalar fields. Here and below we shall find the operator

$$\mathbb{U}_n^{\lambda} \equiv \lambda\mathbb{U}_n + \bar{\lambda} \tag{27.9}$$

particularly useful, where $\mathbb{U}_n$ is the classical temporal displacement operator defined previously.

The vacuum functional is now defined via the DT path integral

$$Z[j] = \int [d\varphi]\exp\{iA[j]\} \equiv \prod_{n=-\infty}^{\infty}\left(\int_{\mathbf{x}}[d\varphi_n]\right)\exp\{iA[j]\}, \tag{27.10}$$

where

$$A[j] \equiv \sum_{n=-\infty}^{\infty} F^n[j] = \sum_{n=-\infty}^{\infty} F^n_{(0)}[j] - i\!\!\!\!\!\!\int_{n\lambda\mathbf{x}} V(\varphi_{n\lambda})$$

and the φ_n are functionally integrated over their spatially indexed degrees of freedom. In the above and in that which follows, we use the notation

$$\oint\!\!\!\!\!\!\sum_{n\lambda\mathbf{x}} \equiv T\int_0^1 d\lambda \sum_{n=-\infty}^{\infty} \int d^3\mathbf{x} \tag{27.11}$$

whenever such a particular combination of spatial integration, summation and virtual-path integration occurs. This replaces the four-dimensional integral $\int d^4x \equiv \int dt\, d^3\mathbf{x}$ found in normal relativistic QFT.

We now postulate our quantum dynamics to be governed by the equation

$$\int [d\varphi]\left[\frac{\delta}{\delta\varphi_n(\mathbf{x})}\{F^n + F^{n-1}\} + Tj_n(\mathbf{x})\right]\exp\{iA[j]\} = 0, \tag{27.12}$$

which is equivalent to a vacuum expectation value of the Heisenberg operator equations of motion derived formally from the DT equation (26.2). Upon integrating by parts, we arrive at the more convenient expression

$$Z[j] = \exp\left\{-i\oint\!\!\!\!\!\!\sum_{n\lambda\mathbf{x}} V(\mathcal{D}_{n\lambda\mathbf{x}})\right\} Z_0[j], \tag{27.13}$$

where

$$\begin{aligned} Z_0[j] &\equiv \int [d\varphi]\exp\left\{iT\sum_{n=-\infty}^{\infty}\int d^3\mathbf{x}\,(\mathcal{F}_0^n + j_n\varphi_n)\right\} \\ &= Z_0[0]\exp\left\{-\frac{1}{2}iT^2\sum_{n,m=-\infty}^{\infty}\int d^3\mathbf{x}\, d^3\mathbf{y}\, j_n(\mathbf{x})\Delta_{\mathrm{F}}^{n-m}(\mathbf{x}-\mathbf{y})j_m(\mathbf{y})\right\} \end{aligned} \tag{27.14}$$

and

$$\mathcal{D}_{n\lambda\mathbf{x}} \equiv -\frac{i}{T}\,\mathbb{U}_n^{\lambda}\frac{\delta}{\delta j_n(\mathbf{x})} = -\frac{i}{T}\left\{\lambda\frac{\delta}{\delta j_{n+1}(\mathbf{x})} + \bar{\lambda}\frac{\delta}{\delta j_n(\mathbf{x})}\right\}. \tag{27.15}$$

Turning now to φ^3 theory, we recall that with hindsight the potential $V^{(3)}(\varphi)$ is normally taken to have the form

$$V^{(3)}(\varphi) = \frac{g}{3!}\{\varphi^3 - \Gamma\varphi\}, \tag{27.16}$$

where the (infinite) subtraction constant Γ is formally given by

$$\Gamma = 3i\Delta_{\mathrm{F}}(0). \tag{27.17}$$

This has the role of cancelling off self-interaction loops at vertices in the Feynman-rules expansion programme. We find that, for DT, the same effect is achieved by taking the potential to have the form

$$V^{(3)}(\varphi_{n\lambda}) = \frac{g}{3!}\left\{\varphi_{n\lambda}^3 - \tilde{\Gamma}\varphi_{n\lambda}\right\}, \tag{27.18}$$

where

$$\tilde{\Gamma} = 2i\Delta_F^0(\mathbf{0}) + \frac{1}{2}i[\Delta_F^1(\mathbf{0}) + \Delta_F^{-1}(\mathbf{0})]. \quad (27.19)$$

The first objective is to find a perturbative expansion for $Z[j]$, which we write in the form

$$Z[j] = Z_0[j] + Z_1[j] + Z_2[j] + \cdots, \quad (27.20)$$

where

$$Z_p[j] \equiv -\frac{i}{p}\oint\!\!\!\!\!\!\sum_{n\lambda\mathbf{x}} V^{(3)}(\mathcal{D}_{n\lambda\mathbf{x}})Z_{p-1}[j], \quad p = 1, 2, \ldots. \quad (27.21)$$

Having found $Z[j]$, we then calculate the required vacuum expectation value of time-ordered products of fields by functional differentiation in the standard way. The results lead to a set of rules for a diagrammatic expansion analogous to the Feynman rules in CT theory, with specific differences. The details of the calculations are omitted here because they are routine and tedious, but the results are as follows.

27.3 Feynman rules for discrete time-ordered products

The objective in this subsection is to present the rules for a diagrammatic expansion of scattering amplitudes in the absence of external sources. The latter are used merely to provide an internal handle on the correlation functions of the theory and are set to zero at the end of the day. This programme is carried out in two stages. In this subsection we give the rules for the evaluation of successive terms in a Feynman-diagram type of expansion for the vacuum expectation value of the time-ordered product

$$\langle 0^{\mathrm{out}}|\tilde{T}\hat{\varphi}_1(\mathbf{x}_1)\hat{\varphi}_2(\mathbf{x}_2)\ldots\hat{\varphi}_k(\mathbf{x}_k)|0^{\mathrm{in}}\rangle \quad (27.22)$$

with k DT scalar fields; we shall give the rules for a system with interaction given by (27.18), so the expansion is effectively in powers in the coupling constant g, as follows.

1. First find the ordinary CT Feynman rules in spacetime.
2. Draw all the different diagrams normally discussed in this programme.
3. For a given diagram with V vertices and I internal lines, find its conventional weighting factor ω, such as the well-known factor of $\frac{1}{2}$ for the simple loop in φ^3 theory.
4. Associate with each vertex a factor

$$igT\oint\!\!\!\!\!\!\sum_{m\lambda\mathbf{z}} \equiv igT\int_0^1 d\lambda \sum_{m=-\infty}^{\infty}\int d^3\mathbf{z}. \quad (27.23)$$

5. For each external line running from the external point $(n, \mathbf{x})$ to a vertex with indices $(m, \lambda, \mathbf{z})$ assign a propagator

$$i\ \mathbb{U}^{\lambda}_{m} \Delta_{\mathrm{F}}^{m-n}(\mathbf{z}-\mathbf{x}). \tag{27.24}$$

6. For each internal line running from vertex $(m_1, \lambda_1, \mathbf{z}_1)$ to vertex $(m_2, \lambda_2, \mathbf{z}_2)$ assign a propagator

$$i\ \mathbb{U}^{\lambda_1}_{m_1} \mathbb{U}^{\lambda_2}_{m_2} \Delta_{\mathrm{F}}^{m_2-m_1}(\mathbf{z}_2-\mathbf{z}_1). \tag{27.25}$$

7. Evaluate the λ integrals.

It is in general much more convenient to perform the virtual-path integrations (over the λs) after the diagrams have been written down rather than before the diagrammatic expansion. In many cases the operator $\mathbb{U}^{\lambda}_{m}$ acting on an external propagator can be transferred to act on internal propagators using the rule

$$\sum_{m=-\infty}^{\infty} \left(\mathbb{U}^{\lambda}_{m} f_m\right) g_m = \sum_{m=-\infty}^{\infty} f_m\ \bar{\mathbb{U}}^{\lambda}_{m} g_m, \tag{27.26}$$

for any indexed functions f_n and g_n, where we define

$$\bar{\mathbb{U}}^{\lambda}_{m} g_m \equiv \lambda\ \mathbb{U}^{-1}_{m} g_m + \bar{\lambda} g_m = \lambda g_{m-1} + \bar{\lambda} g_m. \tag{27.27}$$

However, this does not work so conveniently whenever two or more external lines meet at the same vertex.

To illustrate these rules in operation, consider the conventional perturbation-theory expansion of the time-ordered product $\langle 0|T\hat{\varphi}(x_1)\hat{\varphi}(x_2)|0\rangle$ in powers of the coupling constant. The conventional Feynman rules give the expansion

$$\begin{aligned}\langle 0|T\hat{\varphi}(x_1)\hat{\varphi}(x_2)|0\rangle &= i\Delta_{\mathrm{F}}(x_1-x_2)\\ &\quad - \frac{1}{2}g^2 \int d^4z_1\, d^4z_2\, \Delta_{\mathrm{F}}(x_1-z_1)\Delta_{\mathrm{F}}(z_1-z_2)\\ &\qquad \times \Delta_{\mathrm{F}}(z_2-z_1)\Delta_{\mathrm{F}}(z_2-x_2) + O(g^4).\end{aligned} \tag{27.28}$$

The second term on the right-hand side corresponds to the single-loop diagram with $V=2$ and $I=2$ in φ^3 scalar theory and is divergent. Part of the motivation for investigating DT field theory is the hope that the corresponding diagram might be modified in some significant way.

Using the rules outlined above, the analogous expansion in DT gives

$$\begin{aligned}\langle 0|\tilde{T}\hat{\varphi}_{n_1}(\mathbf{x}_1)\hat{\varphi}_{n_2}(\mathbf{x}_2)|0\rangle &= i\Delta_{\mathrm{F}}^{n_1-n_2}(\mathbf{x}_1-\mathbf{x}_2)\\ &\quad - \frac{1}{2}g^2 \sumint_{m_1\lambda_1\mathbf{z}_1} \sumint_{m_2\lambda_2\mathbf{z}_2} \left\{\mathbb{U}^{\lambda_1}_{m_1}\Delta_{\mathrm{F}}^{m_1-n_1}(\mathbf{z}_1-\mathbf{x}_1)\right\}\\ &\quad \times \left\{\mathbb{U}^{\lambda_1}_{m_1}\mathbb{U}^{\lambda_2}_{m_2}\Delta_{\mathrm{F}}^{m_2-m_1}(\mathbf{z}_2-\mathbf{z}_1)\right\}\\ &\quad \times \left\{\mathbb{U}^{\lambda_2}_{m_2}\mathbb{U}^{\lambda_1}_{m_1}\Delta_{\mathrm{F}}^{m_1-m_2}(\mathbf{z}_1-\mathbf{z}_2)\right\}\\ &\quad \times \left\{\mathbb{U}^{\lambda_2}_{m_2}\Delta_{\mathrm{F}}^{m_2-n_2}(\mathbf{z}_2-\mathbf{x}_2)\right\} + O(g^4).\end{aligned} \tag{27.29}$$

For this particular process the second term on the right-hand side can be rewritten using the rule (27.26) to give

$$\begin{aligned}\langle 0|\tilde{T}\hat{\varphi}_{n_1}(\mathbf{x}_1)\hat{\varphi}_{n_2}(\mathbf{x}_2)|0\rangle &= i\Delta_{\mathrm{F}}^{n_1-n_2}(\mathbf{x}_1-\mathbf{x}_2)\\ &\quad -\frac{1}{2}g^2 \sum\!\!\!\!\!\!\int_{m_1\lambda_1\mathbf{z}_1} \sum\!\!\!\!\!\!\int_{m_2\lambda_2\mathbf{z}_2} \Delta_{\mathrm{F}}^{m_1-n_1}(\mathbf{z}_1-\mathbf{x}_1)\\ &\quad \times\left\{\bar{\mathbb{U}}_{m_1}^{\lambda_1}\bar{\mathbb{U}}_{m_2}^{\lambda_2}\left[\mathbb{U}_{m_1}^{\lambda_1}\mathbb{U}_{m_2}^{\lambda_2}\Delta_{\mathrm{F}}^{m_2-m_1}(\mathbf{z}_2-\mathbf{z}_1)\right]^2\right\}\\ &\quad \times\Delta_{\mathrm{F}}^{m_2-n_2}(\mathbf{z}_2-\mathbf{x}_2)+O(g^4),\end{aligned} \qquad (27.30)$$

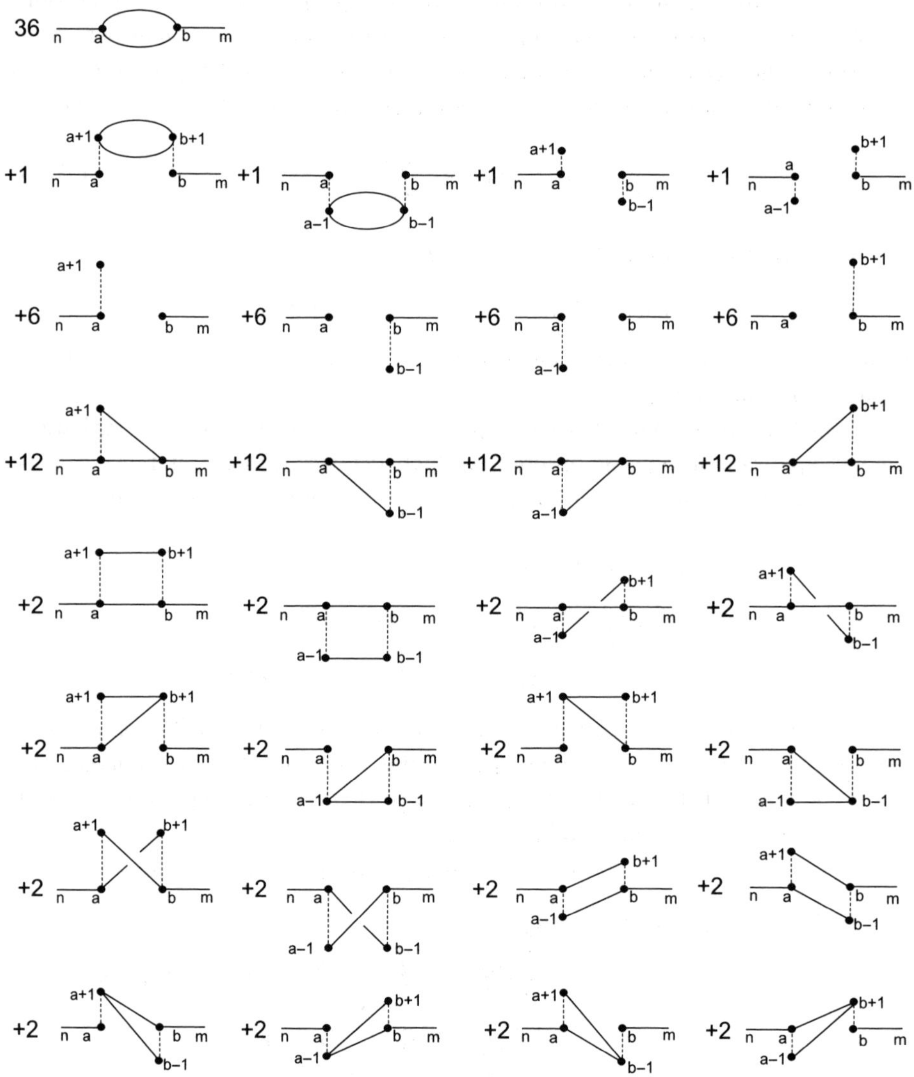

Figure 27.1. Contributions to the propagator.

using the symmetry

$$\Delta_{\mathrm{F}}^{n}(\mathbf{x}) = \Delta_{\mathrm{F}}^{-n}(-\mathbf{x}). \tag{27.31}$$

The integrals over μ and λ can be integrated at this stage to give a multitude of subdiagrams distinguished by different split times, which is the ultimate effect of the discretization process. The various subdiagrams contributing to the loop diagram are shown in Figure 27.1, each with a numerical factor. The sum over all numerical factors for this diagram should add up to 144. The full amplitude corresponding to the loop diagram is the sum of each of these subdiagrams, times the numerical factor for each subdiagram, divided by 288, taking into account the original weighting factor of one half. Symmetry arguments can be used to reduce this number of subdiagrams to twelve.

The above rules are relevant to vacuum expectation values of DT-ordered products of field operators, which shows that DT mechanics can be much more complex than CT mechanics. For particle scattering matrix elements the rules become simpler, as will be discussed next.

27.4 The two–two box scattering diagram

Consider two *incoming* scalar particles with three-momenta **a** and **b**, respectively, scattering via the box scattering diagram shown in Figure 27.2 into two *outgoing* particles with three-momenta **c** and **d**, respectively. Each of these particles is associated with a θ parameter as given by (27.2), which lies within the physical particle interval $[0, \pi)$. Negative values of such a parameter correspond to waves moving backwards in DT and would be interpreted in the usual way as antiparticles in the Feynman–Stueckelberg interpretation. Both positive and negative values occur in the DT Feynman propagators, just as in conventional field theory.

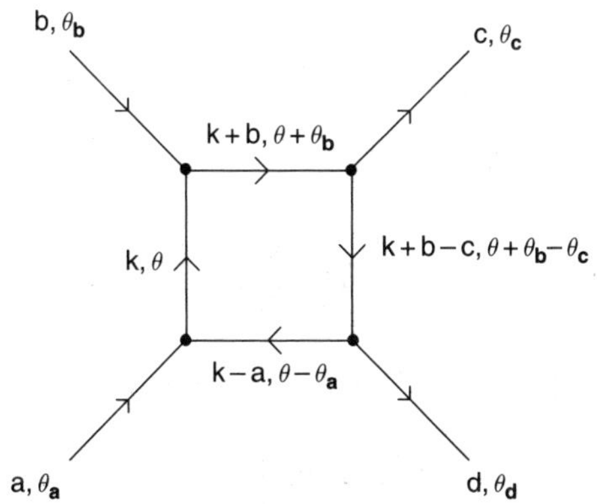

Figure 27.2. The box diagram for two–two scattering.

The details are given by Norton and Jaroszkiewicz (1998a). We expand the four-point function according to the rules outlined above for the box diagram Figure 27.2, viz.

$$\begin{aligned}&\langle 0^{\text{out}}|\tilde{T}\hat{\varphi}_{n_1}(\mathbf{x}_1)\hat{\varphi}_{n_2}(\mathbf{x}_2)\hat{\varphi}_{n_3}(\mathbf{x}_3)\hat{\varphi}_{n_4}(\mathbf{x}_4)|0^{\text{in}}e|_{\text{BOX}}\\ &\quad=(igT)^4\left(\prod_{j=1}^{4}\sum\!\!\!\!\!\!\int_{m_j\lambda_j\mathbf{z}_j} i\,\mathbb{U}_{m_j}^{\lambda_j}\Delta_{\text{F}}^{m_j-n_j}(\mathbf{z}_j-\mathbf{x}_j)\right)\\ &\quad\times\left[\mathbb{U}_{m_2}^{\lambda_2}\mathbb{U}_{m_1}^{\lambda_1}\Delta_{\text{F}}^{m_2-m_1}(\mathbf{z}_2-\mathbf{z}_1)\right]\left[\mathbb{U}_{m_3}^{\lambda_3}\mathbb{U}_{m_2}^{\lambda_2}\Delta_{\text{F}}^{m_3-m_2}(\mathbf{z}_3-\mathbf{z}_2)\right]\\ &\quad\times\left[\mathbb{U}_{m_4}^{\lambda_4}\mathbb{U}_{m_3}^{\lambda_3}\Delta_{\text{F}}^{m_4-m_3}(\mathbf{z}_4-\mathbf{z}_3)\right]\left[\mathbb{U}_{m_1}^{\lambda_1}\mathbb{U}_{m_4}^{\lambda_4}\Delta_{\text{F}}^{m_1-m_4}(\mathbf{z}_1-\mathbf{z}_4)\right].\end{aligned}\tag{27.32}$$

The next step is to do the $\mathbf{x}_j$ integrals, converting the two-point function on each external leg of the diagram to its momentum-space form, using

$$\tilde{\Delta}_{\text{F}}^{n}(\mathbf{p})=\int d\mathbf{x}\,e^{i\mathbf{p}\cdot\mathbf{x}}\Delta_{\text{F}}^{n}(\mathbf{x}).\tag{27.33}$$

Then we use the result

$$\overrightarrow{K_{n,\mathbf{p}}}\tilde{\Delta}_{\text{F}}^{n}(\mathbf{p})=-\delta_n,\tag{27.34}$$

taking care to bring the operators and summations into the brackets whenever the $\mathbb{U}_m^{\lambda}$ operators occur. This effectively amputates the external legs of the diagram. Then we can immediately carry out the summations over the external integers n_i and arrive at the simplified form

$$\begin{aligned}S_{\text{if}}&=(gT)^4\left(\prod_{j=1}^{4}\sum_{m_j=-\infty}^{\infty}\int_0^1 d\lambda_j\int d^3\mathbf{z}_j\right)e^{i\mathbf{a}\cdot\mathbf{z}_1+i\mathbf{b}\cdot\mathbf{z}_2-i\mathbf{c}\cdot\mathbf{z}_3-i\mathbf{d}\cdot\mathbf{z}_4}\\ &\quad\times\left\{\mathbb{U}_{m_1}^{\lambda_1}e^{-im_1\theta_{\mathbf{a}}}\right\}\left\{\mathbb{U}_{m_2}^{\lambda_2}e^{-im_2\theta_{\mathbf{b}}}\right\}\left\{\mathbb{U}_{m_3}^{\lambda_3}e^{im_3\theta_{\mathbf{c}}}\right\}\left\{\mathbb{U}_{m_4}^{\lambda_4}e^{im_4\theta_{\mathbf{d}}}\right\}\\ &\quad\times\left[\mathbb{U}_{m_2}^{\lambda_2}\mathbb{U}_{m_1}^{\lambda_1}\Delta_{\text{F}}^{m_2-m_1}(\mathbf{z}_2-\mathbf{z}_1)\right]\left[\mathbb{U}_{m_3}^{\lambda_3}\mathbb{U}_{m_2}^{\lambda_2}\Delta_{\text{F}}^{m_3-m_2}(\mathbf{z}_3-\mathbf{z}_2)\right]\\ &\quad\times\left[\mathbb{U}_{m_4}^{\lambda_4}\mathbb{U}_{m_3}^{\lambda_3}\Delta_{\text{F}}^{m_4-m_3}(\mathbf{z}_4-\mathbf{z}_3)\right]\left[\mathbb{U}_{m_1}^{\lambda_1}\mathbb{U}_{m_4}^{\lambda_4}\Delta_{\text{F}}^{m_1-m_4}(\mathbf{z}_1-\mathbf{z}_4)\right].\end{aligned}\tag{27.35}$$

Now we use the representation of the propagator

$$\Delta_{\text{F}}^{n}(\mathbf{x})\equiv\frac{1}{(2\pi)^4}\int d^3\mathbf{k}\int_{-\pi}^{\pi}\frac{d\theta}{T}e^{-in\theta+i\mathbf{k}\cdot\mathbf{x}}\tilde{\Delta}_{\text{F}}(\mathbf{k},\theta)\tag{27.36}$$

and evaluate the $\mathbf{z}_i$ integrals to find

$$\begin{aligned}S_{\text{if}}&=g^4(2\pi)^3\delta^3(\mathbf{a}+\mathbf{b}-\mathbf{c}-\mathbf{d})\left(\prod_{j=1}^{4}\sum_{m_j=-\infty}^{\infty}\int_0^1 d\lambda_j\frac{1}{2\pi}\int_{-\pi}^{\pi}d\theta_j\right)\\ &\quad\times\int\frac{d^3\mathbf{k}}{(2\pi)^3}\times\left\{\mathbb{U}_{m_1}^{\lambda_1}e^{-im_1\theta_{\mathbf{a}}}\right\}\left\{\mathbb{U}_{m_2}^{\lambda_2}e^{-im_2\theta_{\mathbf{b}}}\right\}\left\{\mathbb{U}_{m_3}^{\lambda_3}e^{im_3\theta_{\mathbf{c}}}\right\}\left\{\mathbb{U}_{m_4}^{\lambda_4}e^{im_4\theta_{\mathbf{d}}}\right\}\\ &\quad\times\left[\mathbb{U}_{m_2}^{\lambda_2}\mathbb{U}_{m_1}^{\lambda_1}e^{-i(m_2-m_1)\theta_1}\tilde{\Delta}_{\text{F}}(\mathbf{k},\theta_1)\right]\\ &\quad\times\left[\mathbb{U}_{m_3}^{\lambda_3}\mathbb{U}_{m_2}^{\lambda_2}e^{-i(m_3-m_2)}\tilde{\Delta}_{\text{F}}(\mathbf{k}+\mathbf{b},\theta_2)\right]\end{aligned}$$

$$\times \left[\mathbb{U}^{\lambda_4}_{m_4}\mathbb{U}^{\lambda_3}_{m_3} e^{-i(m_4-m_3)}\tilde{\Delta}_{\rm F}(\mathbf{k}+\mathbf{b}-\mathbf{c},\theta_3)\right]$$
$$\times \left[\mathbb{U}^{\lambda_1}_{m_1}\mathbb{U}^{\lambda_4}_{m_4} e^{-i(m_1-m_4)}\tilde{\Delta}_{\rm F}(\mathbf{k}-\mathbf{a},\theta_4)\right]. \tag{27.37}$$

Here we see the appearance of overall linear momentum conservation, as expected. Next we use the result $\mathbb{U}^{\lambda}_{m}e^{im\theta} = e^{im\theta}f_\lambda(\theta)$, where $f_\lambda(\theta) \equiv \lambda e^{i\theta} + \bar{\lambda}$, to find

$$S_{\rm if} = g^4(2\pi)^3\delta^3(\mathbf{a}+\mathbf{b}-\mathbf{c}-\mathbf{d})\left(\prod_{j=1}^{4}\sum_{m_j=-\infty}^{\infty}\int_0^1 d\lambda_j\,\frac{1}{2\pi}\int_{-\pi}^{\pi}d\theta_j\right)$$
$$\times \int\frac{d^3\mathbf{k}}{(2\pi)^3}e^{-im_1\theta_\mathbf{a}}f^*_{\lambda_1}(\theta_\mathbf{a})e^{-im_2\theta_\mathbf{b}}f^*_{\lambda_2}(\theta_\mathbf{b})e^{im_3\theta_\mathbf{c}}f_{\lambda_3}(\theta_\mathbf{c})e^{im_4\theta_\mathbf{d}}f_{\lambda_4}(\theta_\mathbf{d})$$
$$\times\, e^{i(m_1-m_2)\theta_1}f_{\lambda_1}(\theta_1)f^*_{\lambda_2}(\theta_1)\tilde{\Delta}_{\rm F}(\mathbf{k},\theta_1)$$
$$\times\, e^{i(m_2-m_3)\theta_2}f_{\lambda_2}(\theta_2)f^*_{\lambda_3}(\theta_2)\tilde{\Delta}_{\rm F}(\mathbf{k}+\mathbf{b},\theta_2)$$
$$\times\, e^{i(m_3-m_4)\theta_3}f_{\lambda_3}(\theta_3)f^*_{\lambda_4}(\theta_3)\tilde{\Delta}_{\rm F}(\mathbf{k}+\mathbf{b}-\mathbf{c},\theta_3)$$
$$\times\, e^{i(m_4-m_1)\theta_4}f_{\lambda_4}(\theta_4)f^*_{\lambda_1}(\theta_4)\tilde{\Delta}_{\rm F}(\mathbf{k}-\mathbf{a},\theta_4). \tag{27.38}$$

We are now able to do the summations over the m_i. We notice that each summation gives a Fourier-series representation of the periodic Dirac delta, viz.,

$$\sum_{m=-\infty}^{\infty}e^{imx} = 2\pi\sum_{m=-\infty}^{\infty}\delta(x+2m\pi) \equiv 2\pi\delta_{\rm P}(x), \tag{27.39}$$

so we find

$$S_{\rm if} = g^4(2\pi)^4\delta_P(\theta_\mathbf{a}+\theta_\mathbf{b}-\theta_\mathbf{c}-\theta_\mathbf{d})\delta^3(\mathbf{a}+\mathbf{b}-\mathbf{c}-\mathbf{d})\left(\prod_{j=1}^{4}\int_0^1 d\lambda_j\right)$$
$$\times\int\frac{d^3\mathbf{k}}{(2\pi)^4}\int_{-\pi}^{\pi}d\theta\, f^*_{\lambda_1}(\theta_\mathbf{a})f^*_{\lambda_2}(\theta_\mathbf{b})f_{\lambda_3}(\theta_\mathbf{c})f_{\lambda_4}(\theta_\mathbf{d})$$
$$\times\, f_{\lambda_1}(\theta)f^*_{\lambda_2}(\theta)\tilde{\Delta}_{\rm F}(\mathbf{k},\theta)f_{\lambda_2}(\theta+\theta_\mathbf{b})f^*_{\lambda_3}(\theta+\theta_\mathbf{b})\tilde{\Delta}_{\rm F}(\mathbf{k}+\mathbf{b},\theta+\theta_\mathbf{b})$$
$$\times\, f_{\lambda_3}(\theta+\theta_\mathbf{b}-\theta_\mathbf{c})f^*_{\lambda_4}(\theta+\theta_\mathbf{b}-\theta_\mathbf{c})\tilde{\Delta}_{\rm F}(\mathbf{k}+\mathbf{b}-\mathbf{c},\theta+\theta_\mathbf{b}-\theta_\mathbf{c})$$
$$\times\, f_{\lambda_4}(\theta-\theta_\mathbf{a})f^*_{\lambda_1}(\theta-\theta_\mathbf{a})\tilde{\Delta}_{\rm F}(\mathbf{k}-\mathbf{a},\theta-\theta_\mathbf{a}). \tag{27.40}$$

The crucial significance of this step is that we see the appearance of a conservation rule for the parameters θ. This is despite the non-existence of a Hamiltonian in our formulation and the fact that we have not constructed an appropriate invariant for the fully interacting system.

We may go further and do the λ_i integrals. We define the vertex function

$$V(\theta_\mathbf{a},\theta_\mathbf{b}) \equiv \int_0^1 d\lambda\, f^*_\lambda(\theta_\mathbf{a})f^*_\lambda(\theta_\mathbf{b})f_\lambda(\theta_\mathbf{a}+\theta_\mathbf{b})$$
$$= \frac{\cos(\theta_\mathbf{a}+\theta_\mathbf{b})+\cos\theta_\mathbf{a}+\cos\theta_\mathbf{b}+3}{6} \tag{27.41}$$

and so get the final result

$$
\begin{aligned}
S_{\text{if}} &= g^4(2\pi)^4\delta_{\text{P}}(\theta_{\mathbf{a}}+\theta_{\mathbf{b}}-\theta_{\mathbf{c}}-\theta_{\mathbf{d}})\delta^3(\mathbf{a}+\mathbf{b}-\mathbf{c}-\mathbf{d}) \\
&\times \int \frac{d^3\mathbf{k}}{(2\pi)^4}\int_{-\pi}^{\pi} d\theta\, V(\theta_{\mathbf{a}},-\theta)V(\theta_{\mathbf{b}},\theta)V(\theta+\theta_{\mathbf{b}},-\theta_{\mathbf{c}})V(-\theta_{\mathbf{d}},\theta_{\mathbf{a}}-\theta) \\
&\times \tilde{\Delta}_{\text{F}}(\mathbf{k},\theta)\tilde{\Delta}_{\text{F}}(\mathbf{k}+\mathbf{b},\ \theta+\theta_{\mathbf{b}})\tilde{\Delta}_{\text{F}}(\mathbf{k}+\mathbf{b}-\mathbf{c},\ \theta+\theta_{\mathbf{b}}-\theta_{\mathbf{c}}) \\
&\times \tilde{\Delta}_{\text{F}}(\mathbf{k}-\mathbf{a},\ \theta-\theta_{\mathbf{a}}).
\end{aligned}
\tag{27.42}
$$

A diagrammatic representation of the above shows that θ conservation occurs at every vertex.

27.5 The vertex functions

The vertex functions $V(\theta_1,\theta_2)$ represent a degree of softening at each vertex arising from our temporal point splitting via the system function. At each vertex the sum of the *incoming* θ parameters is always zero, including inside loops, so the vertex function always depends on two parameters only. If we had a φ^4 interaction, we would expect that the vertex function will depend on three parameters, and so on. The vertex function has a minimum value of one quarter and attains its maximum value of unity when the θ parameters are each zero. This corresponds to the CT limit $T \to 0$.

27.6 The propagators

The propagators used in the final amplitude are readily found using the basic definition

$$\tilde{\Delta}_{\text{F}}(\mathbf{p},\theta) \equiv T\sum_{n=-\infty}^{\infty} e^{in\theta}\tilde{\Delta}^n_{\text{F}}(\mathbf{p}) \tag{27.43}$$

and the equation

$$(\mathbb{U}_n - 2\eta_E + \mathbb{U}_n^{-1})\tilde{\Delta}^n_{\text{F}}(\mathbf{p}) = -\Gamma_E\delta_n. \tag{27.44}$$

Then

$$2(\cos\theta-\eta_E)\tilde{\Delta}_{\text{F}}(\mathbf{p},\theta) = -T\Gamma_E. \tag{27.45}$$

Now we need to choose the correct solution for Feynman scattering boundary conditions. This is done by referring to the Feynman $-i\epsilon$ prescription, which corresponds to the replacement of E^2 in the above by $E^2 - i\epsilon$. This in turn corresponds to the replacement $\eta_E \to \eta_E + i\epsilon$. Hence we arrive at the desired solution

$$\tilde{\Delta}_{\text{F}}(\mathbf{p},\theta) = \frac{-T\Gamma_E}{2(\cos\theta-\eta_E-i\epsilon)}, \tag{27.46}$$

which holds both for the elliptic region $-1 < \eta_E < 1$ and for the hyperbolic region $-2 < \eta_E < -1$. It may be verified that the indexed propagators (26.94) are given by the integrals

$$\begin{aligned}\tilde{\Delta}^n_{\mathrm{F}}(\mathbf{p}) &= \frac{1}{2\pi}\int_{-\theta}^{\theta} d\theta\, e^{-in\theta}\frac{-\Gamma_E}{2(\cos\theta - \eta_E - i\epsilon)} \\ &= \frac{\Gamma_E}{2\pi i}\oint \frac{dz}{z^n(z^2 - 2(\eta_E + i\epsilon)z + 1)},\end{aligned} \tag{27.47}$$

the contour of integration being the unit circle in the anticlockwise sense. We find, for example,

$$\tilde{\Delta}^n_{\mathrm{F}}(\mathbf{p}) = \frac{\Gamma_E}{2i\sin\theta_E}e^{-i|n|\theta_E} \tag{27.48}$$

in the case of the elliptic regime, $T^2E^2 < 12$, and

$$\tilde{\Delta}^n_{\mathrm{F}}(\mathbf{p}) = \frac{(-1)^{n+1}\Gamma_E}{2\sinh\gamma_E}e^{-|n|\gamma_E} \tag{27.49}$$

in the hyperbolic regime, $T^2E^2 > 12$. Here we make the parametrization

$$\cos\zeta \equiv \eta_E = \frac{6 - 2T^2E^2}{6 + T^2E^2}, \tag{27.50}$$

where ζ is a complex parameter running just below the real axis from the origin to π (when ζ is written as θ_E) and then from π to $\pi - i\ln(2+\sqrt{3})$ (when ζ is written in the form $\pi - i\gamma_E$).

If in (27.46) we introduce the variable p_0 related to θ by the rule

$$\cos\theta \equiv \frac{6 - 2p_0^2T^2}{6 + p_0^2T^2}, \quad \text{sign } \theta = \text{sign } p_0, \tag{27.51}$$

then we find

$$\tilde{\Delta}_{\mathrm{F}}(\mathbf{p},\theta) = \frac{1}{p_0^2 - \mathbf{p}^2 - m^2 + i\epsilon} + \frac{T^2p_0^2}{6(p_0^2 - \mathbf{p}^2 - m^2 + i\epsilon)}, \tag{27.52}$$

an exact result. From this we see the emergence of Lorentz symmetry as an approximate symmetry of the mechanics. If p_0 in the above is taken to represent the zeroth component of a four-vector, with the components of $\mathbf{p}$ representing the remaining components, then we readily see that the first term on the right-hand side of (27.52) is Lorentz invariant. The second term is not Lorentz invariant, but we note that it is proportional to T^2. If, as we expect, T represents an extremely small scale, such as the Planck time or less, then it is clear that Lorentz symmetry should emerge as an extremely good approximate symmetry of our mechanics.

The significance of these results is that not only is spatial momentum conserved during a scattering process, as expected from the DT Noether theorem, but also the sum of the θ parameters of the incoming particles is conserved. This is the DT analogue of energy conservation, since in the limit $T \to 0$ we note

$$\lim_{T\to 0} \frac{\theta_{\mathbf{p}}}{T} = E_{\mathbf{p}} = \sqrt{\mathbf{p}\cdot\mathbf{p} + \mu^2}. \tag{27.53}$$

The θ conservation rule is unexpected at first sight in that we have not discussed any Logan invariant for the full interacting system function. It appears that the analogue of energy conservation occurs here because of the way in which we have set up our incoming and outgoing states and allowed the scattering process to take place over infinite time. The result would probably not hold for scattering over finite time intervals, which would be the analogue of the time–energy uncertainty relation in conventional quantum theory. In essence, the LSZ scattering postulates relate the Logan invariant for in states to the Logan invariant for the out states in such a way that knowledge of the Logan invariant for the intermediate time appears not to be required.

That there is an exact conservation rule for *something* in our DT scattering processes regardless of the magnitude of T is an indicator of the existence of some Logan invariant. The surprise is that it turns out to be the sum of the incoming θ parameters, suggesting that our parametrization of the harmonic oscillator discussed originally was a fortuitously good one.

The scattering amplitude found above for Figure 27.2 reduces to the correct CT amplitude in the limit $T \to 0$.

27.7 Rules for scattering amplitudes

We are now in a position to use our experience with the box diagram Figure 27.2 to write down the general rules for scattering diagrams. Consider a scattering process with a incoming particles with momenta $\mathbf{p}_1, \mathbf{p}_2, \ldots, \mathbf{p}_a$, respectively, and b outgoing particles with momenta $\mathbf{q}_1, \mathbf{q}_2, \ldots, \mathbf{q}_b$, respectively. Make a diagrammatic expansion in the traditional manner of Feynman. For each diagram do the following.

1. At each vertex, conserve linear momentum and θ parameters, i.e., the algebraic sum of *incoming* momenta is zero and the algebraic sum of the *incoming* θ parameters is zero.
2. Associate with each vertex a factor

$$igTV(\theta_1, \theta_2), \tag{27.54}$$

   where θ_1 and θ_2 are any two of the three *incoming* θ parameters,
3. Associate with each internal line carrying momentum $\mathbf{k}$ and θ parameter a factor,

$$iT^{-1}\tilde{\Delta}_{\mathrm{F}}(\mathbf{k},\theta). \tag{27.55}$$

4. Assign, for each loop integral, a factor

$$\int \frac{d^3\mathbf{k}}{(2\pi)^4} \int_{-\pi}^{\pi} d\theta. \tag{27.56}$$

5. Assign an overall momentum–θ-parameter conservation factor

$$(2\pi)^4 \delta_{\mathrm{P}}(\theta_{p_1} + \cdots + \theta_{pa} - \theta_{q_1} - \cdots - \theta_{qb})\delta^3(\mathbf{p}_1 + \cdots + \mathbf{p}_a - \mathbf{q}_1 - \cdots - \mathbf{q}_b). \tag{27.57}$$

6. Assign a weight factor ω for each diagram, exactly as for the standard Feynman rules.

Part VI

Further developments

28

Space, time and gravitation

In this chapter we are going to discuss some aspects touching upon the discretization of time in the context of Einstein's theories of special relativity (SR) and general relativity (GR).

28.1 Snyder's quantized spacetime

At first sight SR seems the wrong theory for temporal discretization. In Newtonian mechanics, time is absolute, which means that simultaneity is absolute: it is the same for all inertial observers. Therefore, if absolute time is discretized for one inertial observer, it is discretized in exactly the same way for all other inertial observers. In that respect, Newtonian discrete time mechanics does not require a preferred inertial frame. Simultaneity is not absolute in SR, however, which suggests that discretizing time will break Lorentz covariance: the use of a preferred inertial frame in which to temporally discretize seems inevitable.

This conclusion is based on classical thinking. The example of quantized angular momentum demonstrates that it is possible to reconcile having a continuous parameter space with a discrete spectrum of observable values. Suppose that an observer is conducting a Stern–Gerlach (SG) experiment, firing a beam of electrons through an apparatus containing a strong inhomogeneous magnetic field (Gerlach and Stern, 1922a). In such an experiment, the observer will first fix the orientation of the apparatus relative to the source. Then it will be found that an electron passing through the apparatus will be detected finally in only one of two possible angular momentum states. This is modelled in quantum mechanics (QM) by assuming that the Hilbert space of outcome states is a quantum bit, i.e., a quantized bit, commonly referred to as a qubit.

The continuum has not been eliminated, however: it resides not in these two states per se but in the freedom of the observer to rotate the apparatus. Then, no matter how the apparatus is rotated, there will be only two possible electron-spin outcomes in each run of the experiment. What will be affected by any rotation of axes of the apparatus will be the probabilities of the two possible outcomes, but that is another story involving another aspect of QM.

The moral is that it may be possible to reconcile discrete time with the continuum of inertial frames. The interpretation is the same as with the SG experiment: discreteness is an attribute of observational context.

Two comments come to mind here.

1. In the early days of Schrödinger wave mechanics, the quantization of angular momentum was referred to as 'spatial quantization'.
2. We stressed right at the start of this book that any discussion of time should take into account observers. Now, close to the end of the book, this point is being reinforced.

We are now in position to discuss *quantized spacetime*, a class of model for relativistic spacetime in which the standard spacetime algebra is extended in such a way as to introduce discreteness in a natural way. These models are generally associated with the name of Hartland S. Snyder (1913–1962), a remarkable physicist whose two landmark papers on the subject continue to inspire fundamentalist theorists (Snyder, 1947a, 1947b). We shall use standard international units here; i.e., we do not set $c = \hbar = 1$.

To understand Snyder's approach, which dealt with pure Lorentz transformations, and that of Yang, who extended it to allow for translations in spacetime (Yang, 1947), we first review some standard concepts of SR. The standard model of relativistic spacetime is that of a four-dimensional spacetime continuum known as Minkowski space, denoted by $\mathcal{M}^4$.

Although $\mathcal{M}^4$ can be considered as a particularly bland GR spacetime, i.e., as a four-dimensional differentiable manifold with the topology of $\mathbb{R}^4$ and a zero-curvature Lorentz signature metric, it can also be treated as a four-dimensional affine vector space with a Lorentzian metric. The particular merit of this latter description is that it readily allows us to find a single coordinate frame that covers the whole spacetime. Inertial frames are such global covers of $\mathcal{M}^4$ adapted to free-particle geodesics.

The standard choice in SR is to use a standard inertial frame $\mathcal{F}$ with coordinates $\{x^\mu : \mu = 0, 1, 2, 3\}$ with $x^0 \equiv ct$, where c is the speed of light. In such a frame the infinitesimal Minkowski line element ds is given by

$$ds^2 = \eta_{\mu\nu}\, dx^\mu\, dx^\nu = (dx^0)^2 - (dx^1)^2 - (dx^2)^2 - (dx^3)^2, \tag{28.1}$$

where the dx^μ are real infinitesimals, $\eta_{\mu\nu}$ is the $\mu\nu$ element of the matrix of metric components given by

$$[\eta_{\mu\nu}] \equiv \begin{bmatrix} 1 & 0 & 0 & 0 \\ 0 & -1 & 0 & 0 \\ 0 & 0 & -1 & 0 \\ 0 & 0 & 0 & -1 \end{bmatrix} \tag{28.2}$$

and the summation convention is employed. Equivalently, the metric tensor g is given in differential geometrical notation by

$$\mathsf{g} = \eta_{\mu\nu}\, \tilde{dx}^{\mu} \otimes \tilde{dx}^{\nu}, \tag{28.3}$$

where the cotangent-space basis one-forms $\{\tilde{dx}^{\mu}\}$ and the tangent-space basis vectors $\{\partial_{\nu}\}$ satisfy the relations $\tilde{dx}^{\mu}(\partial_{\nu}) = \delta^{\mu}_{\nu}$.

One of the principles of SR is that the laws of physics are invariant with respect to standard Poincaré transformations of the form

$$x^{\mu} \to x'^{\mu} = \Lambda^{\mu}{}_{\nu} x^{\nu} + a^{\mu}, \tag{28.4}$$

where $[\Lambda^{\mu}{}_{\nu}]$ is a Lorentz transformation parameter matrix and the $\{a^{\mu}\}$ are translation parameters. A Lorentz transformation can rotate the spatial coordinate axes and/or boost the inertial frame by some subluminal velocity.

In standard transformation theory (Goldstein *et al.*, 2002), infinitesimal transformations are represented by the action of so-called generators of infinitesimal transformations, one generator for each parameter of the transformation. A general Poincaré transformation has a total of ten parameters: four translation parameters, three spatial rotation parameters and three velocity boost parameters. The associated ten generators can be represented in classical Lie-group transformation theory by the differential operators (Hamermesh, 1962)

$$X_{\mu} \mapsto \partial_{\mu}, \quad X_{\mu\nu} \mapsto x_{\nu}\, \partial_{\mu} - x_{\mu}\, \partial_{\nu}. \tag{28.5}$$

In relativistic quantum wave mechanics, these ten classical generators are replaced by the ten operators:

$$\hat{p}_{\mu} \mapsto i\hbar X_{\mu}, \quad \hat{M}_{\mu\nu} \mapsto i\hbar X_{\mu\nu}. \tag{28.6}$$

We shall focus on the generators of spacetime translations, the $\hat{p}_{\mu}$. They satisfy the commutation rule

$$[\hat{p}_{\mu}, \hat{p}_{\nu}] = 0, \tag{28.7}$$

which has the interpretation that $\mathcal{M}^4$ is a flat spacetime.

If it is imagined that the spacetime coordinates are represented by operators, viz., $x^{\mu} \mapsto \hat{x}^{\mu}$, then the phase-space coordinate operators $\{\hat{x}^{\mu}, \hat{p}^{\mu} : \mu = 0, 1, 2, 3\}$ satisfy the algebra

$$[\hat{x}^{\mu}, \hat{x}^{\nu}] = 0, \quad [\hat{p}^{\mu}, \hat{x}^{\nu}] = i\hbar\eta^{\mu\nu}, \quad [\hat{p}^{\mu}, \hat{p}^{\nu}] = 0, \tag{28.8}$$

where $\hat{p}^{\mu} \equiv \eta^{\mu\nu}\hat{p}_{\nu}$. As a check, we take $\mu \to i, \nu \to j$, where $1 \leqslant i, j \leqslant 3$ and recover the standard QM commutation rule $[\hat{p}^{i}, \hat{x}^{j}] = i\hbar\eta^{ij} = -i\hbar\delta_{ij}$, where δ_{ij} is the Euclidean Kronecker delta, taking the value $+1$ when $i = j$ and the value 0 otherwise. So far, all this is standard SR QM.

Snyder proposed replacing the algebra (28.8) with

$$[\hat{x}^\mu, \hat{x}^\nu] = \frac{ia^2}{\hbar}\{\hat{x}^\mu \hat{p}^\nu - \hat{x}^\nu \hat{p}^\mu\},$$
$$[\hat{p}^\mu, \hat{x}^\nu] = i\hbar\eta^{\mu\nu} - i\frac{a^2}{\hbar}\hat{p}^\mu \hat{p}^\nu, \tag{28.9}$$
$$[\hat{p}^\mu, \hat{p}^\nu] = 0,$$

where a is a fundamental length parameter analogous to our chronon (Snyder, 1947a, 1947b). In the limit $a \to 0$, the algebra (28.9) reduces to (28.8). If we wish to replace the fundamental length a by a fundamental time T_S (the Snyder chronon), we simply define $a \equiv Tc$, where c is the speed of light.

Our metric convention and notation differ marginally from those used by Snyder, but are equivalent. Snyder introduced five new dimensionless coordinates θ^μ: $\mu = 0, 1, 2, 3$, and θ^4 for a new spacetime, which we shall call Snyder space, $\mathcal{S}^5$. The conventional spacetime position operators $\hat{x}^\mu$ are represented as vector fields over $T\mathcal{S}^5$ as follows:

$$\hat{x}^\mu \mapsto ia\theta^\mu \frac{\partial}{\partial\theta^4} + ia\theta^4 \frac{\partial}{\partial\theta_\mu}, \quad \hat{p}^\mu \mapsto -\frac{\hbar}{a}\frac{\theta^\mu}{\theta^4}, \quad \mu = 0, 1, 2, 3. \tag{28.10}$$

Then it can be verified that this representation satisfies the Snyder algebra (28.9).

On expanding out, we find for the coordinate operators

$$\hat{x}^0 \equiv c\hat{t} \equiv ia\left(\theta^0 \frac{\partial}{\partial\theta^4} + \theta^4 \frac{\partial}{\partial\theta^0}\right),$$
$$\hat{x}^1 \equiv \hat{x} \equiv ia\left(\theta^1 \frac{\partial}{\partial\theta^4} - \theta^4 \frac{\partial}{\partial\theta^1}\right),$$
$$\hat{x}^2 \equiv \hat{y} \equiv ia\left(\theta^2 \frac{\partial}{\partial\theta^4} - \theta^4 \frac{\partial}{\partial\theta^2}\right), \tag{28.11}$$
$$\hat{x}^3 \equiv \hat{z} \equiv ia\left(\theta^3 \frac{\partial}{\partial\theta^4} - \theta^4 \frac{\partial}{\partial\theta^3}\right).$$

The line element in $\mathcal{S}^5$ is given by

$$ds_S^2 = (d\theta^0)^2 - (d\theta^1)^2 - (d\theta^2)^2 - (d\theta^3)^2 - (d\theta^4)^2. \tag{28.12}$$

Reading off the relevant metric tensor components and using them to raise and lower indices, we note that $\theta^0 = +\theta_0$ and $\theta^i = -\theta_i$, $i = 1, 2, 3, 4$. These signs are a crucial element of the discussion and it should be carefully noted in (28.11) that the first operator, $\hat{x}^0$, differs in the sign of the last term on the right-hand side from the other three operators.

Snyder stated that the $\hat{x}^i$ operators have spectra ma, where m is a positive, negative, or zero integer, whilst $\hat{x}^0$ has a continuous spectrum from $-\infty$ to $+\infty$. To see this heuristically, consider the eigenvalue equation

$$ia\left(u\frac{\partial}{\partial v} - v\frac{\partial}{\partial u}\right)\varphi(u, v) = \lambda\varphi(u, v), \tag{28.13}$$

for some real coordinates u and v. Now change coordinates from (u, v) to $[r, \theta]$, where

$$u = r\cos\theta, \quad v = r\sin\theta. \tag{28.14}$$

Then

$$\partial_\theta = (\partial_\theta u)\partial_u + (\partial_\theta v)\partial_v = -r\sin\theta\,\partial_u + r\cos\theta\,\partial_v = u\,\partial_v - v\,\partial_u,$$

so the eigenvalue equation (28.13) becomes

$$ia\,\partial_\theta\tilde{\varphi}(r,\theta) = \lambda\tilde{\varphi}(r,\theta), \tag{28.15}$$

where $\tilde{\varphi}(r,\theta) \equiv \varphi(u,v)$. This equation is readily solved by the ansatz

$$\tilde{\varphi}(r,\theta) = R(r)e^{-i\lambda\theta/a}. \tag{28.16}$$

Assuming that φ is a single-valued function of u and v, the periodicity of the coordinate θ leads to the quantization condition $\lambda = ma$, where m is some integer. A similar argument is used to deduce that orbital angular momentum in Schrödinger wave mechanics is quantized. Comparing the form of the operators $\hat{x}$, $\hat{y}$ and $\hat{z}$ in (28.11) with (28.13) leads to the conclusion that the spatial coordinate operators $\hat{x}$, $\hat{y}$ and $\hat{z}$ have discrete eigenvalues, which fact corresponds to spatial quantization.

On the other hand, suppose that we had the eigenvalue equation

$$ia\left(u\frac{\partial}{\partial v} + v\frac{\partial}{\partial u}\right)\varphi(u,v) = \lambda\varphi(u,v), \tag{28.17}$$

i.e., with a sign change in the last term on the left-hand side. This would lead us to make the coordinate transformation $u = r\cosh\theta, v = r\sinh\theta$, which gives

$$\partial_\theta = (\partial_\theta u)\partial_u + (\partial_\theta v)\partial_v = r\sinh\theta\,\partial_u + r\cosh\theta\,\partial_v = u\,\partial_v + v\,\partial_u. \tag{28.18}$$

Hence the eigenvalue equation (28.17) becomes

$$ia\,\partial_\theta\tilde{\varphi}(r,\theta) = \lambda\tilde{\varphi}(r,\theta), \tag{28.19}$$

which has solution $\tilde{\varphi}(r,\theta) = R(r)e^{-i\lambda\theta/a}$. Now, however, there is no condition on λ arising from periodicity and therefore no quantization. By inspection of (28.11), we deduce that $\hat{x}^0$ does not have a discrete spectrum.

28.1.1 Commentary

1. Snyder's spacetime model comes remarkably close to discretizing time, but it is clear that it does not do so. Indeed, it is very hard to see how to fix the model so that time is discretized. The issue rests on the crucial sign change in the Minkowski spacetime metric. On the one hand, it is required in order for lightcones to exist, as our spreadsheet mechanics suggested in Chapter 7, and for the field equations to be hyperbolic, as our discussion of Tegmark's

analysis in Chapter 1 showed (Tegmark, 1997). On the other hand, this sign change prevents the spectrum of $\hat{x}^0$ from being discrete. A formal way out of this dilemma might be to make time periodic, perhaps on a vast scale. Then the Snyder space manifold $\mathcal{S}^5$ would be replaced by a periodic-in-time manifold, akin to a torus, and then the periodicity argument used for the spacelike operators $\hat{x}^i$ might apply. However, such a hypothesis would require substantial reinterpretation of fundamental physics, particularly regarding QM and the expansion and contraction models of the Universe (Smolin, 2013).

2. Snyder (1947b) thanked Pauli for pointing out that the $\{\theta^\mu, \theta\}$ can be regarded as homogeneous (projective) coordinates of a real four-dimensional space of constant curvature, a de Sitter space.
3. R. M. Mir-Kasimov has explored the consequences for quantum field theory of interpreting Snyder's algebra in terms of curved momentum space (Mir-Kasimov, 1991).
4. In the second of his two remarkable papers (Snyder, 1947a), Snyder undertook the logical development of the discussion initiated in the first paper (Snyder, 1947b): he investigated the impact on Maxwell's electromagnetic theory of taking the electric and magnetic fields to be functions of non-commuting spacetime coordinates. This was a remarkable and prescient investigation, which has been resurrected in recent years in attempts to formulate the standard model of particle physics in fundamentalist terms by theorists such as Connes (Chamseddine and Connes, 1996). However, there remain problems due to the signature of relativistic metrics.

28.2 Discrete time quantum fields on Robertson–Walker spacetimes

In this section, we move from SR into a background GR spacetime that has a preferred frame of cosmological significance. We consider the temporal discretization of a scalar field evolving over a GR spacetime with a Friedmann–Robertson–Walker (FRW) metric (Cho *et al.*, 1997). Our approach is based on the formalism discussed in Section 17.6 and Appendix B. We take $c = \hbar = 1$.

The latest cosmological evidence is that the Universe is well modelled by an FRW GR structure that is spatially flat over three-dimensional spacelike hypersurfaces (WMAP, 2013). Therefore, we take cosmic coordinates (t, x, y, z), where t is cosmic time, with the line element

$$ds^2 = dt^2 - a(t)^2(dx^2 + dy^2 + dz^2), \tag{28.20}$$

where $a(t)$ is the cosmic scale factor, and consider the action integral for a real scalar field φ of mass m on this background:

$$I = \int dt\, d^3\mathbf{x}\, a^3 \frac{1}{2}\left\{\dot{\varphi}^2 - \frac{1}{a^2}(\nabla\varphi)^2 - (m^2 + \xi R)\varphi^2\right\}, \tag{28.21}$$

where $\xi = 0$ for minimal coupling and $\xi = 1$ for conformal coupling to the Ricci curvature scalar R. The Ricci scalar R is given by

$$R(t) = \frac{6\ddot{a}a + \dot{a}^2}{a^2} \tag{28.22}$$

and is therefore time-dependent.

The conjugate momentum

$$\pi \equiv \frac{\partial \mathcal{L}}{\partial \dot{\varphi}} = a^3 \dot{\varphi} \tag{28.23}$$

has canonical equal-time commutator

$$[\pi(\mathbf{x}, t), \varphi(\mathbf{y}, t)] = -i\delta^3(\mathbf{x} - \mathbf{y}), \tag{28.24}$$

which may be rewritten in the form

$$\lim_{t' \to t} [\dot{\varphi}(\mathbf{x}, t'), \varphi(\mathbf{y}, t)] = -ia^{-3}\delta^3(\mathbf{x} - \mathbf{y}). \tag{28.25}$$

We now take advantage of the empirically supported spatial homogeneity of the line element (28.20), to which the coordinates $\mathbf{x} \equiv (x, y, z)$ are well adapted. The spatial Fourier transform

$$\varphi(\mathbf{x}, t) = \int \frac{d^3\mathbf{k}}{(2\pi)^3} \varphi_{\mathbf{k}}(t) e^{i\mathbf{k}\cdot\mathbf{x}} \tag{28.26}$$

gives the Lagrangian

$$L = \int \frac{d^3\mathbf{k}}{(2\pi)^3} a^3 \frac{1}{2} \{\dot{\varphi}_{\mathbf{k}} \dot{\varphi}_{-\mathbf{k}} - \omega_{\mathbf{k}}^2 \varphi_{\mathbf{k}} \varphi_{-\mathbf{k}}\}, \tag{28.27}$$

where

$$\omega_{\mathbf{k}}^2 = \frac{\mathbf{k}^2}{a^2} + m^2 + \xi R. \tag{28.28}$$

Reality of the scalar field gives

$$\varphi^*(\mathbf{x}, t) = \varphi(\mathbf{x}, t), \qquad \varphi_{\mathbf{k}}^*(t) = \varphi_{-\mathbf{k}}(t). \tag{28.29}$$

Now we write

$$\varphi_{\mathbf{k}} = \varphi_{\mathbf{k}}^{(+)} + i\varphi_{\mathbf{k}}^{(-)}, \quad \varphi_{-\mathbf{k}} = \varphi_{\mathbf{k}}^{(+)} - i\varphi_{\mathbf{k}}^{(-)}, \tag{28.30}$$

where $\varphi_{\mathbf{k}}^{(+)}$ and $\varphi_{\mathbf{k}}^{(-)}$ are real. These are not independent, since $\varphi_{\mathbf{k}}^{(\pm)} = \pm\varphi_{-\mathbf{k}}^{(\pm)}$. This is taken into account by multiplying the integrand in (28.27) by two and integrating over only half of $\mathbf{k}$-space. We may, for example, choose $k_3 > 0$, and call this volume V^+, giving the integral

$$L = \sum_{\sigma=+,-} \int_{V^+} \frac{d^3\mathbf{k}}{(2\pi)^3} \left\{ \left(\dot{\varphi}_{\mathbf{k}}^{(\sigma)}\right)^2 - \omega_{\mathbf{k}}^2 \left(\varphi_{\mathbf{k}}^{(\sigma)}\right)^2 \right\}. \tag{28.31}$$

Now we turn to temporal discretization via virtual paths as before. We choose the virtual paths given by

$$\varphi_{n\mathbf{k}}^{(\sigma)}(\lambda) \equiv \lambda\varphi_{n+1\mathbf{k}}^{(\sigma)} + (1-\lambda)\varphi_{n\mathbf{k}}^{(\sigma)} \tag{28.32}$$

and integrate from $\lambda = 0$ to $\lambda = 1$. The system function turns out to be

$$\mathcal{F}_{\mathbf{k}}^{(\sigma)n} = \frac{1}{2}\alpha_{n\mathbf{k}}\left[\left(\varphi_{n+1\mathbf{k}}^{(\sigma)}\right)^2 + \left(\varphi_{n\mathbf{k}}^{(\sigma)}\right)^2\right] - \beta_{n\mathbf{k}}\varphi_{n+1\mathbf{k}}^{(\sigma)}\varphi_{n\mathbf{k}}^{(\sigma)}, \tag{28.33}$$

where the index σ is $\pm$, $\mathbf{k}$ is in the half space V^+ and

$$\alpha_{n\mathbf{k}} \equiv a_n^3\left[\frac{6-2T^2\omega_{nk}^2}{6T}\right], \quad \beta_{n\mathbf{k}} \equiv \frac{a_n^3}{3T}[6+T^2\omega_{n\mathbf{k}}^2]. \tag{28.34}$$

The variable $\pi_{n\mathbf{k}}^{(\sigma)}$ conjugate to $\varphi_{n\mathbf{k}}^{(\sigma)}$ is defined by

$$\pi_{n\mathbf{k}}^{(\sigma)} \equiv -\frac{\partial}{\partial\varphi_{n\mathbf{k}}^{(\sigma)}}\mathcal{F}_{\mathbf{k}}^{(\sigma)n} = -\alpha_{n\mathbf{k}}\varphi_{n\mathbf{k}}^{(\sigma)} + \beta_{n\mathbf{k}}\varphi_{n+1\mathbf{k}}^{(\sigma)}. \tag{28.35}$$

Quantization is defined via the commutators

$$\left[\hat{\pi}_{n\mathbf{k}}^{(\sigma)}, \hat{\varphi}_{n\mathbf{q}}^{(\sigma')}\right] = -i\delta_{\sigma\sigma'}\delta_{\mathbf{k}-\mathbf{q}}, \quad \mathbf{k},\mathbf{q} \in V^+, \tag{28.36}$$

where $\delta_{\mathbf{k}} \equiv (2\pi)^3\delta^3(\mathbf{k})$. Then

$$\left[\hat{\varphi}_{n+1\mathbf{k}}^{(\sigma)}, \hat{\varphi}_{n\mathbf{q}}^{(\sigma')}\right] = -\frac{i}{\beta_{n\mathbf{k}}}\delta_{\sigma\sigma'}\delta_{\mathbf{k}-\mathbf{q}}, \quad \mathbf{k}, \mathbf{q} \in V^+. \tag{28.37}$$

Hence

$$\left[\hat{\varphi}_{n+1\mathbf{k}}, \hat{\varphi}_{n\mathbf{q}}^{+}\right] = -\frac{i}{a_n^3}\frac{6T}{6+T^2\omega_{n\mathbf{k}}^2}(2\pi)^3\delta^3(\mathbf{k}-\mathbf{q}). \tag{28.38}$$

Finally, we arrive at the field commutators

$$[\hat{\varphi}_{n+1}(\mathbf{x}), \hat{\varphi}_n(\mathbf{y})] = -\frac{i}{a_n^3}\int\frac{d^3\mathbf{k}}{(2\pi)^3}\frac{6T}{6+T^2\omega_{n\mathbf{k}}^2}e^{i\mathbf{k}\cdot(\mathbf{x}-\mathbf{y})}. \tag{28.39}$$

In the limit $T \to 0$ we recover the canonical result (28.25), provided that we may write

$$\lim_{T\to 0}\hat{\varphi}_{n+1}(\mathbf{x}) \sim \hat{\varphi}_n(\mathbf{x}) + \dot{\varphi}_n(\mathbf{x})T + O(T^2). \tag{28.40}$$

28.2.1 Quadratic invariants

Given the Heisenberg equation of motion

$$\hat{\varphi}_{n+1\mathbf{k}} \underset{c}{=} 2\eta_{n\mathbf{k}}\hat{\varphi}_{n\mathbf{k}} - \gamma_{n\mathbf{k}}\hat{\varphi}_{n-1\mathbf{k}}, \tag{28.41}$$

where

$$2\eta_{n\mathbf{k}} = \frac{\alpha_{n\mathbf{k}} + \alpha_{n-1\mathbf{k}}}{\beta_{n\mathbf{k}}}, \quad \gamma_{n\mathbf{k}} = \frac{\beta_{n-1\mathbf{k}}}{\beta_{n\mathbf{k}}}, \tag{28.42}$$

we may construct a quadratic invariant momentum density

$$\mathcal{C}^n_{\mathbf{k}} = \frac{1}{2} i\beta_{n\mathbf{k}} \{ \hat{\varphi}^+_{n+1\mathbf{k}} \hat{\varphi}_{n\mathbf{k}} - \hat{\varphi}^+_{n\mathbf{k}} \hat{\varphi}_{n+1\mathbf{k}} \}, \tag{28.43}$$

which is conserved, viz.,

$$\mathcal{C}^n_{\mathbf{k}} \underset{\mathrm{H}}{=} \mathcal{C}^{n-1}_{\mathbf{k}}. \tag{28.44}$$

This may also be written in the form

$$\mathcal{C}^n_{\mathbf{k}} = \beta_{n\mathbf{k}} \left\{ \hat{\varphi}^{(-)}_{n+1\mathbf{k}} \hat{\varphi}^{(+)}_{n\mathbf{k}} - \hat{\varphi}^{(-)}_{n\mathbf{k}} \hat{\varphi}^{(+)}_{n+1\mathbf{k}} \right\}. \tag{28.45}$$

With the basic commutators (28.37) we find

$$\begin{aligned} \left[\mathcal{C}^n_{\mathbf{k}}, \hat{\varphi}^{(\pm)}_{n\mathbf{q}} \right] &= \pm i \hat{\varphi}^{(\mp)}_{n\mathbf{k}} \delta_{\mathbf{k}-\mathbf{q}}, \\ \left[\mathcal{C}^n_{\mathbf{k}}, \hat{\varphi}^{(\pm)}_{n+1\mathbf{q}} \right] &= \pm i \hat{\varphi}^{(\mp)}_{n+1\mathbf{k}} \delta_{\mathbf{k}-\mathbf{q}}. \end{aligned} \tag{28.46}$$

Hence we find

$$[\mathcal{C}^n_{\mathbf{k}}, \hat{\varphi}_{n\mathbf{q}}] = \hat{\varphi}_{n\mathbf{k}} \delta_{\mathbf{k}-\mathbf{q}}, \quad [\mathcal{C}^n_{\mathbf{k}}, \hat{\varphi}_{n+1\mathbf{q}}] = \hat{\varphi}_{n+1\mathbf{k}} \delta_{\mathbf{k}-\mathbf{q}}. \tag{28.47}$$

If now we construct the integrated invariant

$$\hat{C}^n \equiv \int \frac{d^3\mathbf{k}}{(2\pi)^3} C(\mathbf{k}) \mathcal{C}^n_{\mathbf{k}}, \tag{28.48}$$

where $C(\mathbf{k})$ is an arbitrary real function then we find

$$\left[\hat{C}^n, \hat{\varphi}_{n\mathbf{q}} \right] = C(\mathbf{q}) \hat{\varphi}_{n\mathbf{q}}, \quad \left[\hat{C}^n, \hat{\varphi}_{n+1\mathbf{q}} \right] = C(\mathbf{q}) \hat{\varphi}_{n+1\mathbf{q}}. \tag{28.49}$$

These results allow us to construct various quantities of interest.

28.3 Regge calculus

Although our primary interest has been in fundamental concepts, there are some branches of numerical simulation that deserve comment. One of these is Tullio Regge's 'general relativity without coordinates' approach to GR (Regge, 1961).

GR is a notoriously difficult theory to deal with at all levels. As a classical field theory, the Einstein field equations (6.2) are generally impossible to solve for an arbitrary energy-momentum tensor $T_{\mu\nu}$, and numerical solutions are commonly attempted. Regge contributed to this programme in what can fairly be regarded as a fundamentalist approach. His method is to approximate curved spacetimes by higher-dimensional versions of polyhedra.

There is a degree of similarity between Regge's approach and our virtual-path approach, with two important differences: (i) we discretize one dimension, whereas Regge discretized all four spacetime dimensions; and (ii) Regge's aim was ultimately to take the continuum limit of a discretized model of spacetime. In his classic paper (Regge, 1961), Regge first discussed two-dimensional surfaces in terms of triangulations, i.e., the approximation of a surface by a number of simpler, straight-sided figures such as triangles. This is a variant of the 'method of exhaustion' principle referred to in Section 3.1, whereby Greek mathematicians approximated figures such as circles by polygons with ever increasing numbers of sides.

A fundamental feature of triangulation over curved surfaces is that the sum of the angles at the vertices need not equal 2π. In such approximation schemes, the relevant departure from the Euclidean-plane value of 2π is represented by $2\pi - \varepsilon$, where ε is called the deficiency.

Having set the scene in two dimensions, Regge moved on to matter-free GR. It is well known that the Einstein field equations in this case can be obtained from an action principle that is based on the Einstein–Hilbert action (Hilbert, 1915)

$$A[g] \equiv \int R\sqrt{-g}\, d^4x, \tag{28.50}$$

where R is the Ricci scalar curvature, and we have ignored inessential constants. Regge showed that a triangulation process, which is relatively complicated considering the nature of the theory, resulted in the above action integral $A[g]$ being replaced by the limit of an action sum of the form

$$A[g] \sim \lim_{\text{Regge}} \sum_n \varepsilon_n L^n, \tag{28.51}$$

where L^n is a function on the nth element of the triangulation, ε_n is the relevant deficiency, and the limit requires a careful balance of numbers of elements in the triangulation with the local curvature, as explained by Regge.

What is of interest to us is that Regge showed that, if l_m is the 'length' of side m of the triangulation and is considered to be a dynamical variable (the shade of Lee's mechanics appears at this point), then the GR field equations are given in this approximation by

$$\sum_n \varepsilon_n \frac{\partial L^n}{\partial l_m} \underset{c}{=} 0, \tag{28.52}$$

which bears a remarkable similarity, with important differences of course, to the DT equations of motion we have encountered several times in this book.

29
Causality and observation

29.1 Introduction

Throughout this book we have commented on the need to have a wider vision of physics as more than merely developing and solving equations of motion for states of systems under observation (SUOs). It is our premise that any attempt to discuss time needs to address issues such as the status of observers, apparatus, and suchlike, relative to those states of SUOs. This echoes the following quote from Wheeler:

> Stronger than the anthropic principle is what I might call the participatory principle. According to it we could not even imagine a universe that did not somewhere and for some stretch of time contain observers because the very building materials of the universe are these acts of observer-participancy. You wouldn't have the stuff out of which to build the universe otherwise. This participatory principle takes for its foundation the absolutely central point of the quantum: No elementary phenomenon is a phenomenon until it is an observed (or registered) phenomenon.
>
> (Wheeler, 1979)

We agree with Wheeler to this extent: there seems little, if any, point in speculating about truly unobservable parts of the Universe. If, however, there is a possibility that physical signals or effects from currently unobserved parts of the Universe might one day reach us, then that would in essence make those parts observable, albeit indirectly, and therefore suitable for theoretical analysis and speculation.

In this chapter we discuss some aspects of the discreteness that underpins observations. First we review classical causal-set theory, an approach to discrete spacetime that emphasizes the patterns of causal relationships running throughout the observed Universe. We follow this with an attempt to encode the patters of causality within quantum mechanics (QM). This seems a natural scenario given the possibility that multi-register quantum states can have complex patterns of separability and entanglement, as reviewed in Appendix D. Finally, we discuss an approach to the description of laboratory physics that models the possibility of observers constructing and destroying apparatus during the course of multi-detector observations in a discrete time setting.

29.2 Causal sets

In the opening chapter of this book we referred to Reichenbach's 'star' notation used to discuss the causal relationship between two events A and B, and pointed out that his definition of causality depended on counterfactuality (Reichenbach, 1958). But, in the real world, counterfactuality is a metaphysical concept: the past cannot be experimented with. Rerunning an experiment with altered initial conditions is *not* a genuine counterfactual experiment. Nevertheless, such experiments are interpreted as such by theorists conditioned in classical ways of thinking.

With causality in mind, some authors (Bombelli *et al.*, 1987; Brightwell and Gregory, 1991; Ridout and Sorkin, 2000; Markopoulou, 2000; Requardt, 1999) have discussed the idea that spacetime could be discussed in terms of classical causal sets. A causal set $\mathcal{C}$ is a set $\mathcal{C} \equiv \{x, y, \ldots\}$ of objects (or *events*), some pairs of which have a particular binary relationship amongst themselves, denoted by the symbol $\prec$, which is taken here to be a mathematical representation of a temporal ordering. Not all pairs in a causal set need have such an ordering. For any two different events x and y, if neither $x \prec y$ nor $y \prec x$ holds, x and y are said to be relatively spacelike, causally independent or incomparable (Howson, 1972). The objects in $\mathcal{C}$ are generally assumed to be the ultimate description of spacetime, which in the causal-set hypothesis is often postulated to be discrete (Ridout and Sorkin, 2000). Minkowski spacetime is an example of a causal set with a continuum of events (Brightwell and Gregory, 1991), with the possibility of extending the relationship $\prec$ to include the concept of *null* or *lightlike* relationships. In our approach to quantum causal-set theory, which will be discussed in the next section, any interpretation in terms of spacetime is to be regarded as an emergent one and not necessary in principle.

For comparable events x, y, z of the causal set $\mathcal{C}$, the following relations hold:

$$\begin{aligned} \forall x, y, z \in \mathcal{C}, &\quad x \prec y \text{ and } y \prec z \Rightarrow x \prec z \quad \text{(transitivity)}, \\ \forall x, y \in \mathcal{C}, &\quad x \prec y \Rightarrow y \not\prec x \qquad \text{(asymmetry)}, \\ \forall x \in \mathcal{C}, &\quad x \not\prec x \qquad \text{(irreflexivity)}. \end{aligned} \tag{29.1}$$

Familial concepts such as parents, offspring and siblings are sometimes invoked to discuss causal sets. For example, x is a direct cause or parent of y if $x \prec y$ and there is no event z in $\mathcal{C}$ such that $x \prec z \prec y$. Event y is known as the effect or offspring of x, and we denote this relationship by the notation $x \lhd y$. A given event may be the parent of more than one event, in which case the offspring are called siblings. Siblings of the same parent are necessarily incomparable, and a given event may be the offspring of more than one parent.

We shall restrict our attention to discrete causal sets, i.e., those consisting of a countable number of events, because the causal-set structure we find in quantum register cosmology is necessarily discrete. The importance of discrete causal-set theory is that, in the large-scale limit of very many events, discrete causal sets may yield all the properties of continuous spacetimes, such as metrics, manifold structure and even variable spatial dimensionality, all of which should be determined by the dynamics (Bombelli *et al.*, 1987). This is the converse of the usual procedure of using the properties of the manifold and metric to determine the lightcones of the spacetime, from which the causal order may in turn be inferred.

Concepts of proper time, spatial distance and pseudo-Riemannian metric may be developed in discrete causal-set theory by considering the lengths of paths between events (Bombelli *et al.*, 1987; Brightwell and Gregory, 1991). For example, the notion of 'temporal distance', which is equivalent to proper time in relativity, may be defined as follows.

A causal or timelike path is a set $\{a_1, a_2, \ldots, a_n\}$ of events such that, for $1 \leqslant i < n$, $a_i \lhd a_{i+1}$. We define the proper time along such a path to be $n-1$. There may be more than one timelike path between two events, if $n > 1$, and their proper times may differ. This corresponds to the path dependence of proper time in relativity, which is the origin of the so-called 'twin paradox'.

For incomparable events, the binary relation $\prec$ may be used in a more elaborate and indirect way to provide an analogous definition of spatial distance, in much the same way as light signals may be used in special relativity to determine distances between spacelike separated events. This implements Riemann's notion that the origin of distance may be a counting process (Bombelli *et al.*, 1987).

In an analogous way, concepts of 'volume', 'area', and 'dimension of space' may be defined in terms of counting numbers of causally related events within a specified distance. For example, given an event x in the causal set $\mathcal{C}$, define the 'sphere' $S_1(x)$ to be the subset of $\mathcal{C}$ consisting of events y for which x is a parent. Next, define $S_2(x)$ to be the subset of $\mathcal{C}$ consisting of events for which the elements of $S_1(x)$ are the parents. This process may be continued indefinitely to generate what is essentially the discrete spacetime equivalent of the forwards lightcone with vertex at x in relativity. Suppose now that, for sufficiently large integer n, we found that $S_n(x)$ had the approximate behaviour

$$S_n(x) \sim a + b \ln n, \tag{29.2}$$

where a and b were constants. On comparing the formulae for 'areas' of Euclidean manifolds of various (integral) dimensions, we would be justified in asserting that, over the scale concerned, the causal set appeared to behave like a space of dimension $b+1$. An attractive feature of causal sets is the possibility that different spatial dimensions might emerge on different scales (Bombelli *et al.*, 1987), whereas in conventional theory, higher dimensions generally have to be put in by hand.

29.3 Quantum causal sets

This section discusses how a classical observer could describe an SUO with many degrees of freedom in terms of a succession of quantum register states (Eakins and Jaroszkiewicz, 2005). In this description, time is marked off by the observer in discrete steps called stages, with each stage Ω_n being labelled by an integer n called lab time. Associated with each discrete time will be a collection of detectors, which the observer could look at, if they so decided. Associated with the full set of detectors at time n will be a quantum register Hilbert space $\mathcal{H}_n$, and the states in it will be amenable to the entanglement–separation analysis discussed in Appendix D.

To illustrate how causal-set structure could emerge from such a scenario, consider an experiment that starts at time $n = 0$ with the SUO described by a state Ψ_0 in quantum register $\mathcal{H}_0$. At the next stage, the observer may, or might not, have looked at any of the detectors. If they have, then there will be state reduction, or wavefunction collapse, which in our terms is synonymous with information extraction.

However, suppose that the observer does not attempt to extract information, but, rather, alters the detectors. Then the Hilbert space at time $n = 1$ is now $\mathcal{H}_1$ and the state of the SUO is some element in $\mathcal{H}_1$, denoted by $\ominus_1$. This process can be repeated until the experiment is concluded.

Taking the notation described in Appendix D, consider a sequence of quantum register states given by

$$\Psi^{123456} \to \psi_1^{23\cdot 456} \to \theta_{16}^{23\cdot 35} \to \eta_4^{12\cdot 356} \to \phi^{12\cdot 34\cdot 56} \to \dots, \qquad (29.3)$$

where Ψ^{123456} is in $\mathcal{H}_0$ at time $n = 0$, $\psi_1^{23\cdot 456}$ is in $\mathcal{H}_1$ at time $n = 1$, and so on. Reading off the entanglement–separation information as described in Appendix D, we can represent this process diagrammatically, as shown in Figure 29.1.

29.4 Discrete time and the evolving observer

We finish our investigation into discrete time by a discussion of how an observer could model the processes of observation (Jaroszkiewicz, 2010). This could include the construction, modification and destruction of apparatus.

By their very nature, such processes involve discrete time. This is not the regular discreteness associated with a fundamental Type-1 chronon, but has to do with the discreteness of acquisition of information. The question 'when does a detector click?' would be replaced in this scheme by the question 'when does

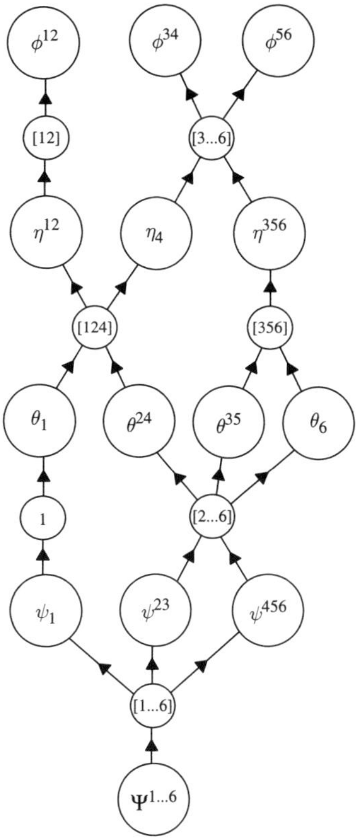

Figure 29.1. An example of quantum causal-set structure that is based on entanglements and separations.

the observer have reason to believe that a detector has clicked?'. It seems to us that a better strategy than discussing states of SUOs per se would be to discuss how an observer perceives those states. In other words, our focus is on observers, not the things they observe.

A typical experiment described by our formalism involves an observer interacting with a discrete time-dependent number r_n of elementary signal detectors (ESDs) at a countable number of times t_n, where the integer n runs from some initial integer M to some final integer $N > M$. An ESD is not necessarily a physical detector, but an *opportunity for detection*: a time and a place where information could be acquired if the observer so wished. For example, each of the two slits in the famous double-slit experiment is an ESD, but the success of the experiment relies on the observer actually *not* placing a detector at any of the slits.

There is no need to assume that $t_{n+1} - t_n$ always has the same value, or that r_n is independent of n. The formalism allows for the creation and destruction of ESDs, which is something that happens in the real world.

29.4.1 The laboratory and the universal register

Observers do not just exist without a great deal of support structure. It is reasonable therefore to assume that an observer exists in a suitable physical environment, referred to as the *laboratory*, Λ. This will have the facilities for the construction or introduction of apparatus consisting of a number of ESDs. At any given discrete time n, the observer will associate a state known as the *lab state* with the collection of ESDs at that time. This state could be a pure state or a mixed state. We shall restrict our attention here to pure lab states for reasons of space.

A lab state carries information as to whether various ESDs exist in the first place and, if so, whether they are functioning normally and either in their ground or signal states, or whether they are faulty.

The power-set approach to ESDs allows us to think of an absence of an ESD in Λ as an observable fact that is representable mathematically. The state corresponding to an absent or non-functioning ESD is represented by the element $|\emptyset)$ of its associated power set. Therefore, we can represent the complete absence of any ESDs whatsoever by an infinite collection of such elements. This corresponds to an observer without any apparatus, i.e., an empty laboratory. We denote this lab state by the symbol $|\Omega)$ and call it the *information void*, or just the *void*. It represents a potential for existence, relative to a given observer.

If the observer's laboratory Λ is in its void state $|\Omega)$, that does not mean that the laboratory Λ and the observer do not exist, or that there are no SUOs in Λ. It means simply that the observer has no current means of acquiring any information. An empty laboratory devoid of detectors is a physically meaningful concept, but one with no interesting empirical content.

The information void can be thought of as one element in an infinite-rank quantum register called the universal register and denoted $\mho$. We write

$$|\Omega) \equiv \prod_i^\infty |\emptyset_i) \in \mho, \tag{29.4}$$

where the index i could in principle be discrete, continuous or a combination thereof. Real observers can only ever deal with finite numbers of ESDs in practice and we can generally ignore all non-existent potential ESDs.

In our notation, the ordering of states in a tensor product is not significant, since labels keep track of the various terms. An arbitrary classical lab state $|\Psi)$ in the universal register $\mho$ will be of the form $\prod_i^\infty |s_i)$, where $|s_i)$ is one of the four elements of the power set of the ith bit, viz., $\mathcal{P}(\mathsf{B}^i)$ (Jaroszkiewicz, 2010).

Operators acting on universal register states will be denoted in open-face font and act as follows. If O_i is a bit operator acting on elements of $\mathcal{P}(\mathsf{B}^i)$, then $\mathbb{O} \equiv \prod_i^\infty O_i$ acts on an arbitrary classical state $|\Psi) \equiv \prod_i^\infty |s_i)$ according to the rule

$$\mathbb{O}|\Psi) \equiv \prod_i^\infty O_i|s_i). \tag{29.5}$$

For every classical register state $|\Psi) \equiv \prod_i^\infty |s_i)$ there will be a corresponding dual register state $(\Psi| \equiv \prod_i^\infty (s_i|$, where $(s_i|$ is dual to $|s_i)$. Classical register states including the void satisfy the orthonormality condition

$$(\Phi|\Psi) \equiv \left\{\prod_i^\infty (r_i|\right\} \prod_j^\infty |s_j) = \prod_i^\infty (r_i|s_i) = \prod_i^\infty \delta_{r_i s_i}. \tag{29.6}$$

Classical register states $|\Phi)$ and $|\Psi)$ that differ in at least one bit power-set element therefore satisfy the rule $(\Phi|\Psi) = (\Psi|\Phi) = 0$.

29.4.2 Contextual vacua

In conventional classical mechanics or Schrödinger–Dirac quantum mechanics, empty space is generally not represented by any specific mathematical object. In quantum field theory, however, empty space is represented by the *vacuum*, a normalized vector in an infinite-dimensional Hilbert space. It has physical properties such as zero total momentum, zero total electric charge, etc., which, despite their being bland, are physically significant attributes nevertheless.

In our approach we encounter an analogous concept. Starting with the void $|\Omega)$, we represent the construction of a collection of ESDs in the laboratory Λ by the application of a corresponding number of construction operators C_i to their respective empty states $|\emptyset_i)$. For example, a lab state consisting of a single ESD i in its ground state is given by $|\Psi) = \mathbb{C}_i|\Omega) = \left\{\prod_{j\neq i}^\infty |\emptyset_j)\right\} \times |0_i)$, where $\mathbb{C}_i$ is the register operator $\mathbb{C}_i \equiv \left\{\prod_{j\neq i}^\infty I_j\right\} \times C_i$. More generally, a state consisting of a number r of ESDs each in its ground state is given by

$$|\Psi^r) = \mathbb{C}_1\mathbb{C}_2 \ldots \mathbb{C}_r|\Omega), \tag{29.7}$$

where, without loss of generality, we label the ESDs involved from 1 to r. Such a state will be said to be a *rank-r ground state*, or *contextual vacuum state*.

We can now draw an analogy between the vacuum of quantum field theory and the rank-r ground states in our formalism. The physical three-dimensional space of conventional physics would correspond to a ground state of extremely large rank, *if* physical space were relevant to the experiment. This would be the case for discussions involving particle scattering or gravitation, for example. For many experiments, however, such as the Stern–Gerlach experiment and quantum-optics networks, physical space would be considered part of the relative external context and therefore could be ignored for the purposes of those experiments. It all depends on what the observer is trying to do.

In the real world there is more than one observer, so a theory of observation should take account of that fact. That is readily done in our theory. For example, the ground state for two or more distinct observers for which some commonality of time had been established would be represented by elements in $\mho$ of the form

$$|\Psi^1, \Psi^2) \equiv \mathbb{C}_1^1\mathbb{C}_2^1 \ldots \mathbb{C}_{r_1}^1\mathbb{C}_1^2\mathbb{C}_2^2 \ldots \mathbb{C}_{r_2}^2|\Omega), \tag{29.8}$$

and so on, where superscripts refer to the different observers. If the subsequent dynamics were such that the ESDs of observer 1 never sent signals to those of observer 2 and vice versa, then, to all intents and purposes, we could discuss each observer as if they were alone. If, on the other hand, some signals did pass between them, then that would be equivalent to having only one observer.

The formalism developed allows a description of experiments that is relatively faithful to actuality (Jaroszkiewicz, 2010). Experiments do not exist before a certain lab time. Then the apparatus has to be constructed, which means that quantum states of an SUO cannot exist before that lab time. Thereafter, the details of the quantum registers involved are determined by the observer. Finally, the apparatus will be decommissioned and, to all intents and purposes, the void state is recovered.

All of these processes take place in discrete time, relative to the observer conducting the experiment.

30
Concluding remarks

The programme of research underpinning this book had a single objective, which was to explore the viability of the hypothesis that time is discrete rather than continuous. Early on in the writing, it became apparent that there were many theorists from diverse branches of science who have dealt with discrete time mechanics in one form or another. The subject matter of this book therefore is not as esoteric as might be imagined, but can be eminently practical in its applications. Discrete time is also of great interest from a fundamentalist perspective.

One of the most pressing factors driving fundamentalist research is the current state of play concerning the divergences in quantum field theories. Numerous *ad hoc* fixes have been explored, ranging from the non-integral spacetime dimensions associated with dimensional regularization to lattice gauge theory and string theory. In this respect, discrete time is but one avenue to explore.

Our programme of investigation into discrete time is in no way complete. During the writing of this book, our review of non-standard analysis raised the possibility that the chronon might be best described in terms of infinitesimals. That would then bypass the need to think of an observable scale for any chronon, because infinitesimals have by definition a 'size' of zero.

Another feature of our investigation that emerged during the writing of this book is the need to clarify the architecture underpinning any theoretical discussion. Systems under observation should not be discussed without reference to observers, because time is synonymous with information acquisition, and that necessarily involves observers. Since information is always acquired discretely, the implication is that time, from the observer's perspective, is really discrete.

Appendix A

Coherent states

Coherent states arise naturally in the theory of continuous time quantized harmonic oscillators. We review this topic now for a one-dimensional harmonic oscillator with Hamiltonian

$$H(p,q) = \frac{1}{2}p^2 + \frac{1}{2}q^2, \tag{A.1}$$

which is a real, nowhere-negative function over p–q phase space. This means that the quantum harmonic oscillator Hamiltonian is a non-negative operator, i.e., none of its eigenstates has a negative eigenvalue. There exists a unique lowest-energy eigenstate called the *ground state*, denoted by $|0)$. We shall use *round* brackets to denote coherent states not normalized to unity. We take $\hbar = 1$, so the only non-zero canonical commutator is $[\hat{p}, \hat{q}] = -i$.

The ladder operators a and $a^\dagger$ are defined by

$$a \equiv \frac{1}{\sqrt{2}}\{\hat{q} + i\hat{p}\}, \quad a^\dagger \equiv \frac{1}{\sqrt{2}}\{\hat{q} - i\hat{p}\}, \tag{A.2}$$

with commutator $[a, a^\dagger] = 1$. The operator a annihilates the ground state, viz., $a|0\rangle = 0$. There is a discrete set of normalized excited states of the ground state, given by

$$|n\rangle \equiv \frac{1}{\sqrt{n!}}(a^+)^n|0), \qquad n \in \mathbb{N}, \tag{A.3}$$

such that $\langle n|m\rangle = \delta_{n,m}$. The excited states form a complete set, with resolution of the identity

$$\sum_{n=0}^{\infty} |n\rangle\langle n| = \mathbb{I}, \tag{A.4}$$

where $\mathbb{I}$ is the identity operator in the oscillator Hilbert space.

A coherent state is characterized by a complex number z and written $|z)$. The excited states satisfy the eigenvalue equations $a|z) = z|z)$ and $(z|a^\dagger = z^*(z|$, $z \in \mathbb{C}$. If we write

$$|z) = \sum_{n=0}^{\infty} C_n(z)|n\rangle \tag{A.5}$$

then we find $|z) = C_0(z)\exp\{za^\dagger\}|0)$. We choose the normalization $(z|w) = \exp\{z^* w\}$.

No two coherent states are orthogonal, forming what is called an *overcomplete* set. This means that they can be used as a basis to expand arbitrary vectors in the Hilbert space. The resolution of the identity is given by

$$\frac{1}{\pi}\int\int e^{-|z|^2}|z\rangle\langle z|d^2z = \mathbb{I}. \tag{A.6}$$

Appendix B

The time-dependent oscillator

The one-dimensional time-dependent oscillator (Lewis and Riesenfeld, 1969; Cho *et al.*, 1997) has time-dependent Hamiltonian

$$H(t) = \frac{1}{2}R(t)^2p^2 + \frac{1}{2}S(t)^2x^2, \tag{B.1}$$

where all quantities are taken to be real and R and S are externally prescribed functions of time such that R never vanishes. The equation of motion obtained from the above Hamiltonian is

$$\ddot{x} - \frac{2\dot{R}}{R}\dot{x} + R^2S^2x \underset{c}{=} 0. \tag{B.2}$$

We jump immediately to quantum mechanics. Working in the Heisenberg picture, the equal-time commutator is given by

$$[\hat{p}(t), \hat{x}(t)] = -i, \tag{B.3}$$

where we take $\hbar = 1$. Now define the operator

$$a(t) \equiv i\{z^*\hat{p} - R^{-2}\dot{z}^*\hat{x}\}, \tag{B.4}$$

where z is a complex-valued function of time. Then

$$\frac{d}{dt}a \equiv \frac{\partial}{\partial t}a + i\left[\hat{H}, a\right] = 0, \tag{B.5}$$

provided that z satisfies the equation of motion (B.2). Similar remarks apply to a^+. Their commutator is given by

$$[a, a^+] = iR^{-2}[\dot{z}z^* - z\dot{z}^*], \tag{B.6}$$

and, since a and a^+ are temporally invariant operators, we deduce that the function

$$f(z) \equiv iR^{-2}[\dot{z}z^* - z\dot{z}^*] \tag{B.7}$$

is a constant of the motion for any solution z to the equation of motion (B.2).

The constancy of the function $f(z)$ is a consequence of Abel's identity (Abel, 1829; Arfken, 1985), which states that, if z_1 and z_2 are two solutions to the differential equation

$$\frac{d^2}{dt^2}z(t) + p(t)\frac{d}{dt}y(t) + q(t)y(t) = 0, \tag{B.8}$$

where p and q do not diverge in the interval $[t_0, t_1]$, then the Wronskian $W(z_1, z_2; t)$ defined by

$$W(z_1, z_2; t) \equiv z_1(t)\dot{z}_2(t) - \dot{z}_1(t)z_2(t) \tag{B.9}$$

is given by

$$W(z_1, z_2; t_1) = W(z_1, z_2; t_0)\exp\left\{-\int_{t_0}^{t_1} p(t)dt\right\}. \tag{B.10}$$

In the present case, $p(t) = -d\ln\{R^2\}/dt$, which, with $z_1 \equiv z$ and $z_2 \equiv z^*$, leads upon integration to the conclusion that $f(z)$ is a constant.

We may choose boundary conditions on z such that $f(z) = 1$ and then the commutator takes the standard form

$$[a, a^+] = 1. \tag{B.11}$$

Upon inverting the relationship (B.4) we find

$$\begin{aligned} \hat{p} &= R^{-2}(\dot{z}a + \dot{z}^* a^+), \\ \hat{x} &= za + z^* a^+. \end{aligned} \tag{B.12}$$

If now we choose boundary conditions such that

$$\dot{z}(0) = -iR(0)S(0)z(0) \tag{B.13}$$

then the Hamiltonian at time zero is given by

$$\hat{H}(0) = R(0)S(0)\left[a^+ a + \frac{1}{2}\right]. \tag{B.14}$$

Appendix C
Quaternions

Given the Pauli algebra $\tau^i\tau^j = \delta_{ij}I_2 + i\epsilon_{ijk}\tau^k$, where I_2 is the 2×2 identity matrix, define the quaternionic symbols $q^i \equiv -i\tau^i$, $q^0 \equiv I_2$. These satisfy the quaternionic multiplication rule

$$q^i q^j = -\delta_{ij}q^0 + \epsilon_{ijk}q^k. \tag{C.1}$$

Now define upper and lower quaternion indices with conjugation as follows:

$$\begin{aligned} q^\mu &\equiv (q^0, q^i), \qquad & q_\mu &\equiv (q_0, q_i), \\ q_0 &\equiv q^0, \qquad & q_i &\equiv -q^i = (q^i)^*. \end{aligned} \tag{C.2}$$

Then conjugation is equivalent to the following quaternionic index-raising and -lowering action: $(q^\mu)^* = q_\mu$, $(q_\mu)^* = q^\mu$. The product rule (C.1) may be written in the compact form

$$q^\mu q^\nu = c^{\mu\nu}{}_\lambda q^\lambda, \tag{C.3}$$

where

$$\begin{aligned} c^{00}{}_0 &= 1, \quad c^{0i}{}_0 = c^{i0}{}_0 = 0, \quad c^{ij}{}_0 = -\delta_{ij}, \\ c^{00}{}_i &= 0, \quad c^{0i}{}_j = c^{i0}{}_j = \delta_{ij}, \quad c^{ij}{}_k = \epsilon_{ijk}. \end{aligned} \tag{C.4}$$

Note that $c^{\mu\nu}{}_0 = \eta^{\mu\nu} = \eta_{\mu\nu}$, where $\eta^{\mu\nu}$ are the components of the Lorentz metric tensor. The product of three quaternions is given by $q^\mu q^\nu q^\alpha = c^{\mu\nu}{}_\lambda c^{\lambda\alpha}{}_\beta q^\beta$, and similarly for higher powers of the quaternions. The Pauli matrices are 2×2 matrices and this permits us to define a linear mapping Tr, called the *trace,* on the quaternions to the reals defined by

$$\begin{aligned} \mathrm{Tr}\ q^\mu &= 2\delta^\mu_0, \quad \mathrm{Tr}(q^\mu q^\nu) = 2\eta^{\mu\nu}, \\ \mathrm{Tr}(q^0 q^i) &= 0, \qquad \mathrm{Tr}(q^i q^j) = -2\delta_{ij}. \end{aligned} \tag{C.5}$$

The trace operation and the product rule (C.3) permit all SU(2) expressions to be simplified in terms of the quaternions rather than 2×2 matrices.

Appendix D

Quantum registers

A quantum register is the tensor product of two or more Hilbert spaces, each known as a subregister (of that register). The number of subregisters in a given tensor product is the rank of that quantum register, relative to that product.

Rank is not the same as dimension. Rank is contextual. For instance, a four-dimensional Hilbert space $\mathcal{H}$ has rank 1, if no further contextual information is given. On the other hand, the tensor product $\mathcal{Q}^{(1)} \otimes \mathcal{Q}^{(2)}$ of two qubits (quantum bits, the two-dimensional quantum analogues of classical bits) is a Hilbert space of dimension four isomorphic to $\mathcal{H}$, but of rank 2. It is physical, contextual information that determines rank, the mathematical isomorphism between $\mathcal{H}$ and $\mathcal{Q}^{(1)} \otimes \mathcal{Q}^{(2)}$ as Hilbert spaces not being relevant in this context.

Our interest here is in changes of rank. These will occur at discrete times, reflecting the dynamics of observation relative to a given observer. It is on this point that fundamental issues of interpretation in quantum mechanics arise. We have in mind those interpretations such as the relative-state formulation of Everett which assert that observers and SUOs are described by the same wavefunction.

Changes of rank are meaningful only relative to the observer involved, constituting an example of contextuality in physics (Eakins and Jaroszkiewicz, 2003). An example of a physical process in which rank changes is any particle decay: before the decay, a particle will be in some spin state $|j, m\rangle$ in a rank-1 Hilbert space, whilst after the decay the SUO can be described in terms of elements of a Hilbert space of rank 2 or more, depending on the number of decay products.

Given p and q are integers such that $p < q$, we use the lower-index notation

$$\mathcal{H}_{[p \ldots q]} \equiv \mathcal{H}_p \otimes \mathcal{H}_{p+1} \otimes \cdots \otimes \mathcal{H}_q \tag{D.1}$$

to denote the decomposition of a register of rank $q+1-p$ into the tensor product of subregisters $\mathcal{H}_p, \mathcal{H}_{p+1}, \ldots, \mathcal{H}_q$. For example, $\mathcal{H}_{[3..6]} \equiv \mathcal{H}_3 \otimes \mathcal{H}_4 \otimes \mathcal{H}_5 \otimes \mathcal{H}_6$. Although the symbol $\otimes$ in this notation denotes a tensor product, the ordering of subregisters in such a tensor product is **not** significant, contrary to standard convention. For example, $\mathcal{H}_1 \otimes \mathcal{H}_2$ means the same as $\mathcal{H}_2 \otimes \mathcal{H}_1$. What is important is the identity of each subregister, i.e., labels are significant. The point here is that the observer can identify each subregister with states of a specific set of detectors in their apparatus, and these detectors can be labelled and identified classically.

Hilbert-space tensor products contain two sorts of vectors. The first sort is the separable vectors, which are products of the form $\phi_p \phi_{p+1} \dots \phi_q$, where ϕ_p is a vector in subregister $\mathcal{H}_p$, ϕ_{p+1} is in $\mathcal{H}_{p+1}$ and so on. We shall henceforth leave out the tensor product symbol $\otimes$ whenever we discuss products of states and operators.

The other sort of vector is the entangled vectors, which are not separable. Entangled states are always expressible as vector sums of separable states. Note that $\Psi_1 \Phi_2 + \Phi_2 \Theta_1$ is separable and equal to $(\Psi_1 + \Theta_1)\Phi_2$, whereas $\Psi_1 \Phi_2 + \eta_2 \Theta_1$ is not separable if Ψ_1 and Θ_1 are linearly independent and Φ_2 and η_2 are linearly independent.

The possibility of finding both separable and entangled states in a quantum register is crucial to our programme of quantum register cosmology, because this models the observed fact that the physical Universe appears to consist of many separate subsystems, such as atoms and molecules, as well as entangled states. States of separate subsystems in the Universe will correspond to separate factors in the state of the Universe at any given time. Moreover, such factors need not be small in terms of dimensionality. For example, a physics experiment would typically involve an isolated system under observation S, inside some region of space, surrounded by equipment A. Beyond that would be the local environment E, and further still beyond that, the rest of the Universe U. At a given time, the state of the Universe Ψ might take the separable form

$$\Psi = \Psi_{\mathrm{S}} \Psi_{\mathrm{A}} \Psi_{\mathrm{E}} \Psi_{\mathrm{U}}, \tag{D.2}$$

where Ψ_{S} represents the state of S, and so on. Each of these factors could itself be an entangled state in a quantum register of extremely large rank.

D.1 Splits

A split is any convenient way of grouping the subregisters in a quantum register into two or more factor registers, each of which is itself a tensor product of subregisters and therefore a vector space. For large rank-quantum registers, very many different splits will be possible. For example, we may write the register $\mathcal{H}_{[123]}$ in five different ways:

$$\mathcal{H}_{[123]} = \mathcal{H}_1 \otimes \mathcal{H}_{[23]} = \mathcal{H}_3 \otimes \mathcal{H}_{[12]} = \mathcal{H}_2 \otimes \mathcal{H}_{[13]} = \mathcal{H}_1 \otimes \mathcal{H}_2 \otimes \mathcal{H}_3. \tag{D.3}$$

Although splits of the same register are isomorphic as vector spaces, they differ in the information content required to define each of them in a physically meaningful way relative to an observer.

D.2 Partitions

If d_i is the dimension of subregister $\mathcal{H}_i$, then the quantum register $\mathcal{H}_{[1...N]}$ is a vector space of dimension $d = d_1 d_2 \ldots d_N$. This quantum register contains both entangled and separable states. In order to filter out these two types of constituents, we define separations and entanglements as follows.

D.3 Separations

For any two subregisters $\mathcal{H}_i$ and $\mathcal{H}_j$ of the quantum register $\mathcal{H}_{[1..N]}$, such that $i \neq j$, the separation $\mathcal{H}_{ij}$ is the subset of $\mathcal{H}_{[ij]}$ consisting of all separable elements, i.e.,

$$\mathcal{H}_{ij} \equiv \{\phi_i \psi_j : \phi_i \in \mathcal{H}_i,\ \psi_j \in \mathcal{H}_j\}. \tag{D.4}$$

By definition, $\mathcal{H}_{ij}$ includes the zero vector $0_{[ij]}$ of $\mathcal{H}_{[ij]}$, because this vector can always be written trivially in the separable form $0_{[ij]} = 0_i 0_j$, where 0_i is the zero vector in $\mathcal{H}_i$ and similarly for 0_j. Note that $\mathcal{H}_{ij} = \mathcal{H}_{ji}$.

The separation $\mathcal{H}_{ij}$ is a rank-2 separation, and this terminology readily generalizes to higher-rank separations as follows. Pick an integer k in the interval $[1, N]$ and then select k different elements $i_1, i_2, \ldots, i_k$ of this interval. Then the rank-k separation $\mathcal{H}_{i_1 i_2 \ldots i_k}$ is the subset of $\mathcal{H}_{[i_1 \ldots i_k]} \equiv \mathcal{H}_{i_1} \otimes \mathcal{H}_{i_2} \otimes \ldots \otimes \mathcal{H}_{i_k}$ given by

$$\mathcal{H}_{i_1 i_2 \ldots i_k} \equiv \{\psi_{i_1} \psi_{i_2} \ldots \psi_{i_k} : \psi_{i_a} \in \mathcal{H}_{i_a},\ 1 \leqslant a \leqslant k\}. \tag{D.5}$$

Every element of a rank-k separation has k factors. A rank-1 separation of a subregister is by definition equal to that subregister, so we may write $\mathcal{H}_i = \mathcal{H}_{[i]}$.

D.4 Rank-2 entanglements

The entanglements are defined in terms of complements. Starting with the lowest-rank entanglement possible, the rank-2 entanglement $\mathcal{H}^{ij}$ is the complement of $\mathcal{H}_{ij}$ in $\mathcal{H}_{[ij]}$, i.e., $\mathcal{H}^{ij} \equiv \mathcal{H}_{[ij]} - \mathcal{H}_{ij} = (\mathcal{H}_{[ij]} \cap \mathcal{H}_{ij})^{c}$. Hence $\mathcal{H}_{[ij]} = \mathcal{H}_{ij} \cup \mathcal{H}^{ij}$, the separation $\mathcal{H}_{ij}$ and the entanglement $\mathcal{H}^{ij}$ being disjoint. In particular, $\mathcal{H}^{ij}$ does not contain the zero vector, which is in $\mathcal{H}_{ij}$. $\mathcal{H}^{ij}$ contains all the entangled states in $\mathcal{H}_{[ij]}$.

An important point about the decomposition $\mathcal{H}_{[ij]} = \mathcal{H}_{ij} \cup \mathcal{H}^{ij}$ is that neither $\mathcal{H}_{ij}$ nor $\mathcal{H}^{ij}$ is a vector space, and this is true of separations and entanglements generally.

D.5 Separation products

The generalization of the above to larger-rank tensor products is straightforward but requires notation for the concept of the separation product. Given arbitrary subsets $\mathcal{H}'_i$ and $\mathcal{H}'_j$ of $\mathcal{H}_i$ and $\mathcal{H}_j$, respectively, where $i \neq j$, we define the separation product $\mathcal{H}'_i \cdot \mathcal{H}'_j$ to be the subset of $\mathcal{H}_{[ij]}$ given by

$$\mathcal{H}'_i \cdot \mathcal{H}'_j \equiv \{\psi\phi : \psi \in \mathcal{H}'_i,\ \phi \in \mathcal{H}'_j\}. \tag{D.6}$$

In this notation the rank-2 separation $\mathcal{H}_{ij}$ can be written $\mathcal{H}_{ij} = \mathcal{H}_i \cdot \mathcal{H}_j$.

The separation product is associative, commutative and cumulative, i.e.,

$$\begin{aligned}(\mathcal{H}_i \cdot \mathcal{H}_j) \cdot \mathcal{H}_k &= \mathcal{H}_i \cdot (\mathcal{H}_j \cdot \mathcal{H}_k) \equiv \mathcal{H}_{ijk},\\ \mathcal{H}_i \cdot \mathcal{H}_j &= \mathcal{H}_j \cdot \mathcal{H}_i,\\ \mathcal{H}_{ij} \cdot \mathcal{H}_k &= \mathcal{H}_{ijk},\end{aligned} \tag{D.7}$$

and so on. The separation product can also be defined to include entanglements. For example,

$$\mathcal{H}^{ij} \cdot \mathcal{H}_k = \{\phi^{ij}\psi_k : \phi^{ij} \in \mathcal{H}^{ij},\ \psi_k \in \mathcal{H}_k\}. \tag{D.8}$$

A further notational simplification is to use a single $\mathcal{H}$ symbol, using the vertical position of indices to indicate separations and entanglements and incorporating the separation-product symbol $\cdot$ with indices directly. For example, the following are equivalent ways of writing the same separation product of entanglements and separations:

$$\mathcal{H}^{15\cdot 97}_{23468} \equiv \mathcal{H}^{15\cdot 97}_{28\cdot 4\cdot 36} \equiv \mathcal{H}^{15} \cdot \mathcal{H}^{97} \cdot \mathcal{H}_{28} \cdot \mathcal{H}_4 \cdot \mathcal{H}_{36}. \tag{D.9}$$

Associativity of the separation product applies both to separations and to entanglements. For example, we may write

$$\mathcal{H}_{ij} \cdot \mathcal{H}_{klm} \cdot \mathcal{H}^{rs} = \mathcal{H}^{rs}_{ij\cdot klm} = \mathcal{H}^{rs}_{ijklm}, \tag{D.10}$$

but note that, whilst

$$\mathcal{H}_{ij} \cdot \mathcal{H}_{kl} = \mathcal{H}_{ij\cdot kl} = \mathcal{H}_{ijkl}, \tag{D.11}$$

we have

$$\mathcal{H}^{ij} \cdot \mathcal{H}^{kl} \equiv \mathcal{H}^{ij\cdot kl} \neq \mathcal{H}^{ijkl}. \tag{D.12}$$

With the above notation for separation products rank-3 and higher entanglements such as $\mathcal{H}^{ijkl}$ are likewise defined in terms of complements, in the same way as that in which $\mathcal{H}^{ij}$ is defined. For example, consider $\mathcal{H}_{[abc]} \equiv \mathcal{H}_a \otimes \mathcal{H}_b \otimes \mathcal{H}_c$. We define

$$\mathcal{H}^{abc} \equiv \mathcal{H}_{[abc]} - \mathcal{H}_{abc} \cup \mathcal{H}^{bc}_a \cup \mathcal{H}^{ac}_b \cup \mathcal{H}^{ab}_c. \tag{D.13}$$

Likewise, given $\mathcal{H}_{[abcd]} \equiv \mathcal{H}_a \otimes \mathcal{H}_b \otimes \mathcal{H}_c \otimes \mathcal{H}_d$, then

$$\begin{aligned}\mathcal{H}^{abcd} \equiv \mathcal{H}_{[abcd]} &- \mathcal{H}_{abcd} \cup \mathcal{H}^{cd}_{ab} \cup \mathcal{H}^{bd}_{ac} \cup \mathcal{H}^{bc}_{ad} \cup \mathcal{H}^{ad}_{bc} \cup \mathcal{H}^{ac}_{bd} \cup \mathcal{H}^{ab}_{cd}\\ &\cup \mathcal{H}^{bcd}_a \cup \mathcal{H}^{acd}_b \cup \mathcal{H}^{abd}_c \cup \mathcal{H}^{abc}_d \cup \mathcal{H}^{ab\cdot cd} \cup \mathcal{H}^{ac\cdot bd} \cup \mathcal{H}^{ad\cdot bc}.\end{aligned} \tag{D.14}$$

Sets such as $\mathcal{H}^{abc}$ are called rank-3 entanglements, and so on. In general, higher-rank entanglements such as $\mathcal{H}^{abcd}$ in the above require a great deal of filtering out of separations from the original tensor-product Hilbert space in order for their definition to be possible, which accounts partly for the fact that entanglements are generally conceptually more complex than separations.

D.6 Natural lattices and partitions

The decomposition of a quantum register into a union of subsets, each the union of separations and entanglements, is unique and will be called the contextual lattice $L(\mathcal{H})$ of $\mathcal{H}$, the context being the factorization of $\mathcal{H}$ into specific subregisters. Each element of a contextual lattice will be called a *partition.* In the above examples, the natural lattice of $\mathcal{H}_{[abc[}$ has five partitions whilst that of $\mathcal{H}_{[abcd]}$ has fifteen. In general, partitions are separation products of entanglements and separations of various ranks, such that, for each partition, the sum of the ranks of its factors equals the rank of the quantum register. The individual separations and entanglements making up a partition will be called *blocks.*

Examples are

$$\begin{aligned} \mathcal{H}_{[1]} &= \mathcal{H}_1, \\ \mathcal{H}_{[12]} &= \mathcal{H}_{12} \cup \mathcal{H}^{12}, \\ \mathcal{H}_{[123]} &= \mathcal{H}_{123} \cup \mathcal{H}_1^{23} \cup \mathcal{H}_2^{13} \cup \mathcal{H}_3^{12} \cup \mathcal{H}^{123}, \\ \mathcal{H}_{[1234]} &= \mathcal{H}_{1234} \cup \mathcal{H}_{12}^{34} \cup \mathcal{H}_{13}^{24} \cup \mathcal{H}_{14}^{23} \cup \mathcal{H}_{23}^{14} \cup \mathcal{H}_{24}^{13} \cup \mathcal{H}_{34}^{12} \cup \mathcal{H}_1^{234} \\ &\quad \cup \mathcal{H}_2^{134} \cup \mathcal{H}_3^{124} \cup \mathcal{H}_4^{123} \cup \mathcal{H}^{12\cdot 34} \cup \mathcal{H}^{13\cdot 24} \cup \mathcal{H}^{14\cdot 23} \cup \mathcal{H}^{1234}. \end{aligned} \tag{D.15}$$

If the number of partitions in the contextual lattice for a rank-r quantum register is denoted by $\#r$, then we find the sequence $\#1 = 1$, $\#2 = 2$, $\#3 = 5$, $\#4 = 15$, $\#5 = 52$, and so on. Eakins has shown that $\#r$ is the rth Bell–Ramanujan number (Eakins, 2004).

Although the number of partitions in the natural lattice of a rank-r quantum register is the same as the number of splits and given by the Bell numbers, splits and partitions cannot coincide for $r > 1$. Every factor subregister in a split is a vector space, whereas no partition is a vector space for $r > 1$. Both splits and partitions are essential ingredients in our approach to causal set structure.

A useful extension of the notation is to label the various vector elements of entanglements and separations. For example, $\psi_{123}^{456\cdot 78}$ denotes an element in the partition $\mathcal{H}_{123}^{456\cdot 78}$ and so on. The notation also permits us to rewrite $\psi_{123}^{456\cdot 78}$ in the factorized form

$$\psi_{123}^{456\cdot 78} = \psi_1 \psi_2 \psi_3 \psi^{456} \psi^{78}, \tag{D.16}$$

where $\psi_1 \in \mathcal{H}_1$, $\psi_2 \in \mathcal{H}_2$, $\psi_3 \in \mathcal{H}_3$, $\psi^{456} \in \mathcal{H}^{456}$ and $\psi^{78} \in \mathcal{H}^{78}$.

References

Abel, N. H. 1829. Précis d'une théorie des fonctions elliptiques. *J. Reine Angew. Math.*, **4**, 309–348.

Abers, E., and Lee, B. W. 1973. Gauge theories. *Phys. Rep.*, **9C**(1), 1–141.

Abraham, R., and Marsden, J. E. 2008. *Foundations of Mechanics*. Second edn. Providence, RI: AMS Chelsea Publishing.

Arfken, G. 1985. *Mathematical Methods for Physicists*. Third edn. New York: Academic Press Inc.

Aristotle 1930. *Physica (The Physics)*. Oxford: Clarendon Press.

Arthur, R. 1988. Continuous creation, continuous time: a refutation of the alleged discontinuity of Cartesian time. *J. Hist. Phil.*, **26**(3), 349–375.

Barrow, J. D. 1992. *Pi in the Sky*. Oxford: Clarendon Press.

Bartholomæus Anglicus 1240. *De Proprietatibus Rerum (On the Properties of Things)*.

Bender, C. M., Cooper, F., Gutschick, V. P., and Nieto, M. M. 1985a. Simple approach to tunneling using the method of finite elements. *Phys. Rev. D*, **32**(6), 1486–1490.

Bender, C. M., Milton, K. A., Sharp, D. H., Simmons. L. M. Jr, and Strong, R. 1985b. Discrete-time quantum mechanics. *Phys. Rev. D*, **32**(6), 1476–1485.

Bender, C. M., Mead, L. R., and Milton, K. A. 1993. Discrete time quantum mechanics. hep-ph/9305246, 1–52.

Benza, V., and Caldirola, P. 1981. De Sitter microuniverse associated to the electron. *Nuovo Cimento*, **62A**(3), 175–185.

Bishop, E. 1977. Book review of *Elementary Calculus*, by H. J. Keisler. *Bull. Am. Math. Soc.*, **83**, 205–208.

Bjorken, J. D., and Drell, S. D. 1964. *Relativistic Quantum Mechanics*. New York: McGraw-Hill.

Bjorken, J. D., and Drell, S. D. 1965. *Relativistic Quantum Fields*. New York: McGraw-Hill Inc.

Blaszczyk, P., Katz, M. G., and Sherry, D. 2013. Ten misconceptions from the history of analysis and their debunking. *Foundations Sci.*, **18**, 43–74.

Bohr, N. 1920. Über die Serienspektra der Elemente. *Z. Phys.*, **2**(5), 423–478.

Bombelli, L., Lee, J., Meyer, D., and Sorkin, R. 1987. Space-time as a causal set. *Phys. Rev. Lett.*, **59**(5), 521–524.

Born, M. 1926. Zur Quantenmechanik der Stoßvorgänge. *Z. Phys.*, **38**, 803–827.

Brightwell, G., and Gregory, R. 1991. Structure of random discrete spacetime. *Phys. Rev. Lett.*, **66**(3), 260–263.

Brown, L. M. 2005. *Feynman's Thesis, A New Approach to Quantum Theory*. Singapore: World Scientific.

Brown, R. 1828. A brief account of microscopical observations made in the months of June, July and August, 1827, on the particles contained in the pollen of plants;

and on the general existence of active molecules in organic and inorganic bodies. *Phil. Mag.*, **4**, 161–173.

Bureau International des Poids et Mesures (BIPM) 2006. *The International System of Units (SI)*. Eight edn. Sèvres: BIPM.

Cadzow, J. A. 1970. Discrete calculus of variations. *Int. J. Control*, **11**(3), 393–407.

Caldirola, P. 1978. The chronon in the quantum theory of the electron and the existence of heavy leptons. *Nuovo Cimento*, **45**(4), 549–579.

Candlin, D. J. 1956. On sums over trajactories for systems with Fermi statistics. *Nuovo Cimento*, **4**(2), 231–239.

Casalbuoni, R. 1976a. The classical mechanics for Bose–Fermi systems. *Nuovo Cimento A Series 11*, **33**(3), 389–431.

Casalbuoni, R. 1976b. On the quantization of systems with anticommuting variables. *Nuovo Cimento A Series 11*, **33**(1), 115–125.

Chamseddine, A., and Connes, A. 1996. Universal formula for noncommutative geometry actions: unification of gravity and the standard model. *Phys. Rev. Lett.*, **24**, 4868–4871.

Cho, K. H., Ji, J. Y., Kim, S. P., Lee, C. H., and Ryu, J. Y. 1997. Heisenberg-picture approach to the evolution of the scalar fields in an expanding universe. *Phys. Rev. D*, **56**, 4916–4921.

Cohen, I. B. 1999. *A Guide to Newton's Principia*. Berkeley, CA: University of California Press.

Colosi, D., and Rovelli, C. 2009. What is a particle? *Classical Quantum Gravity*, **26**, 025002 (22 pp.).

Cornell, J. (ed.). 1989. *Bubbles, Voids, and Bumps in Time: The New Cosmology*. Cambridge: Cambridge University Press.

Cornish, F. H. J. 1984. The hydrogen atom and the four-dimensional harmonic oscillator. *J. Phys. A: Math. Gen.*, **17**, 323–327.

Davisson, C., and Germer, L. H. 1927. Diffraction of electrons by a crystal of nickel. *Phys. Rev.*, **6**, 705–740.

de Broglie, L. 1924. *Recherches sur la théorie des quanta*. Ph.D. thesis, Faculty of Sciences at Paris University.

DeWitt, B. S. 1965. *Dynamical Theory of Groups and Fields*. London: Blackie and Son Limited.

Dirac, P. A. M. 1925. The fundamental equations of quantum mechanics. *Proc. Roy. Soc. A*, **109**, 642–653.

Dirac, P. A. M. 1928. The quantum theory of the electron. *Proc. Roy. Soc. A*, **117**(778), 610–624.

Dirac, P. A. M. 1933. The Lagrangian in quantum mechanics. *Phys. Z. Sowjetunion*, **3**(1), 64–72.

Dirac, P. A. M. 1938. Classical theory of radiating electrons. *Proc. Roy. Soc. A*, **167**, 148–169.

Dirac, P. A. M. 1958. *The Principles of Quantum Mechanics*. Oxford: Clarendon Press.

Dirac, P. A. M. 1964. *Lectures on Quantum Mechanics*. New York: Belfer Graduate School of Science, Yeshiva University.

Dray, T., Manogue, C. A., and Tucker, R. W. 1991. Particle production from signature change. *Gen. Rel. Grav.*, **23**(8), 967–971.

Eakins, J. 2004. *Classical and Quantum Causality in Quantum Field Theory, or, 'The Quantum Universe'*. Ph.D. thesis, University of Nottingham.

Eakins, J., and Jaroszkiewicz, G. 2003. Factorization and entanglement in quantum systems. *J. Phys. A: Math. Gen.*, **36**, 517–526.

Eakins, J., and Jaroszkiewicz, G. 2005. A quantum computational approach to the quantum Universe, Reimer, A. (ed.), *New Developments in Quantum Cosmology Research*. New York: Nova Science Publishers, Inc., pages 1–51.

Eden, R. J., Landshoff, P. V., Olive, D. I., and Polkinghorne, J. C. 1966. *The Analytic S-Matrix*. Cambridge: Cambridge University Press.

Einstein, A. 1905a. Über einen die Erzeugung und Verwandlung des Lichtes betreffenden heuristischen Gesichtspunkt. *Ann. Phys.*, **17**, 132–148. (English Translation: 1965. Concerning an heuristic point of view toward the emission and transformation of light. *Am. J. Phys.* **33**(5), 1–16.)

Einstein, A. 1905b. Über die von der molekularkinetischen Theorie der Wärme geforderte Bewegung von in ruhenden Flüssigkeiten suspendierten Teilchen. *Ann. Phys.*, **17**, 549–560.

Einstein, A. 1913. *Letter to Ernst Mach.* Reprinted with commentary in Misner, C., Thorne, K., and Wheeler, J., 1973. *Gravitation.* San Francisco, CA: W. H. Freeman.

Einstein, A. 1915. Die Feldgleichungen der Gravitation. *Sitzungsber. Preuß. Akad. Wiss. Berlin*, 844–847.

Encyclopædia Britannica 1993. Time. In *Encyclopædia Britannica*, 15th edn, Vol. 28. London: Encyclopædia Britannica, Inc., pages 662–673.

Encyclopædia Britannica. 2000. CD Rom edn. britannica.co.uk.

Farias, R. H. A., and Recami, E. 2010. Introduction of a quantum of time ("chronon") and its consequences for quantum mechanics. *Adv. Imaging Electron Phys.*, **163**, 33–115.

Fermi, E. 1926. Sulla quantizzazione del gas perfetto monoatomico. *Rend. Lincei*, **3**, 145–149. (English translation by A. Zannoni: On the quantization of the monoatomic ideal gas. arXiv:cond-mat/9912229.)

Feynman, R. P. 1948. Space-time approach to non-relativistic quantum mechanics. *Rev. Mod. Phys.*, **20**(2), 367–387.

Feynman, R. P. 1982. Simulating physics with computers. *Int. J. Theor. Phys.*, **21**(6/7), 467–488.

Feynman, R. P., and Hibbs, A. R. 1965. *Quantum Mechanics and Path Integrals*. New York: McGraw-Hill.

Finkelstein, D. 1969. Space-time code. *Phys. Rev.*, **184**(4), 1969–1971.

Finkelstein, D. 1972a. Space-time code. II. *Phys. Rev. D*, **5**(2), 320–328.

Finkelstein, D. 1972b. Space-time code. III. *Phys. Rev. D*, **5**(12), 2923–2931.

Finkelstein, D. 1974. Space-time code. IV. *Phys. Rev. D*, **9**(8), 2219–2231.

Finkelstein, D., Frye, G., and Susskind, L. 1974. Space-time code. V. *Phys. Rev. D*, **9**(8), 2231–2236.

FitzGerald, G. F. 1889. The ether and the Earth's atmosphere. *Science*, **13**, 390.

Fredkin, E. 1990. Digital mechanics: an informational process based on reversible universal CA. *Physica*, **D45**, 254–270.

Fredkin, E. 2001. A physicist's model of computation. *Digital Mechanics*, 11 November, 1–13.

Fredkin, E., and Toffoli, T. 1982. Conservative logic. *Int. J. Theor. Phys.*, **21**(3–4), 219–253.

Gagnon, P. 2012. Is the moon full? Just ask the LHC operators. http://www.quantumdiaries.org/2012/06/07/is-the-moon-full-just-ask-the-lhc-operators/.

Galapon, E. A. 2002. Pauli's theorem and quantum canonical pairs: the consistency of a bounded, self-adjoint time operator canonically conjugate to a Hamiltonian with non-empty point spectrum. *Proc. R. Soc. Lond. A*, **458**, 451–472.

Gardner, M. 1970. The fantastic combinations of John Conway's new solitaire game "life". *Scient. Am.*, **223**, 120–123.

Gasiorowicz, S. 1967. *Elementary Particle Physics*. New York: John Wiley and Sons.

Gelfand, I. M., and Shilov, G. E. 1964. *Generalised Functions*. New York: Academic Press.

Gerlach, W., and Stern, O. 1922a. Der experimentelle Nachweis des magnetischen Moments des Silberatoms *Z. Phys.*, **8**, 110–111.

Gerlach, W., and Stern, O. 1922b. Der experimentelle Nachweis der Richtungsquantelung im Magnetfeld. *Z. Phys.*, **9**, 349–355.

Gödel, K. 1949. An example of a new type of cosmological solutions of Einstein's field equations of gravity. *Rev. Mod. Phys.*, **21**(3), 447–450.

Goldstein, H. 1964. *Classical Mechanics*. New York: Addison-Wesley.

Goldstein, H., Poole, C., and Safko, J. 2002. *Classical Mechanics*. Third edn. New York: Addison-Wesley.

Hamermesh, M. 1962. *Group Theory and Its Applications to Physical Problems*. New York: Addison-Wesley.

Hartley, L. P. 1953. *The Go-Between*. London: Hamish Hamilton.

Heisenberg, W. 1925. Über quantentheoretische Umdeutung kinematischer und mechanischer Beziehungen. *Z. Phys. A*, **33**(1), 879–893.

Heisenberg, W. 1927. Über den anschaulichen Inhalt der quantentheoretischen Kinematik und Mechanik. *Z. Phys*, **43**, 172–198. (Reprinted English translation: The physical content of quantum kinematics and mechanics, in Wheeler, J. A., and Zurek, W. H. (eds.) 1983. *Quantum Theory of Measurement*. Princeton, NJ: Princeton University Press.)

Hilbert, D. 1915. Die Grundlagen der Physik. *Nachr. Königl. Gesell. Wiss. Göttingen, Math.-Phys.*, **3**, 395–407.

Hildebrandt, S. 2012. Mountains are younger than we think. *ScienceNordic*. Available online at sciencenordic.com.

Howson, A. G. 1972. *A Handbook of Terms Used in Algebra and Analysis*. Cambridge: Cambridge University Press.

Ikeda, M., and Maeda, S. 1978. On symmetries in a discrete model of mechanical systems. *Math. Japonica*, **23**(2), 231–244.

Jackson, F. H. 1910. On q-definite integrals. *Q. J. Pure Appl. Math.*, **41**, 193–203.

Jaroszkiewicz, G. 1994a. Conserved quantities in classical and quantised discrete time systems. Nottingham University Mathematics Department preprint. (Talk given at International Workshop on Finite dimensional Integrable systems, Dubna, Russia, 1994.)

Jaroszkiewicz, G. 1994b. A q-ball simulation of bag scattering. *J. Phys. G: Nucl. Part. Phys.*, **21**, 501–516.

Jaroszkiewicz, G. A. 1995a. Deformed mechanics and the discrete time quantised anharmonic oscillator, in Lukierski, J., Popowicz, Z., and Sobczyk, J. (eds.), *Proceedings of XXX Karpacz Winter School of Theoretical Physics Poland, on Quantum Groups, Formalism and Applications*, Wrocław, 1994. Warsaw: Wydawnictwo Naukowe PWN.

Jaroszkiewicz, G. 1995b. Hilbert space condition on conserved quantities in second-order discrete time classical mechanics. *J. Phys. A: Math. Gen.*, **28**, L197–L199.

Jaroszkiewicz, G. 1999. Discrete spacetime: classical causality, prediction, retrodiction and the mathematical arrow of time. (Talk at First Interdisciplinary Workshop on

Studies on the Structure of Time: from Physics to Psycho(patho)logy, Palermo, (1999).)

Jaroszkiewicz, G. 2002. *Analysis of the Relationship Between Real and Imaginary Time in Physics.* Dordrecht: Kluwer Academic Publishers.

Jaroszkiewicz, G. 2008. Quantized detector networks: a review of recent developments. *Int. J. Mod. Phys. B*, **22**(3), 123–188.

Jaroszkiewicz, G. 2010. Towards a dynamical theory of observation. *Proc. Roy. Soc. A*, **466**(2124), 3715–3739.

Jaroszkiewicz, G., and Nikolaev, V. 2001. Temporal discretization of the Skyrme model. arXiv:hep-th/0110052, 1–12.

Jaroszkiewicz, G., and Norton, K. 1997a. Principles of discrete time mechanics: I. Particle systems. *J. Phys. A: Math. Gen.*, **30**(9), 3115–3144.

Jaroszkiewicz, G., and Norton, K. 1997b. Principles of discrete time mechanics: II. Classical field theory. *J. of Phys. A: Math. Gen.*, **30**(9), 3145–3163.

Jordan, P., and Wigner, E. P. 1928. Über das Paulische Äquivalenzverbot. *Z. Phys.*, **47**, 631–651.

Keisler, H. J. 2012. *Elementary Calculus: An Elementary Approach.* (available online).

Kennard, E. H. 1927. Zur Quantenmechanik einfacher Bewegungstypen. *Z. Phys.*, **44**(4–5), 326–352.

Klimek, M. 1993. Extension of q-deformed analysis and q-deformed models of classical mechanics. *J. Phys. A: Math. Gen.*, **26**, 955–967.

Klimek, M. 1996. The conservation laws and integrals of motion for a certain class of equations in discrete models. *J. Phys. A: Math. Gen.*, **29**, 1747–1758.

Koke, S., Grebing, C., Frei H., *et al.* 2010. Direct frequency comb synthesis with arbitrary offset and shot-noise-limited phase noise. *Nature Photonics*, **4**, 463–465.

Kowalczyński, J. K. 2000. Can we detect tachyons now? *Acta Phys. Slovaca*, **50**, 381–395.

Kowalski, K. 1994. *Methods of Hilbert Spaces in the Theory of Nonlinear Dynamical Systems.* Singapore: World Scientific.

Lee, T. D. 1983. Can time be a discrete dynamical variable? *Phys. Lett. B*, **122**(3–4), 217–220.

Leech, J. W. 1965. *Classical Mechanics.* London: Methuen and Co. Ltd.

Lehmann, H., Symanzik, K., and Zimmermann, W. 1955. Zur Formulierung quantisierter Feldtheorien. *Nuovo Cimento*, **1**(1), 205–225.

Lévi, R. 1927. Théorie de l'action universelle et discontinue. *J. Phys. Radium*, **8**, 182–198.

Lewis, H. R., and Riesenfeld, W. B. 1969. An exact quantum theory of the time-dependent harmonic oscillator and of a charged particle in a time-dependent electromagnetic field. *J. Math. Phys*, **10**, 1458–1473.

Li, A., and Richardson, M. 2009. The speed and lifetime of cosmic-ray muons. MIT Department of Physics, 1–5 (available online).

Logan, J. D. 1973. First integrals in the discrete variational calculus. *Aequat. Math.*, **9**, 210–220.

Lukierski, J., Nowicki, A., and Ruegg, H. 1992. New quantum Poincaré algebra and kappa-deformed field theory. *Phys. Lett.*, **293***B*, 344–352.

Mach, E. 1912. Die Mechanik in ihrer Entwicklung, historisch-kritisch dargestellt. (English translation by T. J. McCormack, 1934. *The Science of Mechanics.* La Salle, IL: Open Court.)

Maeda, S. 1981. Extension of discrete Noether theorem. *Math. Japonica*, **26**(1), 85–90.

Maimonides, M. 1190. *The Guide for the Perplexed.* London: George Routledge, (Translated by M. Freidlander (1904).)

Markopoulou, F. 2000. Quantum causal histories. *Class. Quantum Gravity.* **17**, 2059–2072.

Marsden, J. E., and West, M. 2001. Discrete mechanics and variational integrators. *Acta Numer.*, 1–158.

Martin, J. L. 1959a. The Feynman principle for a Fermi system. *Proc. Roy. Soc. A*, **251**, 543–549.

Martin, J. L. 1959b. Generalized classical dynamics and the "classical analogue" of a Fermi oscillator. *Proc. Roy. Soc. A*, **251**, 536–542.

McAllister, R. W., and Hofstadter, R. 1956. Elastic scattering of 188 MeV electrons from proton and the alpha particle. *Phys. Rev.*, **102**, 851–856.

Meschini, D. 2006. Planck-scale physics: facts and belief. *Foundations Sci.*, **11**, 1233–1821.

Minkowski, H. 1908. Space and time. (A translation of an address delivered at the 80th Assembly of German Natural Scientists and Physicians, at Cologne, 21 September), in Lorentz, H. A., Einstein, A., Minkowski, H., and Weyl, H. *The Principle of Relativity: A Collection of Original Memoirs on the Special and General Theory of Relativity* (1952). New York: Dover Publications, Inc.

Minsky, M. 1982. Cellular vacuum. *Int. J. Theor. Phys.*, **21**(6/7), 537–551.

Mir-Kasimov, R. M. 1991. On the principle of gauge invariance in field theory with curved momentum space. *Phys. Lett. B*, **259**(1–2), 79–83.

Moser, J., and Veselov, A. P. 1991. Discrete versions of some classical integrable systems and factorization of matrix polynomials. *Commun. Math. Phys.*, **139**, 217–243.

Muga, J. G., Sala Mayato, R., and Egusquiza, Í. L. (eds.). 2008. *Time in Quantum Mechanics.* Second edn. Berlin: Springer.

Newton, I. 1687. *The Principia (Philosophiae Naturalis Principia Mathematica).* New translation by I. B. Cohen and Anne Whitman, Berkeley, CA: University of California Press (1999).

Noether, E. 1918. Invariante Variationsprobleme. *Nachr. Königl. Gesell. Wiss. Göttingen, Math.-Phys.*, 235–257.

Norton, K., and Jaroszkiewicz, G. 1998a. Principles of discrete time mechanics: III. Quantum field theory. *J. Phys. A: Math. Gen.*, **31**(3), 977–1000.

Norton, K., and Jaroszkiewicz, G. 1998b. Principles of discrete time mechanics: IV. The Dirac equation, particles and oscillons. *J. Phys. A: Math. Gen.*, **31**(3), 1001–1023.

Oney, S. 2007. The Jackson integral, http://www.stephenoney.com/papers/JacksonIntegral.pdf.

Pauli, W. 1925. Über den Einfluß der Geschwindigkeitsabhängigkeit der Elektronenmasse auf den Zeemaneffekt. *Z. Phys.*, **31**, 373–385.

Pauli, W. 1933. *Die allgemeinen Prinzipien der Wellenmechanik.* Berlin: Springer.

Pauli, W. 1946. Exclusion principle and quantum mechanics. Nobel Lecture (available online).

Peierls, R. E. 1952. The commutation laws of relativistic field theory. *Proc. Roy. Soc. A*, **214**, 143–157.

Peres, A. 1993. *Quantum Theory: Concepts and Methods.* Dordrecht: Kluwer Academic Publishers.

Pinney, E. 1958. *Ordinary Difference-Differential Equations.* Berkeley, CA: University of California Press.

Planck, M. 1900. Über eine Verbesserung der Wienschen Spektralgleichung *Verhandl. Deutsch. Phys. Gesell.*, **2**, 202–204.

Planck, M. 1901. Über das Gesetz der Energieverteilung im Normalspektrum. *Ann. Phys.*, **309**(3), 553–563.

Poincaré, H. 1890. Sur le problème des trois corps et les équations de la dynamique. *Acta Math.*, **13**, 1–270.

Price, H. 1997. *Time's Arrow*. Oxford: Oxford University Press.

Rabei, E., Ajlouni, A., and Ghassib, H. 2006. Quantization of Brownian Motion. *Int. J. Theor. Phys.*, **45**(9), 1613–1623.

Regge, T. 1961. General relativity without coordinates. *Nuovo Cimento*, **19**(3), 558–571.

Reichenbach, H. 1958. *The Philosophy of Time and Space*. New York: Dover Publications.

Requardt, M. 1999. Space-time as an orderparameter manifold in random networks and the emergence of physical points. gr-qc/99023031, 1–40.

Ridout, D. P., and Sorkin, R. D. 2000. A classical sequential growth dynamics for causal sets. *Phys. Rev. D*, **61**, 024002.

Rindler, W. 1969. *Essential Relativity*. New York: Van Nostrand Reinhold Company.

Robinson, A. 1966. *Non-standard Analysis*. Amsterdam: North-Holland Publishing Co.

Roman, P. 1969. *Introduction to Quantum Field Theory*. New York: John Wiley and Sons, Inc.

Rovelli, C. 1996. Relational quantum mechanics. *Int. J. Theor. Phys.*, **35**, 1637–1678.

Rudin, W. 1964. *Principles of Mathematical Analysis*. New York: McGraw-Hill Book Company.

Sambursky, S. 1959. *Physics of the Stoics*. London: Routledge and Kegan Paul.

Scarani, V., Tittel, W., Zbinden, H., and Gisin, N. 2000. The speed of quantum information and the preferred frame: analysis of experimental data. *Phys. Lett. A*, **276**, 1–7.

Schneider, D. P., Hell, P. B., Richards, G. T., *et al.* 2007. The Sloan Digital Sky Survey Quasar Catalog IV. Fifth data release. *Astron. J.*, **134**, 102–117.

Schrödinger, E. 1926. Quantisierung als Eigenwertproblem (erste Mitteilung). *Ann. Phys.*, **79**, 361–376.

Schutz, B. 1980. *Geometrical Methods of Mathematical Physics*. Cambridge: Cambridge University Press.

Schwinger, J. 1959. Field theory commutators. *Phys. Rev. Lett.*, **3**(6), 296–297.

Schwinger, J. 1963. Gauge theories of vector particles, in *Theoretical Physics* (Trieste Seminar 1962). Vienna: IAEA, pages 89–134.

Schwinger, J. 1965. Relativistic quantum field theory. Nobel Prize Lecture, 1–13 (available online).

Schwinger, J. 1969. *Particles and Sources*. New York: Gordon and Breach.

Secada, J. E. K. 1990. Descartes on time and causality. *Philos. Rev.*, **99**(1), 45–72.

Skyrme, T. H. 1961. A non-linear field theory. *Proc. Roy. Soc. A*, **260**, 127–138.

Smolin, L. 2013. *Time Reborn: From the Crisis of Physics to the Future of the Universe*. London: Allen Lane.

Smoluchowski, M. 1906. Zur kinetischen Theorie der Brownschen Molekularbewegung und der Suspensionen. *Ann. Phys.*, **21**, 756–780.

Snyder, H. S. 1947a. The electromagnetic field in quantized space-time. *Phys. Rev.*, **72**(1), 68–71.

Snyder, H. S. 1947b. Quantized space-time. *Phys. Rev.*, **71**(1), 38–41.

Spiegel, M. R. 1965. *Laplace Transforms*. New York: McGraw-Hill Book Company.

Stein, E., and Shakarchi, R. 2003. *Fourier Analysis: An Introduction*. Princeton, NJ: Princeton University Press.

Stern, A., and Desbrun, M. 2006. Discrete geometric mechanics for variational integrators, *Proceedings of in SIGGRAPH '06.* New York: ACM, Pages 75–80.

Streater, R. F., and Wightman, A. S. 1964. *PCT, Spin and Statistics, and All That.* New York: W. A. Benjamin, Inc.

Stueckelberg, E. C. G. 1941. Remarque à propos de la création de paires de particules en théorie de la relativité. *Helv. Phys. Acta*, **14**, 588–594.

Sudarshan, E. C. G., and Mukunda, N. 1983. *Classical Dynamics: A Modern Perspective.* Malabar, FL: Robert E. Krieger Publishing Company.

Susskind, L. 1995. The world as a hologram. *J. Math. Phys.*, **36**(11), 6377–6396.

't Hooft, G. 1993. Dimensional reduction in quantum gravity. arXiv:gr-qc/9310026, 1–13.

Tapia, V. 1988. Second order field theory and nonstandard Lagrangians. *Nuovo Cimento B*, **101**, 183–196.

Taylor, J. G. (ed.) 1987. *Tributes to Paul Dirac, Cambridge, 1985.* Bristol: Adom Hilger.

Tegmark, M. 1997. On the dimensionality of spacetime. *Class. Quantum Gravity.*, **14**, L69–L75.

Tifft, W. G. 1996. Three-dimensional quantized time in cosmology. *Astrophys. Space Sci.*, **244**, 187–210.

Tonomura, A., Endo, J., Matsuda T., Kawasaki, T., and Ezawa, H. 1989. Demonstration of single-electron buildup of an interference pattern. *Am. J. Phys.*, **57**(2), 117–120.

TOTEM Collaboration 2012. Luminosity-independent measurements of total, elastic and inelastic cross-sections at $\sqrt{s} = 7$ TeV. Preprint CERN-PH-EP-2012-353, 1–7.

Weiss, P. 1936. On the quantization of a theory arising from a variational principle for multiple integrals with application to Born's electrodynamics. *Proc. Roy. Soc. A*, **156**, 192–220.

Weiss, P. 1938. *Proc. Roy. Soc. A*, **169**, 102–119.

Wells, H. G. 1895. *The Time Machine.* London: Willian Heinemann.

Wheeler, J. A. 1979. From the Big Bang to the Big Crunch. *Cosmic Search Mag.*, **1**(4). Interview with J. A. Wheeler (available online).

Whitrow, G. J. 1980. *The Natural Philosophy of Time.* Second edn. Oxford: Clarendon Press.

WMAP 2013. Wilkinson Microwave Anisotropy Probe empirical results. http://lambda.gsfc.nasa.gov/product/map/current/parameters.cfm.

Wolfram, S. 1986. *Theory and Applications of Cellular Automata.* Singapore: World Scientific.

Yang, C. N. 1947. On quantized space-time. *Phys. Rev.*, **72**, 874.

Index